The Science of Useful Nature in Central America

T0200714

In this ambitious new study, Sophie Brockmann argues that interactions with landscape and environment were central to the construction of Central American identities in the Age of Enlightenment. She argues that new intellectual connections and novel ways of understanding landscapes had a transformative impact on political culture, as patriotic reformers sought to improve the region's fortunes by applying scientific and 'useful' knowledge gathered from local and global networks to the land. These reformers established networks that extended into the countryside and far beyond Central America's borders. Tracing these networks and following the bureaucrats, priests, labourers, merchants and scholars within them, Brockmann shows how they made a lasting impact by defining a new place for the natural world in narratives of nation and progress.

SOPHIE BROCKMANN is a lecturer in history at De Montfort University, Leicester.

The Science of Useful Nature in Central America

Landscapes, Networks and Practical Enlightenment, 1784–1838

Sophie Brockmann

De Montfort University, Leicester

CAMBRIDGE
UNIVERSITY PRESS

University Printing House, Cambridge CB2 8BS, United Kingdom

One Liberty Plaza, 20th Floor, New York, NY 10006, USA

477 Williamstown Road, Port Melbourne, VIC 3207, Australia

314-321, 3rd Floor, Plot 3, Splendor Forum, Jasola District Centre, New Delhi - 110025, India

103 Penang Road, #05-06/07, Visioncrest Commercial, Singapore 238467

Cambridge University Press is part of the University of Cambridge.

It furthers the University's mission by disseminating knowledge in the pursuit of education, learning and research at the highest international levels of excellence.

www.cambridge.org
Information on this title: www.cambridge.org/9781108431620
DOI: 10.1017/9781108367615

© Sophie Brockmann 2020

First published 2020
First paperback edition 2022

A catalogue record for this publication is available from the British Library

ISBN 978-1-108-42123-2 Hardback
ISBN 978-1-108-43162-0 Paperback

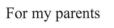
For my parents

Contents

List of Figures		*page* viii
Acknowledgements		ix
Notes on Translations and Names		xi
List of Abbreviations and Acronyms		xii
	Introduction: A Kingdom of Vast Extension	1
1	Landscape, Ruins, and Governance	29
2	Networks of Knowledge and Action	58
3	Making Enlightenment Local	91
4	Useful Geography in Practice	120
5	Transforming Environments	156
6	Independence and Useful Nature	194
	Conclusions	230
Glossary of Colonial Administrative Terms		235
Bibliography		238
Index		262

Figures

0.1 Political map of Central America, c. 1800.   *page* xiii
0.2 Map of Central American harbours and trade routes.   xiv
1.1 Antonio Bernasconi, Map of the site of Palenque, 1785. 40
1.2 Map of the site of Quiché, 1834.   41
4.1 Map of the coast of Zapotitlan and Suchitepequez, eighteenth
     century.   132
4.2 Vista of San Fernando de Guadelupe (Salto de Agua), 1795. 138
4.3 Spatial representation of roads in the *Gazeta de Guatemala*,
     1797.   139
4.4 José Rossi y Rubí and Pedro Garci Aguirre, Map of a new road
     in the province of Suchitepéquez, 1800.   146
5.1 Juan Bautista Jauregui, Plan for a new settlement at Izabal,
     1807.   184
6.1 Miguel Rivera Maestre, 'Carta del Estado de Guatemala en
     Centro-America. Año de 1832'.   224

Acknowledgements

In the research and writing of this book, I have incurred many intellectual debts to other scholars and colleagues, whose friendship, support, and inspiring work I am grateful for. I have benefited from, and been inspired by, the knowledge of Guatemalan and Latin American history which Jordana Dym, Rebecca Earle, and Sylvia Sellers-García so generously shared in person, via email, and through their publications. George Lovell, Christopher Lutz, Héctor Concoha Chet, and Catherine Poupeney Hart welcomed me into the small world of Central Americanists with kind words and indispensable advice. I also thank Daniela Bleichmar, Lina del Castillo, James Delbourgo, Irina Podgorny, Neil Safier, Simon Schaffer, Emma Spary, William Whyte, and above all Nicholas Jardine for their encouragement, help, and advice at crucial points during my PhD and beyond. The members of various conference audiences and seminars helped me to refine my arguments and pointed my research in new directions – especially the members of the Latin America in Global History Leverhulme research network, the Cabinet of Natural History seminar in Cambridge, and the staff and fellows of the John Carter Brown Library, who were always generous with their time and happy to share their knowledge. I must thank especially my friends and colleagues, Heather Dichter and Beatriz Pichel, for their comments on draft chapters, as well as two anonymous Cambridge University Press reviewers for their thorough and helpful comments. All errors, of course, are entirely my own.

I am most grateful for the financial support of several organisations, without whom this book could not have been researched or written: The UK Arts and Humanities Research Council, the John Carter Brown Library at Brown University, the Max Planck Institute for the History of Science in Berlin, St John's College Cambridge, the US National Endowment for the Humanities, the Institute of Latin American Studies in London, and De Montfort University in Leicester who awarded me stipends and travel grants to study for my PhD, visit archives and rare book collections, enjoy the stimulating company of other researchers at these institutions, and finally to transform my doctoral dissertation into this book.

Librarians, archivists, and administrators have been unfailingly helpful and knowledgeable. No historical research is possible without their work, and I owe great thanks to them. Special thanks go to the staff at the Guatemalan National Library and the Archivo General de Centro América, above all Jorge Castellanos and Vicky Gómez, as well as librarians at the Max Planck Institute for the History of Science in Berlin, the John Carter Brown Library in Providence, and the Newberry Library in Chicago. During the years that I worked on this topic, a number of rare treatises from the John Carter Brown Library and the Universidad Francisco Marroquín in Guatemala have been digitised, which allowed me to revisit some texts and gain entirely new perspectives, and I thank the teams responsible. I am most grateful to Tamara Hug at the Department of History and Philosophy of Science in Cambridge, Valerie Andrews at the John Carter Brown Library, and Olga Jiménez at the Institute of Latin American Studies in London for resolving administrative issues (some of which could have easily derailed scholarships and fellowships) with amazing efficiency.

I am indebted to many friends and colleagues who over the years have provided advice and moral support. I must thank especially Alex, Amy, Catherine, Hannah, Jack, Jennie, Liz, and Namrata for their patience and listening skills during the process of writing this book.

And, most importantly, my family: my sister Evelyn, who always believes in me, and my parents, to whom this book is dedicated and who have made everything possible with their generous and unwavering support.

Notes on Translations and Names

All translations from Spanish are my own. In translations of Spanish quotations, idiosyncrasies of the text have been reflected as far as possible while making the translation idiomatic and intelligible.

Modernised and English names have been used in the text for geographical locations for clarity where these place names are widely known, for instance Seville and Mexico City. Guatemala City is generally used for the Spanish capital Nueva Guatemala de la Asunción, although the designation of Nueva Guatemala is used where it is necessary to make a distinction from the old capital, Antigua, or Santiago de Guatemala.

The names of the modern countries of Honduras, Nicaragua, and El Salvador are generally used to refer to the territories more or less encompassed by the intendancies of Comayagua, León, and San Salvador to avoid confusion with cities of that name. A notable exception is the province of Sonsonate, which is today in El Salvador but was part of the Guatemalan provinces in colonial times.

The spelling of geographical locations has been modernised for consistency, that is Trujillo instead of Truxillo or Trugillo, except when in quotations and titles. Spanish names that appear in different spellings across primary documents have also been edited for consistency.

The names of indigenous language groups and related place names, where these are well known, reflect the modern orthography of indigenous languages rather than the Spanish spelling, for instance K'iché instead of Quiché. In quotations and where paraphrasing a Spanish author's words, the original spelling has been kept, for instance Quiché instead of K'iché.

I have translated some Spanish terms for ethnic and social groups. *Casta* terms such as *indio* or *mestizo* are further discussed in the Introduction.

Abbreviations and Acronyms

AFEHC   Asociación para el Fomento de los Estudios Históricos en
          Centroamérica
AGCA   Archivo General de Centro América. Guatemala City, Guatemala
AGI   Archivo General de Indias. Seville, Spain
AMN   Archivo del Museo Naval. Madrid, Spain
BL   The British Library. London, UK
CSIC   Consejo Superior de Investigaciones Científicas, Madrid
          (Publisher)
Gazeta   *Gazeta de Guatemala.* Ignacio Beteta: Guatemala, 1797–1807
HSA   Archive of the Hispanic Society of America. New York, USA
MP   Mapas y Planos (Archivo General de Indias classification)
RGS   Royal Geographical Society Archives. London, UK
TNA   The National Archives. London, UK

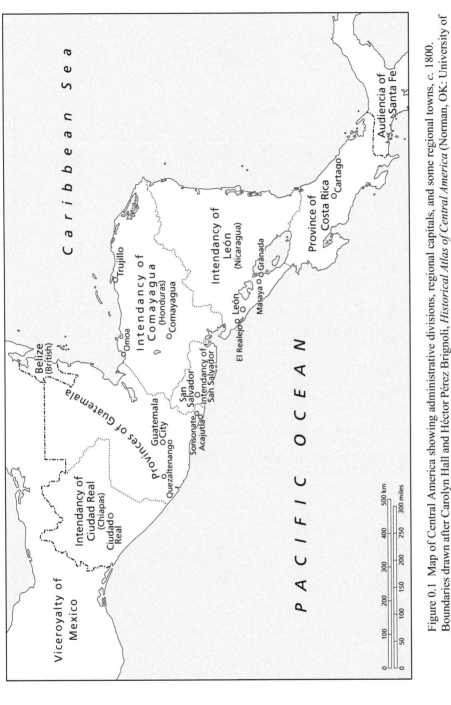

Figure 0.1 Map of Central America showing administrative divisions, regional capitals, and some regional towns, c. 1800. Boundaries drawn after Carolyn Hall and Héctor Pérez Brignoli, *Historical Atlas of Central America* (Norman, OK: University of Oklahoma Press, 2003), 38.

Figure 0.2 Harbours, major trade routes, and some trading settlements of
Central America, *c.* 1800. Trade routes drawn after Carolyn Hall and Héctor
Pérez Brignoli, *Historical Atlas of Central America* (Norman, OK: University
of Oklahoma Press, 2003), 162 and 179.

Introduction: A Kingdom of Vast Extension

The Kingdom of Guatemala in the late eighteenth century was a paradox. It was simultaneously rich and poor, according to its colonial administrators, priests, merchants, and ordinary residents. It was a 'kingdom of vast extension' that should by all accounts be a wealthy province within the Spanish empire, as puzzled observers noted. As the Spanish administrative region spanning modern-day Chiapas, Guatemala, El Salvador, Honduras, Nicaragua, and Costa Rica, it had great natural advantages, from being a place where 'the most exquisite fruits of all climes grow in abundance' to possessing harbours on both the Atlantic and Pacific coasts. Indeed, it might well be 'the best of all the King's possessions'.[1] Yet, as a contributor to Guatemala's newspaper remarked in 1803, 'this kingdom, which should be one of the most prosperous, is one of the most miserable ones in America'. How to square these contradictions, and help fulfil the true potential of the region, was the mission of a group of reformers who came together in patriotic associations in the late eighteenth and early nineteenth centuries. They believed that geographical, climatic, botanical, agricultural, and demographic knowledge held the key to 'enlightened' progress. True patriots would not just be content with gathering such knowledge. Instead, theirs was a practical Enlightenment that would offer prosperity by applying scientific knowledge to the management of landscapes. As this book argues, Central American reformers found the meaning of a homeland not in abstract ideas of idealised national landscapes, but in experiential engagement with them.

By the early nineteenth century, reformers imagined a new region, one that was self-confidently connected to the rest of the world through scientific communication networks, and one whose inhabitants were dedicated to developing its bountiful landscapes into ever more prosperous spaces. Although patriotic identities of the eighteenth century map onto nineteenth-century nationalisms imprecisely at best, the legacy of new visions of Central

---

[1] 'Apuntamientos estadisticos del Br. Talcamábida, sobre la agricultura, industria y comercio de este reyno', *Gazeta de Guatemala* (hereafter: *Gazeta*), Vol. 7, no. 313 (25 July 1803), 297; 'Descripción geográfica', *Gazeta*, Vol. 6, no. 279 (2 October 1802), 245. While Costa Rica was technically a part of this jurisdiction, the Audiencia (High Court) had little effective power there.

1

America created by late-colonial reformers shaped the nation-states that emerged after the region's independence from Spain in 1821.[2] Statesmen would draw on these earlier shifts in thinking about nature and strategies for managing landscapes to help define national identities within a larger Central American Federation, as well as the place of Central America in global networks. This contribution of the natural world to ideas of identity and nationalism has sometimes been placed in the realm of the literary, the learned, the intellectual. Jorge Cañizares-Esguerra has pointed to the legacies of a scientific patriotism in the nineteenth century as offering 'ideological tools that allowed those communities to think of themselves as central to the world', while Mark Thurner has made the argument for the intellectual history of Peru that a 'romantico-scientific' idea of soil and natural productions was a crucial hallmark in the formation of the idea of a 'nation'.[3] Central American scientific Enlightenment must certainly be placed in the context of such intellectual traditions, and the reformers' most ambitious goals of a social and environmental transformation of entire tracts of land indeed remained merely a powerful ideal.

However, exponents of enlightened reform in the eighteenth century insisted that their patriotic worlds were already taking shape in the countryside. They constructed patriotic ideals not just from studying landscapes, but from intervening in them. Their ideals had much in common with other scientific-patriotic traditions of Spanish America embodied by José Antonio de Alzate y Ramírez, Hipólito Unanue, José Celestino Mutis, or Francisco José de Caldas, who argued that scientific knowledge was at its most useful when it found practical application. Central American reformers doubled down on these concepts of utility and applicability forcefully and made it clear that it was the practical effects of knowledge that concerned them. Although they also developed a particular localist epistemology to fend off doubts about the reliability of different kinds of knowledge, information was not an end in itself. Instead, reformers worked within a political-economic framework of controlling landscapes and labourers rather than an abstract learned one. They attempted to intervene directly in agricultural activities, planned new villages,

---

[2] Anthony McFarlane, 'Identity, Enlightenment and Political Dissent in Late Colonial Spanish America', *Transactions of the Royal Historical Society*, 6th Series, 8 (1998): 309–35; Anthony Pagden, 'Identity Formation in Spanish America', in Nicholas Canny and Anthony Pagden (eds.), *Colonial Identity in the Atlantic World, 1500–1800* (Princeton, NJ: Institute for Advanced Study, 1987), 51–94; Luis Monguió, 'Palabras e ideas: "patria" y "nación" en el virreinato del Perú', *Revista iberoamericana*, 44 (1978): 451–70. Gabriel Paquette, 'The Dissolution of the Spanish Atlantic Monarchy', *The Historical Journal* 52, no. 1 (2009): 175–212, is a useful review of historiographic trends.

[3] Jorge Cañizares-Esguerra, *Nature, Empire and Nation: Explorations of the History of Science in the Iberian World* (Stanford University Press, 2006), 127; Mark Thurner, *History's Peru: The Poetics of Colonial and Postcolonial Historiography* (Gainesville, FL: University of Florida Press, 2011), 90–2.

acclimatised plants, designed infrastructure projects, and even tried to improve public health through controlling vegetation. Indeed, the late eighteenth century was a time of particular importance when it came to thinking about landscapes and their potential for change in the Kingdom of Guatemala (also known as the Audiencia de Guatemala), in practical terms as much as in the sense of patriotically imagined abundant soils. The memory of recent natural disasters, such as a major 1773 earthquake, still loomed large, while reformers drew new lessons from other natural disturbances, such as overgrown roads, rains, and locust plagues. The questions that members of the colonial administration asked of topographies, the travels of merchants, the agricultural and natural-historical designs of scholars and 'enthusiasts', as well as the practices of farmers who worked the land, helped to draw up their new programme for imagining progress that was built around experiences and understandings of landscape.

To speak of 'landscapes' in this context may be an anachronistic amalgamation of a number of Spanish concepts that are discussed in this book such as *país, tierras, terreno, montaña* or even *temperamento* or *clima*, but it is a fitting concept because of its multi-layered ability to encompass man-made as well as natural space, the physical environment, and human settlements. In addition, across Spanish America governments and scholars themselves were preoccupied with creating knowledge about these spaces. Many historians have consequently noted the importance of spatial practices to understanding the governance and intellectual culture of Spanish America.[4] Drawing on the work of historical geographers, landscape is here taken to be the 'surface of the land' as perceived and recorded (and therefore constructed as landscape) by bureaucrats, engineers, reformers, farmers, and travellers.[5] Although these perceptions never resembled a nineteenth-century sense of the 'picturesque', any implication of a European

---

[4] Nuria Valverde and Antonio Lafuente, 'Space Production and Spanish Imperial Geopolitics', in Daniela Bleichmar et al. (eds.), *Science in the Spanish and Portuguese Empires, 1500–1800* (Stanford University Press, 2009), 198–215; Ricardo Padrón, *The Spacious Word: Cartography, Literature and Empire in Early Modern Spain* (University of Chicago Press, 2004); Barbara Mundy, *The Mapping of New Spain: Indigenous Cartography and the Maps of the Relaciones Geográficas* (University of Chicago Press, 1996); Raymond Craib, *Cartographic Mexico: A History of State Fixations and Fugitive Landscapes* (Durham, NC: Duke University Press, 2004). Mariselle Meléndez, 'The Cultural Production of Space in Colonial Latin America', in Barney Warf and Santa Arias (eds.), *The Spatial Turn: Interdisciplinary Perspectives* (London: Routledge, 2009), 173–91, at 187.

[5] Denis Cosgrove, *Social Formation and Symbolic Landscape* (London: Croom Helm, 1984), 1; Alan Baker, *Geography and History: Bridging the Divide* (Cambridge University Press, 2003), 109–13 and 128–30; Denis Cosgrove and Stephen Daniels, 'Introduction', in Cosgrove and Daniels (eds.), *The Iconography of Landscape: Essays on the Symbolic Representation, Design and Use of Past Environments* (Cambridge University Press, 1988); Henri Lefebvre, *The Production of Space* (Oxford: Blackwell, 1991), Introduction. Jorge Cañizares-Esguerra has spoken about 'patriotic landscapes' for the case of nineteenth-century Mexico, but uses the term in the aesthetic sense of landscape painting: *Nature, Empire and Nation*, chapter 7.

'gaze' in the use of 'landscape' in this book reflects this history of largely European or Creole authors recording their attempts to make sense of a country. Definitions of landscape as social and cultural construct do not negate the importance of materiality. They rather underscore the importance of conceptualising the land through mapping and bureaucratic practices in eighteenth-century state governance, as well as through civic reform.[6] In the case of Guatemala, George Lovell and Severo Martínez Peláez have drawn influential conclusions about landscape as a historical category. George Lovell is concerned with the 'cultural landscape' of the Cuchumatanes region as a whole, that is, the interaction of history, land, and people. For Martínez Peláez, landscape is a more narrow and necessarily superficial term that appears as a foil to contrast with a true understanding of the land as means of production, but has explanatory power in the erasures of indigenous labour it contains.[7] My argument is influenced by these observations about the relationship between land, governance, and people. However, this book also contends that descriptions of landscapes broadly conceived as they appear in a variety of archival documents of the late eighteenth and early nineteenth centuries reveal further intellectual worlds and practical attitudes alike.

These intellectual and practical worlds came together in the actions of a patriotic association founded by a group of men who saw themselves as enlightened reformers, the *Real Sociedad Económica de Amantes de la Patria de Guatemala*, from 1795 onwards.[8] It became the key forum for exchanging knowledge deemed 'useful' in the sense that it could be applied to Central American landscapes. In bringing new ideas and scientific knowledge to their projects, the reformers drew on a range of sources from local as well as more global correspondents. Their ideologies and practices can therefore be better

---

[6] Charles Withers, *Placing the Enlightenment: Thinking Geographically about the Age of Reason* (Chicago, IL and London: University of Chicago Press, 2007), 12–13; Catarina Madeira Santos, 'Administrative Knowledge in a Colonial Context: Angola in the Eighteenth Century', *The British Journal for the History of Science* 43, no. 4 (2010): 539–56, at 542.

[7] George Lovell, *Conquest and Survival in Colonial Guatemala: A Historical Geography of the Cuchumatán Highlands, 1500–1821* (4th edition, Montreal and Kingston: McGill-Queen's University Press, 2015); Severo Martínez Peláez, *La patria del criollo. Ensayo de interpretación de la realidad colonial guatemalteca* (Mexico City: Fondo de Cultura Económica, 2006), 99–103.

[8] Royal Economic Society of the Lovers of the Homeland of Guatemala. Historians often interchangeably refer to it as *Real Sociedad Económica de Amigos del País* or Royal Economic Society of Friends of the Country of Guatemala, since some official correspondence refers to it that way, but its own statutes and publications are by the name of '*amantes de la patria*'. The classic work is Elisa Luque Alcaide, *La Sociedad Económica de Amigos del País de Guatemala* (Seville: Escuela de Estudios Hispano-Americanos, 1962), which mainly relies on archival documentation from the AGI. José Luis Maldonado Polo, *Las huellas de la razón: la expedición científica de Centroamérica (1795–1803)* (Madrid: Consejo Superior de Investigaciones Científicas, 2001) also outlines many of the Economic Society's activities, and biographical information on the Central American Enlightenment's main proponents.

understood by thinking spatially about the practice of scientific knowledge. This approach draws the established link between science and the affirmation of American identities into dialogue with historical geography and the spatial history of science, and is more broadly influenced by the 'spatial turn' in intellectual history and the history of science over the last three decades.[9] For Guatemala, it relies especially on Sylvia Sellers-García's work for understanding the possibilities of correspondence and conceptions of distance.[10] In Central America, increasingly localist views of spaces developed through the practical application of scientific and other empirical knowledge. Although these projects of useful nature did not define future national boundaries, they helped to re-imagine the relationship between different regions. However, these spatial views were often contradictory and fragmented, especially during the early independence period, as Jordana Dym's conclusions on the construction of political and cultural spaces have demonstrated.[11] The 'patriotism' of these eighteenth-century reformers did not meaningfully prefigure the political territories of the nineteenth century. Instead, reformers' efforts took place in a broader context of imperial governance, where enlightened ideas interacted with local social and economic priorities and routine bureaucratic practices.[12] Theirs was less a coherent intellectual programme than a patchwork of specific responses and solutions to social, economic, or geographical problems that they perceived around them.[13] There are similarities here to the observations of Lina

---

[9] On science and political identity: Anthony McFarlane, 'Science and Sedition in Spanish America: New Granada in the Age of Revolution, 1776–1810', in Susan Manning and Peter France (eds.), *Enlightenment and Emancipation* (Lewisberg, PA: Bucknell University Press, 2006), 97–116; Maria Rachel Fróes da Fonseca, 'La construcción de la patria por el discurso científico: Mexico y Brasil (1770–1830)', *Secuencia* 45 (1999): 5–26; José Luis Peset, *Ciencia y libertad: el papel del científico ante la independencia americana* (Madrid: CSIC, 1987). On spatial turn, for instance: David Livingstone, *Putting Science In its Place: Geographies of Scientific Knowledge* (University of Chicago Press, 2003); Crosbie Smith and Jon Agar (eds.), *Making Space for Science: Territorial Themes in the Shaping of Knowledge* (Basingstoke, New York, and Manchester: Palgrave Macmillan and Manchester University Press, 1998); Kapil Raj, *Relocating Modern Science: Circulation and the Construction of Knowledge in South Asia and Europe, 1650–1900* (Basingstoke and New York: Palgrave Macmillan, 2007).

[10] Sylvia Sellers-García, *Distance and Documents at the Spanish Empire's Periphery* (Stanford University Press, 2014).

[11] Jordana Dym, *From Sovereign Villages to National States* (Albuquerque, NM: University of New Mexico Press, 2006); Jordana Dym, 'Democratizing the Map: The Geo-body and National Cartography in Guatemala, 1821–2010', in James Akerman (ed.), *Decolonizing the Map: Cartography from Colony to Nation* (University of Chicago Press, 2017), 160–204, and other publications on the history of cartography.

[12] Renán Silva, *Los Ilustrados de la Nueva Granada, 1760–1808: genealogía de una comunidad de interpretación* (Medellín: Banco de la República; EAFIT, 2002), 48–9, calls this the 'context of application' of enlightened ideas.

[13] This builds on Sebastian Conrad's definition of 'Enlightenment' as local reaction: Conrad, 'Enlightenment in Global History: A Historiographical Critique', *The American Historical Review* 117, no. 4 (2012): 999–1027.

del Castillo and especially María José Afanador Llach, who have recently described a blending of political-economic thought and geographical knowledge for the case of New Granada and independent Colombia.[14] I argue that reformers' belief in the material consequences of practical interventions rather than just a detached vision of landscape helped to shape imaginations of territory. This case study alongside the Colombian parallels therefore suggests that ideas of the nation-state across Latin America were constructed by a larger range of colonial and independent, political, and scientific influences than has hitherto been recognised. If there was a spatial dimension to patriotism, it was the extent of the places where reform impacted the cultural landscape, what we might call the 'territory of intervention'. The reformers' homeland was an ever-shifting concept, able to accommodate political changes in the geographical territory.

Enlightenment and Reform

Ideas about the improvement and management of nature emerged within the context of two much-debated historical phenomena: the Enlightenment, and the Spanish Empire's so-called Bourbon Reforms. For self-proclaimed Central American reformers, they provided a set of tools as well as values: a belief in progress through applied scientific learning and ideals of good governance, but also an institutional framework to support associations dedicated to the furthering of these causes, including not just the Economic Society but also the Guatemala City merchant association, the *Consulado de Comercio*, which was established in 1793 and often supported similar ideals. Economic Societies, or Patriotic Societies as they were sometimes known, had originated on the Spanish mainland (following models from other continental European countries) and were encouraged by the Crown overseas.[15] As 'quasi-governmental' organisations, to use Gabriel Paquette's term, their objective was a broader revitalisation of commerce and trade as well as promotion of locally specific initiatives covering everything from growing more wheat to establishing schools, which would in turn lead to *felicidad pública*, or 'public

---

[14] María José Afanador Llach, 'Political Economy, Geographical Imagination, and Territory in the Making and Unmaking of New Granada, 1739–1830' (PhD dissertation, The University of Texas at Austin, 2016); Lina del Castillo, *Crafting a Republic for the World: Scientific, Geographic, and Historiographic Inventions of Colombia* (Lincoln, NE: University of Nebraska Press, 2018).

[15] Gabriel Paquette, 'State-Civil Society Cooperation and Conflict in the Spanish Empire: The Intellectual and Political Activities of the Ultramarine Consulados and Economic Societies, c. 1780–1810', *Journal of Latin American Studies* 39 (2007): 263–98; Koen Stapelbroek and Jani Marjanen (eds.), *The Rise of Economic Societies in the Eighteenth Century* (Basingstoke and New York: Palgrave Macmillan, 2012); Robert Shafer, *The Economic Societies in the Spanish World, 1763–1821* (Syracuse University Press, 1958), esp. 204–34 on Guatemala.

happiness'.[16] Since these were voluntary associations, but not 'private' in the sense distinguished, for instance, by Henry Lowood in the context of similar German societies, the reformers' actions had the capacity to be tied not just to their roles as 'patriots' but as officers of the state, as village priests, as bishops.[17] This meant that many of their members' visions for landscape interventions were supported or even shaped by some form of state authority, as well as by more obviously non-gubernatorial forms of power such as being a landowner. Guatemala's Economic Society was quite typical in that some of their grandest projects did not come to fruition: they did not succeed in completely transforming Central America's economic fortunes.[18] However, Society members' work offers an example of remarkable ambition across different fields of knowledge, as well as the establishment of new local and global intellectual networks and print cultures which endured into the nineteenth century. This book argues that we should take seriously many scattered short reports from across the kingdom, mainly in matters of agriculture or natural history, sometimes infrastructure, geography, or medicine, which reported attempts at improvement, progress, and pride in members' achievements. To them, this was Enlightenment in action.

Reformers often used the Spanish term *las ciencias* to refer to the body of structured knowledge that would render Central American nature useful.[19] Although *ciencia* can be a much broader term than the English 'science' (encompassing all branches of knowledge), many of the Economic Society's approaches can be described as embracing 'scientific knowledge' in the English meaning of the word as well. Its members, for instance, turned to natural history, geography, but also some historical archival materials as the basis for understanding landscapes, for applying new industrial methods, and as key to producing knowledge that would be useful and applicable to Central America. In 1815, the Society succinctly explained the ideology behind such useful science. In their opinion, there was a direct correlation between knowledge and wealth. Europe was the

---

[16] Gabriel Paquette, *Enlightenment, Governance, and Reform in Spain and Its Empire, 1759–1808* (Basingstoke: Palgrave Macmillan, 2008), 56–8; Miles Wortman, 'Bourbon Reforms in Central America', *The Americas* 32 (1975): 222–38, at 229.

[17] Henry Lowood, *Patriotism, Profit, and the Promotion of Science in the German Enlightenment: The Economic and Scientific Societies, 1760–1815* (New York: Garland Publishing, 1991), 25.

[18] A parallel can be drawn with what Cañizares-Esguerra sees as the failure of Bourbon Spain's ultimate ambition for its scientific projects, to break Dutch and British monopolies: *Nature, Empire and Nation*, 127.

[19] Joaquín Fernández Pérez, 'La ciencia ilustrada y las Sociedades Económicas de Amigos del País', in Manuel Sellés, José Luis Peset and Antonio Lafuente (eds.), *Carlos III y la ciencia de la Ilustración* (Madrid: Alianza, 1988), 217–32; Lowood, *Patriotism*, 26–7 explains that similar German societies differentiated between 'economic' and 'scientific' societies by the 1790s, but the Guatemalan society embraced all these interests. Joel Mokyr, *Enlightened Economy: An Economic History of Britain, 1700–1850* (New Haven, CT and London: Yale University Press, 2009), 188, notes that 'enlightened agriculture' in Britain also entailed a plethora of practices that were termed 'useful knowledge', but that we might not 'recognise as formal science'.

richest part of the world, and also 'the most enlightened [*la parte mas ilustrada*]'. It followed that studying and applying the sciences would give prosperity to 'even the most sterile lands', and unleash 'the genius of industry' to drive progress.[20] Europe, however, would not be directly copied. Instead, careful selection of methodologies and sources would ensure that such scientific knowledge was relevant to Central America. In Central America, as elsewhere in Latin America, reformers considered a plethora of different branches of knowledge and an eclectic range of sources.

Eighteenth-century scholars believed that only through a holistic approach that took into account geography, political economy, medicine, demography, and natural history could their political visions be achieved. Concepts of scientific learning, utility, and progress were linked throughout European statecraft and scientific thought. Theorists of political economy from Hume to Smith engaged closely with questions of nature and its productions, while botanists at Kew, Paris, and Madrid also espoused principles of the utility of natural history to the broader political economy. From French physiocrats to Charles III's ministers in Madrid, eighteenth-century scholars and politicians looked to the production of useful knowledge through the scientific study of nature as key to exploiting a kingdom's natural wealth. German cameralists even imagined themselves to be leading an entirely new academic discipline that would do away with the boundaries of learned and applied knowledge in their studies of the natural world, considering political economy, agricultural, and technical approaches as one.[21] Political economy in particular was inseparable from agronomy, natural history, and natural philosophy.[22] Central Americans' attitude of selecting a range of models and influences that included, where relevant to them, Adam Smith's writings on slavery, rice-growing

---

[20] *Periódico de la Sociedad Económica de Guatemala*, No. 4 (15 June 1815), 56–8.

[21] Richard Drayton, *Nature's Government. Science, Imperial Britain, and the 'Improvement' of the World* (New Haven, CT and London: Yale University Press, 2000), 67–128; Lissa Roberts, ' "*Le centre de toutes choses*": Constructing and Managing Centralization on the Isle de France', *History of Science* 52, no. 3 (2014): 319–42; Emma Spary. ' "Peaches Which the Patriarchs Lacked": Natural History, Natural Resources, and the Natural Economy in France', *History of Political Economy* 35: Annual Supplement (2003): 14–41; Meyer Reinhold, 'The Quest for "Useful Knowledge" in Eighteenth-Century America', *Proceedings of the American Philosophical Society* 119, no. 2 (1975): 108–32; Sellés, Peset, and Lafuente (eds.), *Carlos III y la ciencia de la Ilustración*; Joel Mokyr, *The Enlightened Economy: An Economic History of Britain, 1700–1850* (New Haven, CT and London: Yale University Press, 2009), 40–62; Andre Wakefield, 'Cameralism: A German Alternative to Mercantilism', in Philip J. Stern and Carl Wennerlind (eds.), *Mercantilism Reimagined: Political Economy in Early Modern Britain and Its Empire* (Oxford University Press, 2014).

[22] Richard Drayton, *Nature's Government. Science, Imperial Britain, and the 'Improvement' of the World* (New Haven, CT and London: Yale University Press, 2000), esp. 70–2; Margaret Schabas, *The Natural Origins of Economics* (Chicago, IL and London: University of Chicago Press, 2005); David Lindenfeld, *The Practical Imagination: The German Sciences of State in the Nineteenth Century* (University of Chicago Press, 1997), esp. chapters 1–2; Lowood, *Patriotism*.

technologies in the Carolinas, or English technologies of spinning and weaving cotton reflected more general practices of emulation and translation within the political economies of Europe, and the interactions of political-economic theory with the practices of empire.[23]

Although Central Americans developed particular definitions for what they considered useful knowledge, these ideas were also part of the wider background of Spain's 'Bourbon Reforms'. In the second half of the eighteenth century, Spain's new rulers of the House of Bourbon tried to implement a series of reforms in their American colonies, through policies for the most part devised by the Spanish ministers José del Campillo y Cosío, Pedro de Campomanes, José Moñino, and José de Gálvez. These reforms have been associated with Spanish 'enlightened absolutism' and top-down centralisation, although in Guatemala, as elsewhere in the empire, the reach of these administrative and fiscal reforms was never quite as far-reaching as their designers intended.[24] Broadly speaking, their aims were a reorganisation of local administration, a stimulation of economic growth, and an attack on the privileges of religious orders, and to some extent the Church. The centrepiece of administrative reform was the establishment of intendancies, a new layer of regional government in the Americas aimed at making government more uniform and more powerful. Other aspects of the reforms promoted new philosophies of political economy (for instance, through the works of political economist Bernardo Ward), support of road-building schemes, tax reforms to increase revenue, and in the 1780s and 90s experimentation with limited free trade. They also included support for the systematic study of nature, manifested most decisively in the scientific expeditions that the Crown sent to the Americas, but also botanical gardens and cabinets of natural history.[25] The most prolific of the

---

[23] Sophus Reinert and Pernille Røge (eds.), *The Political Economy of Empire in the Early Modern World* (Basingstoke: Palgrave Macmillan, 2013), esp. chapter by Gabriel Paquette, 'Views from the South: Images of Britain and Its Empire in Portuguese and Spanish Political Discourse, ca. 1740–1810', 76–104; Sophus Reinert, *Translating Empire: Emulation and the Origins of Political Economy* (Cambridge, MA: Harvard University Press, 2011). Cotton is discussed e.g. in Real Sociedad Económica, *Quinta Junta Pública de la Real Sociedad Economica de Amantes de la Patria de Guatemala* (Guatemala City: Viuda de Sebastian de Arevalo, 1799), Real Sociedad, *Quinta junta pública*, 9–12; Adam Smith and rice cultivation in Chapter 5.

[24] Classic works and recent revisions on the reforms include Paquette, *Enlightenment, Governance, and Reform*; Jeremy Adelman, *Sovereignty and Revolution in the Iberian Atlantic* (Princeton University Press, 2016); Barbara and Stanley Stein, *Apogee of Empire: Spain and New Spain in the Age of Charles III, 1759–1789* (Baltimore, MD: The Johns Hopkins University Press, 2003); Agustín Guimerá (ed.), *El reformismo borbónico: una visión interdisciplinar* (Madrid: CSIC, Alianza, 1996). For Guatemala, see: Jordana Dym and Christophe Belaubre (eds.), *Politics, Economy and Society in Bourbon Central America* (Boulder, CO: University of Colorado Press, 2007), especially chapters by Gonzáles Alzate, Dym, and Palma Murga; Dym, *From Sovereign Villages*, 33–61.

[25] The vast historiography on botanical expeditions includes: Neil Safier, *Measuring the New World: Enlightenment Science and South America* (University of Chicago Press, 2008); Daniela Bleichmar, *Visible Empire: Botanical Expeditions and Visual Culture in the*

expeditions were the botanical expeditions sent to New Spain and New Granada, as well as Alejandro Malaspina's circumnavigation of the world, and Jorge Juan and Antonio Ulloa's participation in La Condamine's geodetic expedition. Expeditions were often influenced by local elites, or included elements of negotiation between metropolitan and local scholars, just as local elites in many places were able to negotiate their own interests within the broader Bourbon project.[26] While Central American elites generally welcomed the visit of some members of the New Spain expedition, principles of enlightened reform existed in Central America before and independently of this, and were influenced by a much wider array of factors, as local as they were oriented towards the overall success of the empire.[27]

Guatemalan reformers were certain that their pursuits were 'enlightened'. While there are few historical terms as disputed as 'Enlightenment', it is a useful framework for connecting these Central American reformers to global history. Historians have defined a multitude of phenomena that fall under this umbrella term, and increasingly embrace forms of Enlightenment that were not included in traditional narratives centred on northern Europe: Enlightenments among Catholics, monarchists, and anti-imperialists; 'practical', 'eclectic', and 'agricultural' Enlightenments; and movements not simply centred on canonical European texts.[28] In the Spanish historical context, Francisco Sánchez-Blanco

---

Eighteenth-Century Hispanic World (University of Chicago Press, 2012); Mauricio Nieto Olarte, *Remedios para el imperio: historia natural y la apropiación del Nuevo Mundo* (Bogotá: Instituto Colombiano de Antropología e Historia, 2000); Francisco Puerto Sarmiento, *La ilusión quebrada: Botánica, sanidad y política científica en la España ilustrada* (Madrid: Consejo Superior de Investigaciones Científicas, 1988); Ingrid Engstrand, 'Of Fish and Men: Spanish Marine Science During the Late Eighteenth Century', *The Pacific Historical Review* 69, no. 1 (2000): 3–30; Paula De Vos, 'Natural History and the Pursuit of Empire in Eighteenth-Century Spain', *Eighteenth-Century Studies*, 40 (2007): 209–39; Juan José Saldaña (ed.), *Science in Latin America. A History* (Austin, TX: University of Texas Press, 2006).

[26] Antonio Lafuente, 'Enlightenment in an Imperial Context: Local Science in the Late-Eighteenth-Century Hispanic World', *Osiris* 2nd Series, 15 (2000): 155–73, at 158–9; Susan Migden Socolow, *The Bureaucrats of Buenos Aires, 1769–1810: Amor al Real Servicio* (Durham, NC and London: Duke University Press, 1987); John Coatsworth, 'The Limits of Colonial Absolutism: The State in Eighteenth-Century Mexico', in Karen Spalding (ed.), *Essays in the Political, Economic, and Social History of Colonial Latin America* (Newark, DE: University of Delaware, 1982), 25–51.

[27] Maldonado Polo, *Huellas de la razón* is a meticulous narrative of the expedition's work in Guatemala. It also includes an account of the Central American Enlightenment that is based on a detailed examination of sources in the AGI, but establishes a more diffusionist model of 'Enlightenment' than my study, e.g. 171–4, 180–6. See also Arturo Taracena Arriola, *La expedición científica al reino de Guatemala* (Guatemala City: Editorial Universitaria de Guatemala, 1983); María Luisa Muñoz Calvo, 'Las actividades de José Mariano Mociño en el Reino de Guatemala (1795–1799)', in José Luis Peset (ed.), *Ciencia, vida y espacio en Iberoamérica*, Vol. I (Madrid: CSIC, 1989), 3–19.

[28] Conrad, 'Enlightenment in Global History'; Roy Porter and Mikuláš Teich, *The Enlightenment in National Context* (Cambridge and New York: Cambridge University Press, 1981); Sankar Muthu, *Enlightenment Against Empire* (Princeton University Press, 2003); Peter

has critiqued the idea that *Ilustración* has come to be so broad a term as to virtually be meaningless, for instance being used to refer to state interventions in anything from infrastructure to grain prices and collecting statistics, rather than a true intellectual process or agitation for social change.[29] And yet, part of what makes the multiplicity of definitions interesting is actors' own insistence that they were performing 'enlightened' actions. While 'Enlightenment' as a historical and philosophical period is of course a retrospectively established category, Central Americans were fond of employing the terms *ilustración* and *las luzes* to describe their own work and that of their colleagues, implying knowledge and actions that would shine a metaphorical light onto darkness, or lead to transformations.[30] Priests, high- and low-ranking government administrators, and intellectuals all earnestly professed that they were implementing Enlightenment when they built roads, summarised information about medicinal plants, read scientific news from Europe, or pontificated about the climate of Central American port-towns. In this case, Enlightenment was also an actors' category.

Self-proclaimed Central American reformers negotiated their own visions of Enlightenment, which was influenced by writers from Montesquieu to Bernardo Ward, but was ultimately guided by a localism that prioritised the applicability of any philosophy or treatise to the context and landscapes of Central America. Classic studies of the Guatemalan Enlightenment have amply demonstrated the familiarity of Central American scholars with 'modern' science and the writings of authors such as Condillac, Montesquieu, and Voltaire.[31] Expanding the analysis to the region's wider intellectual environment and governance, however, to questions of geographical, botanical, and environmental knowledge, a practical Enlightenment only tangentially tied to canonical texts emerges. Partly out of theoretical conviction and partly out of sheer necessity brought on by occasionally unreliable connections with the rest of the world, reformers largely prioritised knowledge that was drawn from, or tested within, Central American landscapes.[32] In this, they were influenced and

---

M. Jones, *Agricultural Enlightenment: Knowledge, Technology, and Nature, 1750–1840* (Oxford University Press, 2016); Paquette, *Enlightenment, Governance, and Reform.*

[29] Francisco Sánchez-Blanco, *La mentalidad ilustrada* (Madrid: Taurus, 1999), Introduction.

[30] A parallel study of a transformative, locally based Enlightenment can be found in the works of Bishop Martínez Compañón: Emily Berquist Soule, *The Bishop's Utopia: Envisioning Improvement in Colonial Peru* (Philadelphia, PA: University of Pennsylvania Press, 2014).

[31] John Tate Lanning, *The Eighteenth-Century Enlightenment in the University of San Carlos de Guatemala* (Ithaca, NY: Cornell University Press, 1956), 63; André Saint-Lu, *Condición colonial y conciencia criolla en Guatemala, 1524–1821* (Guatemala: Editorial Universitaria, 1978), 169–70.

[32] An exception to such localism was the realm of formal medicine, where reformers embraced more universalist ideas of progress: Martha Few, *For All of Humanity: Mesoamerican and Colonial Medicine in Enlightenment Guatemala* (Tucson, AZ: University of Arizona Press, 2015). See also Chapter 3.

informed by the colonial bureaucracy's own information-gathering methods. As Bianca Premo has recently stressed, modes of thinking that were recognisably 'enlightened' could exist in one facet of society even while other aspects of society were rooted in older traditions.[33] Within such an Enlightenment, the old and the new, the imperial-bureaucratic and 'scientific' could also co-exist when it came to the production and interpretation of knowledge. María José Afanador Llach has recently made this argument for spatial knowledge in New Granada, while Margaret Ewalt has also seen a hybrid 'eclectic' Enlightenment in Gumilla's natural history of the Orinoco.[34] The idea of Enlightenment as practical also had further precedents in the Hispanic world. Premo cites a prize-winning essay from Madrid's Economic Society in which an author complained that Enlightenment should not be literary, but practical, rejecting the association with the 'idleness of many intellectuals'.[35] Similar accusations of 'writing panegyrics' instead of applying oneself, or producing 'papers and words' instead of results, can also be seen in Guatemala.[36]

The increasingly accepted view of Enlightenment as an eclectic and fragmented phenomenon, revisionist additions to the histories of European Enlightenments as well as calls for examining Enlightenments within their local social and political circumstances invite comparisons of similar 'strands' of Enlightenment in different geographical locations. Central American reformers were, for instance, part of a global movement focused on agricultural improvement. Central American thought on agriculture and environment, including practical encounters with the countryside, shared approaches with Scottish or British reformers identified by Fredrick Albritton Jonsson or Peter Jones.[37] Central Americans' efforts paralleled those of European or American enlightened reformers practically as well as theoretically when they tried to acclimatise to Guatemala a collection of seeds originally collected for the British Board of Agriculture in Sumatra. Within an Enlightenment that was often paternalistic and patrician in its conceptions, the argument of 'improving' the landscape as a moral one that appeared in the writings of British eighteenth- and nineteenth-century botanists also came to bear in Central America, particularly when extended to debates about indigenous people as caretakers of the

---

[33] Bianca Premo, *The Enlightenment on Trial: Ordinary Litigants and Colonialism in the Spanish Empire* (Oxford University Press, 2017), especially 13–17, 70–7.

[34] Afanador Llach, 'Political Economy, Geographical Imagination, and Territory'. Margaret Ewalt, *Peripheral Wonders: Nature, Knowledge and Enlightenment in the Eighteenth-Century Orinoco* (Lewisburg, PA: Bucknell University Press, 2008).

[35] Premo, *Enlightenment on Trial*, 9.

[36] 'Carta', *Gazeta*, Vol. 1, no. 8 (3 April 1797), 59; 'Junta de Gazeta', *Gazeta*, Vol. 4, no. 145 (24 February 1800), 171.

[37] Fredrik Albritton Jonsson, *Enlightenment's Frontier: The Scottish Highlands and the Origins of Environmentalism* (New Haven, CT: Yale University Press, 2013), esp. 129–34; Jones, *Agricultural Enlightenment*.

land. In different political contexts across Europe and the world, including in Central America, enlightened reformers made similar efforts to 'conquer' nature, often justifying physical control of the land with their supposedly superior scientific approaches.[38]

Central America as a Case Study

What made reformers believe that the Kingdom of Guatemala was an especially fertile ground for their improvement projects was also what frustrated them: Central America's peripherality and relative economic insignificance within a Spanish empire dominated by the viceregal centres of Mexico City and Lima. In Central America, bureaucrats and scholars were often conscious of their own peripherality, even as they schemed to prove to the world that it was unjustified. Historians have emphasised the importance of supposedly 'peripheral' regions to empires, the potential of 'peripheries' to become 'centres' from a local perspective; and historians of science have further emphasised the particular potential of sites of natural history in peripheral places to acquire the significance of a 'centre'.[39] Nevertheless, many residents of Central America were painfully conscious of its peripheral status. Sylvia Sellers-García defends the 'periphery' as a historical category that need not be pejorative, especially for a place like Guatemala, which contemporaries often saw as difficult to reach, trade with, and administer.[40] The director of the Economic Society complained that Guatemala was 'unknown in the civilised world'.[41] Given this sense of remoteness, some Spanish magistrates considered a posting in Guatemala a punishment, or a stepping stone to a better appointment. One prominent reformer complained about being posted to Comayagua, capital of the Honduras intendancy, a 'desolate city' full of 'misery' (a periphery within the periphery), while another referred to Guatemala as a 'purgatory' in an initial reaction to his new posting.[42] And yet, as an isthmus with access to the Pacific

---

[38] Compare, for instance, Drayton, *Nature's Government*; Kavita Philip, *Civilizing Natures: Race, Resources and Modernity in Colonial South India* (Hyderabad: Orient Longman, 2003); David Blackbourn, *The Conquest of Nature: Water, Landscape and the Making of Modern Germany* (London: Jonathan Cape, 2006), 37–46. These ideas are further discussed in Chapter 5.

[39] Christine Daniels and Michael Kennedy (eds.), *Negotiated Empires: Centers and Peripheries in the Americas, 1500–1820* (New York and London: Routledge, 2002). George Lovell and Christopher Lutz, 'Core and Periphery in Colonial Guatemala', in Carol A. Smith (ed.), *Guatemalan Indians and the State: 1540 to 1988* (Austin, TX: University of Texas Press, 1990), 35–51; Bleichmar, *Visible Empire*, 127; John McAleer, '"A Young Slip of Botany": Botanical Networks, the South Atlantic, and Britain's Maritime Worlds, c.1790–1810', *Journal of Global History* 11, no. 1 (2016): 24–43.

[40] Sellers-García, *Distance and Documents*, 3–16.   [41] Real Sociedad, *Quinta junta pública*, 2.

[42] Jorge Cañizares-Esguerra, *How to Write the History of the New World: Histories, Epistemologies, and Identities in the Eighteenth-Century Atlantic World* (Stanford University Press, 2001), 338.

as well as the Atlantic coasts, and neighbouring politically powerful territories such as New Spain (Mexico) and Cuba, the kingdom was geopolitically important, even if Spanish metropolitan attention often stayed focused on the coasts rather than interior of the isthmus. Part of the significance of the Audiencia de Guatemala as a case study lies in this contradiction.

Reformers' hopes for enlightened ideas of progress were amplified by Guatemala's perceived peripherality. Those who felt that Central America had been overlooked by Spanish imperial authorities and foreigners alike used the language of enlightened improvement to prove its worth to other parts of the empire and the wider world. They measured their progress against that of other Spanish American territories, but also against European or North American scientific and agricultural advances, even as they insisted on their own visions of Enlightenment and epistemologies. The marginality of the Audiencia de Guatemala in many networks of trade and political power made Central American reformers especially conscious of the difficulty of claiming a place in the world of knowledge, too. They clung to useful science and the effective exploitation of landscapes as a potentially transformative mechanism for asserting Central America's presence within the empire and a wider world. The issues they tackled in the process, negotiating between the local and the global, the particular and the universal, were familiar to any eighteenth-century scholar. However, for Central Americans, the gulf to be overcome between the particular and the universal seemed especially wide at times. The perception that Guatemalan landscapes were as remote, threatening, and 'exceptionally harsh and mountainous' as they were vast and fertile would make overcoming these challenges with the help of enlightened ideas a particular achievement, and meant that the transformative promise of Enlightenment was especially great in the context of such desolation.

The landscapes of Central America were home to a society that was highly unequal and deeply hierarchical. Modern historians have estimated a population of 920,409 in the Kingdom of Guatemala in 1800. Nueva Guatemala, which counted a population of nearly 26,000 by 1812, was by far the largest city. Indians and Creoles were the majority, with Spaniards making up as few as 10 per cent of the population.[43] Spaniards and the wealthiest Creoles alongside institutions such as the religious orders claimed ownership of much of the land. In line with the conventions of this historical field, the

---

[43] Dym, *From Sovereign Villages*, 266–71, gives extensive population statistics, including the percentage of different *castas*. Bernardo Belzunegui Ormázabal, *Pensamiento económico y reforma agraria en el Reino de Guatemala, 1797–1812* (Guatemala: Comisión Interuniversitaria Guatemalteca de Conmemoración del Quinto Centenario del Descubrimiento de América, 1992), 269–72, gives statistics that seem accurate for Guatemala City, but probably underestimates the total number of inhabitants. The *Consulado de Comercio* estimated one million inhabitants in its 1811 *Apuntamientos sobre Agricultura y Comercio*.

adjective and noun 'Indian' is used when discussing people within the Spanish category of '*indio*'. This was a Spanish legal and social as well as racial construct within the colonial *casta* (caste) system, and cannot be uncritically translated with the modern ethnic term of indigenous or *indígena*. Eighteenth-century authors also used the term *naturales* to refer to the native inhabitants of specific places, often with the implication that these were indigenous people. An extensive literature on the construction of ethnic identity through clothing, racial 'passing', and 'purchasing whiteness' has greatly nuanced and compli-cated our understanding of these terms.[44] In Central America, American-born persons of Spanish descent were often known as *ladino*, and only occasionally *criollo*. I use the English term Creole to translate both, since it is widely used in histories of Latin American science. Spaniards from Spain were known as *peninsulares* or *europeos*, although *español* could also be a *casta* term that encompassed both peninsulars and Creoles. Issues of translation also arise with Spanish terms such as *negro* or *esclavos negros*, free or enslaved persons supposedly entirely of African descent. While Indians featured prominently in the documents of enlightened reformers, there were few discussions of enslaved black labour in the Society's ambit. While there were fewer enslaved blacks in Central America than in other Spanish colonies (probably because *hacienda* owners lacked funds, and had access to cheap Indian labour through the *repartimiento* system), a few *haciendas* including the Dominican order's lucrative sugar plantation at San Gerónimo used enslaved labour, as did the British plantations along the Mosquito Coast and logging camps of Belize. Crown slaves were engaged in harbour-works at Omoa. In addition, there was a substantial free black population working in mines, as artisans, and in black militias.[45] Other Spanish *casta* categories such as *mestizo*, *mulato*, *zambo*, and *pardo* are included in the text in italics.

Elite Spanish and Creole perspectives on landscapes were more likely to be recorded in archival documents, but they were inevitably also shaped by the Indian population who, as the majority of agricultural workers, knew the landscape best, and were more likely to reside in

---

[44] Ann Twinam, *Purchasing Whiteness: Pardos, Mulattos, and the Quest for Social Mobility in the Spanish Indies* (Stanford University Press, 2015). Andrew B. Fisher and Matthew D. O'Hara (eds.), *Imperial Subjects: Race and Identity in Colonial Latin America* (Durham, NC and London: Duke University Press, 2009), especially the introduction by Irene Silverblatt and chapter 3 by David Tavárez. Tamara Walker, *Exquisite Slaves: Race, Clothing, and Status in Colonial Lima* (Cambridge University Press, 2017); Laura E. Matthew, *Memories of Conquest: Becoming Mexicano in Colonial Guatemala* (Chapel Hill, NC: University of North Carolina Press, 2012), especially 244–67.

[45] Robinson Herrera, '"Por Que No Sabemos Firmar": Black Slaves in Early Guatemala', *The Americas* 57, no. 2 (2000): 247–67; Thomas Fiehrer, 'Slaves and Freedmen in Colonial Central America: Rediscovering a Forgotten Black Past', *The Journal of Negro History* 64, no. 1 (1979): 39–57; compare Ben Vinson, *Bearing Arms for His Majesty: The Free-Colored Militia in Colonial Mexico* (Stanford University Press, 2003).

rural villages.[46] Reformers often imagined that Indians could be 'led' and managed as if the labour force were just another natural resource, or alternatively paternalistically schemed to protect them from coercive labour practices. The 'improvement' that reformers imagined rested on an assumption that they would be able to direct and control the management of landscapes, but also by extension the bodies and work of rural and urban labourers, by directing certain harvests or road-building projects to be carried out, or instructing them how to process the results of a flax harvest, for instance. Even rural Indians, however, possessed significant agency within the life of the colony: they were much more likely than city scholars to have technical and practical knowledge of farming methods, local geographies, and plants, and voiced their opinion through informal protests, by exercising their rights within Spanish courts, or threatening legal action or petitioning authorities.[47] Intellectuals and statesmen may have been the most vocal proponents of these new patriotic ideas within political discourses, but the membership and debates of the Economic Society also show that a relatively broad public participated in these constructions of patriotic identities. They included landowners, low-ranking secular officials, and parish priests who might be considered 'elite' in the sense that they were Spanish or Creole literate men in a position of some power, but whose voices were amplified by the Economic Society's fora more than in government archival records.

The opportunities for organising projects of improvement opened up by the Economic Society did not mean that officials suddenly pursued only this path for promoting their own plans to make the kingdom prosperous. After all, Enlightenment in Spanish America worked within the structures of government as often as without, and there was often congruence between the aims of the Bourbon Reforms and the Society's regional goals. Many of the reformers' concerns about useful nature in the 1790s were also already shared by colonial administrators in the 1780s. This meant that regional governors who featured in the pages of the Society's newspaper also utilised the usual pathways of the colonial bureaucracy for their projects. They might promote a project as an individual initiative, yet assumed that the Guatemala City government would authorise a draft of Indian labour if it were needed. Nor did projects and reforms start or end with the Economic Society. In the 1780s and early 1790s, for instance, the Spanish official Antonio López Peñalver y Alcalá pursued a number of projects in the spirit of the Bourbon reforms through the

---

[46] Elite reformers usually assumed that rural labourers were 'Indian' and recorded their identity this way.
[47] On Indian agency through law, see also Brian Owensby, *Empire of Law and Indian Justice in Colonial Mexico* (Stanford University Press, 2008)

usual channels of government. He left for Spain in 1796, just before the heyday of the Economic Society.[48] While in Renán Silva's account of the New Granadan Enlightenment, intellectual ideas and the practice of government are usually distinguished, in Guatemala, the realms of enlightened enthusiasm and bureaucracy were not separate.[49]

The roles of Society members and administrator were sometimes blurred. For instance, when Juan Ortiz de Letona was put in charge of a rare collection of plants, his position in the fiscal administration was evidently one of the reasons which gave him control over the plants which were seen as potentially economically useful.[50] However, when he in turn handed the collection to a member of the Guatemala City administration, Juan Payés y Font, it was much more likely that he did so in acknowledgement of their mutual membership of the Economic Society. These institutional structures suggest that this Central American Enlightenment was less revolutionary than gradual, a successful insertion of some new modes of thinking into established social, intellectual, and political circles. Although significant figures emerged in its first director, the *oidor* Jacobo de Villaurrutia, and enthusiastic contributors Alejandro Ramírez, José Rossi y Rubí, and Juan Ortiz de Letona, the often decentralised pathways of creating and relaying a variety of knowledges that we can see in Central America defies interpretation as either an 'interpretive community' (after Stanley Fish) or an 'intellectual field' (after Pierre Bourdieu). Renán Silva and David Lindenfeld's use of these terms to describe groups of reformers in New Granada and the German lands, respectively, suggests a more elite and scholarly context than the Guatemalan reformers' actions.[51] It may be best to imagine them as a loose network that only relied on Guatemala City as a Latourian 'centre of calculation' in some cases. Although many networks of correspondence and administrative power did converge on that city, Society members also emphasised their ability to take action in the countryside, and often highlighted practical actions over abstracted information. Smaller regional towns and rural projects of progress made these networks multi-sited, and employed a variety of intellectual and practical influences.[52] However, despite its less-than-monolithic nature, the Economic Society is

---

[48] 'Sobre fomento de la Mineria en aquel Reyno', 1793, AGI, Guatemala, 770. Christophe Belaubre, 'López Peñalver y Alcalá, Antonio', *Diccionario AFEHC.*

[49] Silva, *Los Ilustrados*, 578. Afanador Llach, 'Political Economy, Geographical Imagination and Territory' is a better parallel here, although the bureaucrats she describes were not generally self-consciously pursuing 'Enlightenment'.

[50] 'Catálogo de plantas', *Gazeta*, Vol. 5, no. 194 (9 March 1801), 415.

[51] Silva, *Los Ilustrados*, 584; Lindenfeld, *Practical Imagination,* 5.

[52] Bruno Latour, *Science in Action: How to Follow Scientists and Engineers Through Society* (Cambridge, MA: Harvard University Press, 1987); Emma Spary's emphasis on competing centres is useful here: 'Botanical Networks Revisited', in Dauser *et al.* (eds.), *Wissen im Netz. Botanik und Pflanzentransfer in europäischen Korrespondenznetzen des 18. Jahrhunderts* (Berlin: Akademie Verlag, 2008), 113–34.

a useful prism to trace Enlightenment ideas across the period. It formed a loose but distinct group whose members gravitated to the ideologies of enlightened reform, and improvement through learning and science, as well as making these ideas a matter of public debate.

Localist ideas of reform in the Americas have sometimes been described as 'Creole patriotism', but the precise category of 'Creole' in the sense of a person of Spanish origin born in the Americas is of limited use in the context of the Guatemalan Society's members' varied backgrounds.[53] Central America-born Creoles, especially friars and clergy educated at institutions such as the University of San Carlos in Guatemala or seminaries in Comayagua and León certainly shaped the work of the Society. The scholar and reformer of Guatemala's university curriculum, José Antonio Liendo y Goicoechea, was born in Costa Rica, to give just one example. However, many of the 'enlightened elite', from whose ranks the Society's most prolific members were drawn, were originally from Spain or from other parts of the Spanish empire, including the first director of the Economic Society, Jacobo de Villaurrutia (born in Santo Domingo), and the Italian-born José Rossi y Rubí. Postings within the empire could generate new loyalties. Nobody could doubt the commitment of Alejandro Ramírez, co-founder of the *Gazeta de Guatemala*, to the Society's project. Born in Spain, he stayed in Central America for two decades and married a local woman, but left in 1812, having been promoted to the post of superintendent of Puerto Rico.[54] Jordana Dym has also made the argument that the Society's newspaper formed a bridge between Spaniards and Creoles, and some residents of Central America certainly thought these debates of difference absurd.[55] One anonymous writer by the pen-name of *Guatemalofilo* admonished European residents to 'respect the Creoles', reminding them that 'your sons will tomorrow become part of this class which you now so disdain'.[56] Others were openly ambivalent about their identity, like the priest who noted 'I am a European; but I have lived more than 30 years of my life in this country . . . and I am not dissatisfied with the fate which has been given to me in this world.'[57]

---

[53] Silva, *Los Ilustrados,* 575–6, 646 also rejects this term for the Enlightenment in New Granada, pointing to its members' varied origins, but does maintain pride in one's regional origins as one of the group's identifying features.

[54] Catherine Poupeney Hart, 'Parcours journalistiques en régime colonial: José Rossi y Rubí, Alejandro Ramírez et Simón Bergaño', *El argonauta español* 6 (2009): 6–11.

[55] Jordana Dym, 'Conceiving Central America: A Bourbon Republic in the *Gazeta* (1797–1807)', in Gabriel Paquette (ed.), *Enlightened Reform in Southern Europe and its Atlantic Colonies* (Farnham: Ashgate, 2009), 99–118, at 102.

[56] 'Carta', *Gazeta* Vol. 1, no. 8 (3 April 1797), 60–4.

[57] 'Cartas del Cura de N.', *Gazeta,* Vol. 1, no. 26 (31 July 1797), 205. Antonio Croquer y Muñoz of Nunalco (Santiago de Nonalco) is the likely candidate for this pseudonym.

The *patria* that reformers worked towards improving, then, need not have anything to do with place of birth. The '*patria*' in the Economic Society's title was defined as 'Guatemala', that is the Spanish Audiencia or Kingdom of Guatemala, but a 'love of the *patria*' did not refer only to Central American betterment. In imperial Spanish ideology, what was good for Spain was automatically considered beneficial to all colonies, too, and that local improvement in the Spanish territories would benefit the improvement of the whole empire. Governments in Madrid and Guatemala City, the Economic Society, and the Consulado were therefore all assumed to participate in the same rhetoric of reform for the sake of public utility and public happiness. This made it politically possible to pursue a local patriotic agenda that prioritised just one part of the wider empire, a situation that had parallels in local improvement initiatives in the context of other empires in eighteenth-century Hungary, Ireland, and Scotland.[58] A more distinct localism also developed in the Kingdom of Guatemala. Several historians have suggested that this prepared the way for the construction of national identities during independence, but the links between such patriotism and later political ideas of a nation-state are tenuous.[59] For instance, patriotism could also simultaneously exist at the scale of an allegiance to 'American' rather than 'Guatemalan' territory.[60] While even long-term Spanish residents might have subscribed to localist sentiments in their reform activities, membership in the Economic Society did not necessarily signify emotional attachment to Central America, and certainly not the idea of a united Central America: the prolific contributor to the Guatemalan Economic Society Matías de Córdova was a signatory of the declaration of independence of Chiapas from Guatemala in 1823.[61] As Jordana Dym has demonstrated, Central American states did not 'emerge already formed from the process of independence'. Instead, political loyalties were divided between

---

[58] Richard Butterwick, 'Peripheries of the Enlightenment: An Introduction', in Richard Butterwick, Simon Davies, and Gabriel Sánchez Espinosa (eds.), *Peripheries of the Enlightenment* (Oxford: Voltaire Foundation, 2008), 10–13; Fredrik Albritton Jonsson, 'Scottish Tobacco and Rhubarb: The Natural Order of Civil Cameralism in the Scottish Enlightenment', *Eighteenth-Century Studies* 49, no. 2 (2016): 129–47.

[59] Luque Alcaide, *Sociedad Económica*, 170; Adolfo Bonilla Bonilla, 'The Central American Enlightenment: An Interpretation of Political Ideas and Political History' (PhD dissertation, University of Manchester, 1996), 16.

[60] 'Bleichmar, *Visible Empire*, 148; Gabriel Entin, 'El patriotismo americano en el siglo XVIII: Ambigüedades de un discurso político hispánico', in Véronique Hébrard and Geneviéve Verdo (eds.), *Las independencias hispanoamericanas* (Madrid: Collection de la Casa de Velásquez, 2013), 22–3; Monguió, 'Palabras e ideas', 452–7. Thurner, *History's Peru*, 86–8 distinguishes *país* from *patria* as being more oriented towards American soils. The words seem more interchangeable in the Guatemalan context.

[61] Catherine Poupeney Hart, 'Entre gaceta y "espectador": avatares de la prensa antigua en América Central', *Cuadernos de ilustración y romanticismo* 16 (2010): 11; 'Córdova, Matías de', in Flavio Rojas Lima (ed.), *Diccionario histórico-biográfico de Guatemala* (Guatemala City: Asociación de Amigos del País, 2004), 304.

municipality, state, and federation. Nation-states emerged along the lines of provincial divisions that had meant very little to colonial understandings of sovereignty.[62]

Colonial-era patriotism was therefore not geared towards independence, but nor was it territorially easily defined. For instance, as with reformers elsewhere in Spanish America, emotional language was often tied to descriptions of physical terrain and soils in a patriotic sentiment that might be linked to national territory, but the 'local' in Central America that reformers might show allegiance to was fragmented. 'Guatemala' as the capital and its surrounding provinces was a conceptual category for the Economic Society, but other provinces within Central America were less clearly defined, reflecting a fractured political and economic landscape. There were some important instances in the Economic Society's project that did hint at a growing spatialised view of an area or the Audiencia as a whole as a patriotic, economic, and geographical unit. In addition, specific locations could now be added to the catalogue of places that were being transformed by enlightened interventions. The construction of such geographies had many layers of meaning. The idea of landscapes embodying their future promise in archival evidence about their history, for instance, appears in many case studies in this book. Although these geographies did not configure nation-states *per se*, there were increasingly some territorial-patriotic spaces that post-independence reformers would be able to fill with new meanings. The case studies in this book show that whether defined as *patria*, *país*, *estas tierras*, *estos payses*, or simply 'Guatemala' (as in the case of the anonymous author *Guatemalofilo*), members of the Society chose a place within, or version of, this homeland as their target when they contributed suggestions, projects, or practical actions to its work.

If the later recasting of patriotic identities as national had to rest on concepts of 'profound emotional legitimacy', to use Benedict Anderson's term, it was not necessarily a geographical definition of homeland as much as enlightened actions which defined the patriot, a belief in the material consequences of interventions in the cultural landscape rather than 'Central America'.[63] While education and incentives might create 'useful' citizens or subjects, patriotism as a voluntary act deserved the highest praise. In Central America, emotive language was associated with the act of attempting to apply Enlightenment. Director of the Economic Society Jacobo de Villaurrutia, expressed these sentiments particularly clearly when he explained that he was 'talking from his heart' about the necessity of being a patriot and lauded the 'very fertile land' as well as knowledge of 'our soils, our climes'.[64] However, these ideas of

---

[62] Dym, *From Sovereign Villages*, quote from p. 5.
[63] Benedict Anderson, *Imagined Communities: Reflections on the Origin and Spread of Nationalism* (London and New York: Verso, 1991), 13.
[64] Real Sociedad, *Quinta junta pública*, 2–4.

patriotism and Enlightenment were rooted in the ideas of taking specific practical action. Elisa Martín-Valdepeñas Yagüe has highlighted a similar connection between patriotism as the efforts and enlightened actions for the case of the Madrid Economic Society's members, commenting also on the vaguely defined political space in which the Spanish patriots acted.[65] A patriot of the eighteenth century, as Gabriel Entin also explains, was expected to be selfless, pursue utility or progress for the public good, keywords that appear frequently in the Economic Society's works.[66] The Director of the Economic Society Villaurrutia himself contrasted action with words, which were of moderate use, since 'a patriot cannot be excited by speeches'.[67] One reformer was praised for his *actividad*, his willingness to take action, another for his 'active dedication [*zelo activo*]', while yet another was lauded for the 'patriotic eagerness of his dedication' relating specifically to the flax harvest.[68] I argue that this sense of embodied patriotism as defined by action is visible throughout the period. A Guatemalan patriotic space therefore emerged at the intersection of the political-economic logic that sought to create wealth from nature, specific interventions in the landscape that reformers believed to represent progress, and clearly defined localist epistemologies that reformers chose as practical rather than ideological solutions to problems of knowledge.

There were instances, even in generally Crown-loyal Central America, where local improvers were at odds with the Madrid government. Despite a general political and social congruence between elites on the two sides of the Atlantic, elites in the Americas generally conceived of the empire as a looser construct of composite monarchies, not colonies in the classical sense.[69] In the Economic Society's work, we can see the publication of a handful of political articles that may have raised suspicion in Madrid, as well as disagreements and perhaps misunderstandings leading to the suppression of the Society. The Economic Society was suspended indefinitely by a royal decree of November 1799, which took effect when it reached Guatemala in May 1800, and not re-established until 1811, although its newspaper continued to be published. No precise explanation was given for its suppression, but the *Gazeta* had previously published a number of articles that may have rankled Crown authorities: a series of articles that flattered the

[65] Elisa Martín-Valdepeñas Yagüe, 'Del amigo del país al ciudadano útil: una aproximación al discurso patriótico en la Real Sociedad Económica Matritense de Amigos del País en el Antiguo Régimen', *Cuadernos de Historia Moderna, Anejo XI* (2012): 23–47, at 28–32.

[66] Entin, 'El patriotismo americano', 21.

[67] 'Sociedad', *Gazeta*, Vol. 3, no. 106 (27 May 1799), 37.

[68] *Gazeta*, Vol. 1, no. 14 (15 May 1797), 111; Real Sociedad, *Quinta junta*, 8; 'Junta extraordinaria' 88, 24 October 1799, HSA, HC 418/563.

[69] John H. Elliott, 'A Europe of Composite Monarchies', *Past & Present* 137 (1992): 48–71; Cañizares-Esguerra, *How to Write the History*, 63–4. Intellectual tensions between reformers and Spain are discussed in Chapter 3.

governments of Russia and Sweden, rather than Spain; extracts of Francisco de Clavijero's censored work *Storia Antica de Messico*; and an essay critical of Spanish economic policy.[70] The paper's first editor, Alejandro Ramírez y Blanco, made some influential enemies, for instance by promoting the cause of the liberalisation of trade.[71] Another editor, Simon Bergaño y Villegas, wrote articles against ecclesiastical power and in favour of teaching in the vernacular instead of Latin in the *Gazeta de Guatemala*, which brought him into trouble with a conservative archbishop, and penned an article advocating drastic reforms of the colonial system. In 1808, he was accused of inciting an uprising of artisans, as well as showing disloyalty towards the king. After this, he was jailed in Spain for two years, although he was later freed and allowed to return to America, founding several newspapers in Havana.[72] The clashes between the *Gazeta* and Madrid, though on the whole they were far from suggesting any revolutionary political attitudes, allow us to conceive of there being some intellectual distance between enlightened reformers in Guatemala and Spain. Useful knowledge and economic improvement were not always interpreted in the same way on both sides of the Atlantic.[73]

Even during the Napoleonic invasion of Spain in 1808, Guatemalan officials generally remained Crown-loyal. Guatemala was quick to declare its loyalty to the deposed king in 1808, and any separatist movement was largely suppressed after 1811 by the new captain-general, José Bustamante y Guerra, who also did his best to slow down the implementation of the 1812 Cádiz Constitution. The restoration of King Ferdinand VII to the Spanish throne in 1814 largely silenced the liberal faction, and Spanish rule appeared firmly re-established. Nevertheless, the Napoleonic invasion had led many across Spanish America to question the colonies' relationship to Spain and the monarchy. In Guatemala, this was also a period when new political ideas and groups were formed which, especially in Guatemala City, were discontented with the status quo. Across Central America, between 1811 and 1820, old grievances were exacerbated in disagreements between representatives of the provinces and Guatemala City,

---

[70] 'De la Rusia', *Gazeta*, Vol. 1, no. 3 (27 February 1797), 17–8; 'De la Suecia y Dinamarca', *Gazeta*, Vol. 1, no. 10 (17 April 1797), 73–4; 'Extracto de un libro que no se ha escrito', *Gazeta*, Vol. 2, no. 54–7, 64, 74–5, 81 (1798). Shafer, *Economic Societies*, 215; John Browning, 'El despertar de la consciencia nacional en Guatemala', in Luján Muñoz and Zilbermann de Luján (eds.), *Historia general de Guatemala* (Guatemala City: Asociación de Amigos del País, Fundación para la Cultura y el Desarrollo. Guatemala: Asociación de Amigos del País, 1995), 634. Luque Alcaide, *La Sociedad Económica*, 159–60. Maldonado Polo, *Huellas de la razón*, 401–3; Bonilla Bonilla, 'Central American Enlightenment', 202.

[71] José Toribio Medina, *La imprenta en Guatemala, 1660–1821* (2nd ed. Guatemala City: Imprenta Nacional de Guatemala, 1960), 310.

[72] Rojas Lima, *Diccionario historico-biográfico de Guatemala*, 187. Christophe Belaubre, 'Bergaño y Villegas, Simon', *Diccionario AFEHC*.

[73] Bonilla Bonilla, 'Central American Enlightenment', 168; Silva, *Ilustración en el virreinato de Nueva Granada*, 79–147; Dym, 'Conveiving Central America', 111.

with Nicaraguans, Costa Ricans, and Salvadoreans protesting about being disadvantaged economically and treated unfairly in matters of taxation, sowing some of the seeds for the political debates of the independence period. Audiencia President Antonio Gonzáles Mollinedo now also used the relative power vacuum of the Cádiz *Cortes* to re-establish the Society in 1811. Arguing that nobody could deny the Society's utility and that it was only intended to be suspended, not abolished, in the first place, he claimed that Ferdinand VII would surely approve it if given the chance.[74] Yet again, while this did not show the Society as a subversive entity, it demonstrated its appeal to governance and projects of progress designed from within Central America.

On a day-to-day basis, enlightened reformers in the late eighteenth century found far more pressing matters to occupy their minds than plotting revolution. The memory of the destruction of the old capital city Santiago in 1773 and construction of the new city of Nueva Guatemala de la Asunción loomed large in the capital's recent memory. By the 1790s, the process of constructing the new city had barely been completed.[75] Several of the men who supported the creation of an Economic Society did so with explicit reference to the destruction wrought by earthquakes: President José Domás y Valle thought that an Economic Society would be useful, because 'twenty years on from the move [of the capital], nobody has yet taken the first step' in sorting out some fundamental problems of political economy.[76] He referred to the regulation of trades in particular, but it is also clear that Guatemala, to him, was a project in progress. Another essay asking for the establishment of the Society pointed out that the move of the capital after the 1773 earthquake meant that they needed knowledge (*facultades*) now 'more than ever', but the author saw a disappointing lack of it in the population.[77] Beyond that, Guatemala had its share of economic and social challenges in this period. They were not uniform: the local and regional nature of much commerce meant that local economies, such as Quetzaltenango's textile industry in the late eighteenth century, were sometimes unscathed by larger trends.[78] However, a general local perception of deterioration of economic and societal circumstances chimed in with a broader

---

[74] Shafer, *Economic Societies*, 224.

[75] Cristina Zilbermann de Luján, 'Destrucción y traslado de la capital. La Nueva Guatemala de la Asunción', in Luján Muñoz and Zilbermann de Luján (eds.), *Historia general de Guatemala*, Vol. iii, 206.

[76] 'Presidente de Guatemala sobre establecer una Sociedad Económica', 1795. AGI, Estado, 48, N.7, 5v.

[77] 'Discurso sobre las utilidades que puede producir una Sociedad Económica en Guatemala', 1795. AGI, Estado, 48, N.7, 1v.

[78] Jorge Gonzáles Alzate, 'State Reform, Popular Resistance, and Negotiation of Rule in Late Bourbon Guatemala: The Quetzaltenango Aguardiente Monopoly, 1785–1807', in Dym and Belaubre, *Politics, Economy and Society in Bourbon Central America*, 129–55; Wortman, *Government and Society in Central America, 1680–1840* (New York: Columbia University Press, 1982), 15–16, 41–90.

narrative of decline across the empire.[79] In addition, the smallpox epidemics of 1780–1 and 1794–6 and the typhus epidemics throughout the period constituted a humanitarian crisis (and provoked a comprehensive smallpox vaccination campaign, which was led and supported by key Economic Society members).[80] They disrupted what was already perceived as a decline in agriculture and trade. Falling prices for indigo (Central America's main export in the eighteenth century) on the world market and wars with Britain that interrupted most legal transatlantic trade contributed to an economic depression in parts of the kingdom.[81] Locust plagues that affected the entire region in the 1770s and again between 1798 and 1805 further slowed agricultural production, while in 1796 the important cattle industry was affected by a mysterious epidemic that killed thousands of cows.[82] Central America's enlightened reformers were determined to find solutions to some of these problems and turn the narrative of decline into prosperity. Enlightenment action represented a fitting response to urgent problems.

Sources and Chapter Outline

The events of this book cover the period between 1784 (when the government in Guatemala City started investigating a set of ruins found near the Chiapas hamlet of Palenque, and the year that archbishop Francos y Monroy concluded a visit of Guatemala by recommending that an Economic Society for its improvement be established), and 1838 (when the federal republic of independent Central American states fell apart). Most case studies come from the years in which the Economic Society's schemes were most pronounced: between 1796 and 1806, with another flurry of activity in the 1810s. In the wake of independence from Spain in 1821, Central American states banded together in a federation known as the 'United Provinces of Central America'. Only the last chapter of the book covers the period of 1821 to 1838, a time in which the legacy of the colonial-era reformism became clear. This is not a political history of Central America's independence processes, but the last chapter demonstrates the surprising resilience of the colonial era's geographical imagination and information networks through changed political circumstances. The criteria for useful and practical knowledge established by the Economic Society had led to a reconsideration of the meaning and potential of Central American

---

[79] Cañizares-Esguerra, *Nature, Empire and Nation*, 98–106.

[80] Martha Few, *For All of Humanity*.

[81] Belzunegui Ormázabal, *Pensamiento económico y reforma agraria*, 160–203; Maldonado Polo, *Huellas de la razón*, 305–48; Wortman, *Government and Society,* 180–94.

[82] Martha Few, 'Killing Locusts in Colonial Guatemala', in Martha Few and Zeb Tortorici (eds.), *Centering Animals in Latin American History* (Durham, NC: Duke University Press, 2013), 62–92. Cattle: Nueva Guatemala, 2 March 1796. AGI, Guatemala, 887.

agricultural, commercial, and environmental geographies, leaving a lasting impact on the construction of Central America's new nations and their dealings with other states.

This study relies on manuscript and printed sources from archives in Guatemala, Spain, Britain, and the United States, particularly from the Archivo General de Indias in Seville, the Archivo General de Centroamérica in Guatemala City, and a small but notable collection of Economic Society papers, long disappeared from the view of historians, from the Hispanic Society of America in New York.[83] The authors of these documents were members of the Economic Society, the Consulado de Comercio, priests, engineers, and officials of the Crown, but also recorded the voices of other actors including Indian farmers or villagers. Official manuscript reports, private correspondence, printed treatises, and pamphlets reveal some of the communication pathways that formed the basis of the acquisition of Central American scientific knowledge. The correspondence archive of the Economic Society, partially preserved in the Archivo General de Centroamérica, and articles printed in its newspaper, the *Gazeta de Guatemala* (1797–1808), allow for an analysis of the correspondence networks that formed the basis of scientific enquiries in Central America. Sources that are broadly geographical, agricultural, or natural-historical, from instructions for growing plants to road surveys, carry particular weight, since they often contain subjective and experiential accounts of encounters with the landscape. City and rural government officials' reports in turn reflect how questions of day-to-day governance were influenced by perceptions of geography and landscape.

Chapter 1 shows the varied ways in which colonial administrative traditions approached the study of landscape. Through the case study of the discovery of Maya ruins near Palenque in Chiapas in the 1780s, it examines the way officials and scholars within the Guatemala City government understood and recorded information about man-made and natural landscapes. It argues that concerns about Central American environments, including the threat of volcanic eruptions and earthquakes, informed the explorations of this site, illuminating the extent to which concerns about natural factors influenced day-to-day understandings of landscapes and the practice of governance in Central America. These practices in turn established powerful models for the work of later reformers. The methodologies of information-gathering and the debates about the site also highlighted the idea of economic improvement through harnessing existing natural resources. This was central to the economic thought

---

[83] On the latter, see Wendy Kramer, George Lovell, and Christopher Lutz. 'Pillage in the Archives: The Whereabouts of Guatemalan Documentary Treasures', *Latin American Research Review* 48, no. 3 (2013): 153–67. Early papers of the Economic Society are part of the Hiersemann collection this article refers to, alongside documents such as draft maps for bishop Cortés y Larraz's *Visita*, which will be edited and published by George Lovell.

of the Bourbon reforms, which was clearly widespread across Guatemala at this time. Officials, engineers, and scholars 'read' from the landscape: Palenque's landscapes were seen throughout the period as key to questions of its past glory and future potential.

Chapter 2 contrasts these relatively rigid pathways of bureaucratic information-gathering with the novel pathways of communication that the newly founded Economic Society provided. It demonstrates how the Society built up a network of information exchange through correspondence as well as the publication of a newspaper, the *Gazeta de Guatemala*. These networks were designed to extend the reach of the Society from urban contexts into rural ones and had an active purpose: members and their associates were exhorted to grow, collect, and harvest economically useful plants. Reports from members over two decades show that, on a small scale at least, this succeeded, leading to an exchange of useful plant material. The varied social position and geographical locations of the newspaper's subscribers also made its pages an exceptional forum for debate, creating a nascent 'public sphere'. The networks even extended beyond the Audiencia's borders, placing Central America in a context of global economic botany and scholarship. One manifestation of the extension of the Society's practical network was that a member imported a collection of 'exotic' plants and seeds from Sumatra and Jamaica, plants which were then grown and harvested in Central America with some success.

Chapter 3 explores the promises and contradictions inherent in the information drawn from these local and global knowledge networks. There were tensions that were never quite resolved between the production of locally relevant knowledge that rejected theoretical approaches and a global intellectual movement that praised universal knowledge. The Economic Society responded to this by carefully negotiating the sources of knowledge which it received from its networks, especially on the topics of natural history and medical botany, and building up its own epistemologies and definitions of practical Enlightenment that made the local applicability of any information the ultimate test of its value. Frameworks of knowledge with universal aspirations, such as Linnaean taxonomy, were not welcome when local descriptions would be more translateable within Central America. I argue that these stubbornly local conceptualisations of knowledge became problematic when a comparison with other places was required, for instance in the context of attempting to export plants from Guatemala to other places, and in debating the merits of plantain trees with scholars in other parts of the empire.

Chapter 4 shows that tensions about the applicability of knowledge were never more pronounced than when it came to geographical information. The *Gazeta* newspaper used its networks of knowledge to attempt to create a new 'Description of Guatemala' that would not just counter erroneous claims about the Americas peddled by some European philosophers, but also critically examined existing sources and formats of geographical knowledge. They rejected geography as a universal science that related places across the globe to each other, and instead prioritised information from current statistics as well as local historical archives. Individual reformers also contributed even more practical geographies that reported their own experience of travel. Geography and chorography were considered useful only in so far as it would help to increase trade and prosperity. It followed that securing transport connections through road- and harbour works that would allow for an exploitation of natural wealth was the most important application of such knowledge. Reformers attempted to rewrite the geography of the Audiencia's trade routes, and thought of their projects as integrating specific places more firmly into the geography of the region.

Chapter 5 argues that preoccupation with travel, topography, and geography merely formed the basis for even more ambitious projects that did, however, show the limits of what practical patriotism might achieve. When combined with a providential belief in the potential of the land, the application of geographical and botanical knowledge to the countryside meant that spaces which had hitherto been considered 'empty' or 'wild' could be filled with new meaning. Reformers were concerned with the role of people (Indians, but also Europeans, Africans, and Caribbeans of African descent, as well as enslaved people) in managing landscapes. They increasingly discussed questions of what we might call 'biopower' after Foucault, conceiving of labour and the management of the population as a resource. In this, reformers paid particular attention to the possibility that humans might influence environments in more profound ways than just by building roads. They hoped that human errors that had made Caribbean environments 'unhealthy' in the past could be reversed by building better-ventilated settlements, or regulating military barracks to help soldiers behave like agricultural settlers and make this land productive.

Chapter 6 demonstrates that projects for managing the natural world established in the colonial period enjoyed continued relevance after independence from Spain in 1821, as the new states of the federal republic of Central America embarked on renewed efforts to study nature. Focusing on the case study of Guatemala, the new republic maintained 'useful patriotism' as an ideal of citizenship and re-established colonial institutions such as the Economic Society. New transcontinental networks of scientific knowledge, centred on London now in parallel to Guatemala City, expanded the scientific worlds of Guatemalans and resident foreigners in a dramatically different geopolitical context. Despite the economic importance of Britain in Central America at this time, and despite the often negative opinions that British scientists and governments expressed of Spanish American (colonial and independent) knowledge,

British investors and geographers were heavily reliant not just on the new maps of the independent state, but colonial geographical material from Spanish archives too. Colonial-era reform projects, often through a material legacy of plans in an archive, had laid the groundwork for imagining a Central America built on idea of useful knowledge, patriotic dedication, and connected scientific networks around the world.

1   Landscape, Ruins, and Governance

Between 1784 and 1788, officials of the Spanish government in Guatemala City faced a conundrum: how to record, categorise, and study a group of stone buildings, apparently of great antiquity, that had been discovered in the Chiapas rainforest near the hamlet of Palenque? Rather than an abstract academic activity, this question went to the heart of how officials of the colonial bureaucracy related to Central American landscapes, and how local intellectuals and (often indigenous) residents of rural villages interacted with them. Archaeological sites were a novel topic to most of the officials who encountered them, but they did not seek to establish radically new forms of knowledge, and indeed often ignored or misunderstood indigenous or other local historical knowledge. Instead of trying to be historical scholars, they relied on Spanish bureaucratic traditions that provided established empirical methods for recording and interpreting landscapes, but were also undergoing changes in the eighteenth century to increasingly incorporate the Bourbon state's concern with landscapes as a source for natural wealth. Responding to the challenge of explaining the hitherto unknown, the bureaucracy's explorations of the Maya city of Palenque drew on the knowledge of an architect, military engineers, and other colonial officials who were considered suited to archaeological exploration. Archaeological exploration therefore became a showcase of geographical and technical knowledge as well as bureaucratic epistemology.

The explorations provide a framework for understanding the role of geography and natural risk in the day-to-day governance of Spanish America. Engineers and bureaucrats used the language of prospecting for natural resources to establish links between landscape, ruins, and economic prosperity. Their attempts to glean meaning not just from the ruins, but the nature and topography surrounding them, reflected the Bourbon reform era's concern with the land and its productions which permeated questions of governance throughout the late eighteenth century. They cited the landscapes in and around Palenque as potential witnesses to the city's demise, but were also interested in the commercial possibilities for the future that these landscapes might embody. Their approaches are also a window onto Enlightenment-era notions about the potential for improvement based on the landscape, and even a certain geographical determinism. Most likely influenced by earthquakes and volcanic eruptions of the 1770s that had destroyed the old capital city and threatened the new one, the Palenque excavations exemplify the idea that the natural world might shape human history. Officials presented man-made environments alongside natural ones, merging long-established administrative traditions with contemporary concerns about the relationship between land, natural forces, and societies. Everywhere they looked, bureaucrats, bishops, and civic reformers could see ruins: in the old capital destroyed by a volcano, in churches destroyed by earthquakes, in the perceived commercial desolation of certain provinces, and in Palenque's archaeological ruins. The language of improvement and enlightened renewal as tied to the land, and knowledge about the land, provided another narrative.

The language of improvement contained in the Palenque explorations, marked by concerns specific to Guatemalan landscapes, would go on to shape the narrative about the government's and inhabitants' relationship with Central American nature for years to come. In the decades that followed, these practices of communication and correspondence, as well as the conceptualisation of rural landscapes that appeared in these documents, became the templates that enlightened reformers built their strategies of knowledge creation on. While later reformers would counter the bureaucracy's relatively rigid correspondence circuits with a far less structured periodical press, they would continue to rely on or adapt for new purposes these Bourbon reformers' formats of surveying, geographical description, and natural history. The Palenque explorations also show that a strong relationship between geography, history, and the use of archives had been established by the 1780s that remained influential. More self-consciously 'enlightened' and 'patriotic' reformers driven by the aim of recording specifically Central American landscapes would later use these epistemologies and correspondence practices as a blueprint for creating and communicating their own Enlightenment knowledges.

Bureaucracy, Landscape, and Ruins

The methods of information-gathering employed by the colonial state's officials embodied long-established approaches to understanding landscapes, and people's interactions with them. The Spanish state had always been interested in collecting information about its colonies in order to effectively control and govern them from a distance, as well as to exploit natural resources. The most prolific examples of such information-gathering are the botanical expeditions that visited the Americas during the last third of the eighteenth century, but since the sixteenth century, secular and religious administrators had undertaken a variety of systematic explorations that covered the geography, economic productions, life, and customs of particular administrative areas, known as *visitas* (visits), *visitas pastorales* (pastoral visits), or *relaciones geográficas* (geographical reports). Similar information was found in a range of less specifically titled government reports within the loose category of *informes* (reports). By harnessing the knowledge of administrators and priests scattered across small towns who were asked to respond to these questionnaires, the reports also at least indirectly drew on the knowledge of local populations.[1] The breadth of possible topics encompassed, in addition to observations about humans and their towns, all parts of the landscape which could be observed or collected, falling into a broad definition of 'natural history'. That included antiquities.[2] To the representatives of the state in Guatemala City, Maya ruins in the late eighteenth century could appear as just one – albeit puzzling and extraordinary – part of the landscape. For instance, when the military engineer Luis Diez Navarro visited the ruins of Copán in Honduras in the process of surveying for a road in 1758, he employed a style of 'thick description' that we might call, to use Ricardo Padrón's term, 'prose cartography'.[3] A passage about there being 'large statues, five or six feet in height' seemed to exist primarily to add depth to the description of the valley he was travelling in, a valley in which there were indications of a past 'King of the Indian Gentiles'. The engineer recorded the ruins, in the same manner as he did villages or rivers, as part of the landscape.

This integrated view was not accidental, but encouraged by the Spanish bureaucracy's traditions – a testament to the malleability of categories that

---

[1] For *relaciones*, see Pilar Ponce, 'Burocracia colonial y territorio americano: Las relaciones de Indias', in Antonio Lafuente and José Sala Catalá (eds.), *Ciencia colonial en América* (Madrid: Alianza, 1992), 29–44; Francisco Solano, *Relaciones geográficas del arzobispado de Mexico, 1743* (Madrid: CSIC, 1988); Robert West, 'The Relaciones Geográficas of Mexico and Central America, 1740–1792', in Howard Cline (ed.), *Handbook of Middle American Indians* (Austin, TX: University of Texas Press, 1972), Vol. xii, 396–452; Jorge Luján Muñoz, *Relaciones geográficas e históricas del siglo XVIII del reino de Guatemala. Vol. i. Relaciones geográficas e históricas de la década de 1740* (Guatemala: Universidad Del Valle, 2006). For natural history traditions, see Antonio Lafuente and Leoncio López-Ocón, 'Scientific Traditions and Enlightenment Expeditions', in José Saldaña (ed.), *Science in Latin America. A History* (Austin, TX: University of Texas Press, 2006), 123–50; Antonio Barrera-Osorio, *Experiencing Nature: The Spanish American Empire and the Early Scientific Revolution* (Austin, TX: University of Texas Press, 2006), 3; see also Catherine Poupeney Hart, 'Entre historia natural y relación geográfica: el discurso sobre la tierra en el reino de Guatemala', in Ignacio Arellano and Fermín del Pino (eds.), *Lecturas y ediciones de crónicas de Indias* (Madrid: Iberoamericana, 2004), 441–60.

[2] Philip Kohl, Irina Podgorny, and Stefanie Gänger, *Nature and Antiquities: The Making of Archaeology in the Americas* (Tucson, AZ: The University of Arizona Press, 2014). See also Lisa Trever and Joanne Pillsbury, 'Martínez Compañón and His Illustrated "Museum"', in Daniela Bleichmar and Peter Mancall (eds.), *Collecting Across Cultures* (Philadelphia, PA: University of Pennsylvania Press, 2010), 236–53; José Alcina Franch, *Arqueólogos o anticuarios: historia antigua de la arqueología en la América española* (Barcelona: Serbal, 1995), 12; María Calatayud Arinero, 'El real gabinete de historia natural', in Sellés et al. (eds.), *Carlos III y la ciencia de la Ilustración*, 274. For British parallels, see Philippa Levine, *The Amateur and the Professional: Antiquarians, Historians and Archaeologists in Victorian England, 1838–1886* (Cambridge University Press, 1986), 16–20.

[3] Padrón, *Spacious Word*.

helped the state to create seamless records of landscapes, and often its people within. Such flattening of places, cultivated and uncultivated landscapes, and histories, without taking into account contemporary indigenous uses was symptomatic of the 'colonial gaze' of government officials.[4] As Severo Martínez Peláez points out in his discussion of the seventeenth-century chronicle *Recordación Florida*, such a detached perspective signified the construction of an idealised landscape that allowed for the work of Indians in cultivating it to be bracketed out and discussed separately.[5] It was, however, also some acknowledgement within the bureaucracy of a cultural landscape that might indiscriminately be described as natural wilderness by some city-dwellers, but within which detailed records of human action upon it could be found and recorded 'on the ground'. The format in which reports were recorded did not just shape descriptions of the landscape, but also the way that bureaucrats perceived them, marshalling potentially ground-breaking discoveries into a language of economic reform and prosperity that was familiar in its bureaucratic formats, but new in the sense of being tied to the logic of the Bourbon reforms. Reports such as Diez Navarro's, which saw monuments and landscape in a fairly disinterested and 'harmonic' manner, could be less easily upheld when questions about the authorship and significance of ruins became politicised in the nineteenth century, but did not disappear entirely.

The chronology of events and official correspondence around the 'discovery' of Palenque is well documented by historians, including Paz Cabello Carro, Miruna Achim, and Jorge Cañizares-Esguerra.[6] The reports show the array of actors who might be involved in knowledge creation about the Central American landscape, although of course their voices were not recorded equally: a group spanning the elite of Guatemalan society as well as humble

---

[4] Marie Louise Pratt, *Imperial Eyes: Travel Writing and Transculturation* (New York: Routledge, 2008); see pp. 130–1 for Humboldt's contradictory conflation of nature and ruins. See David Arnold, *The Tropics and the Travelling Gaze. India, Landscape and Science* (Seattle, WA: University of Washington Press, 2006), 76–80, for a parallel from British India. Ruins also appeared as part of a (politicised, national) landscape in late nineteenth-century Mexico: Cañizares-Esguerra, *Nature, Empire and Nation,* 142.

[5] Martínez Peláez, *Patria del criollo*, 106.

[6] A thorough discussion of the Spanish reports and additional archival sources in the context of the epistemologies of Spanish history-writing is Cañizares-Esguerra, *How to Write the History*, 321–34; also Achim, *From Idols to Antiquity: Forging the National Museum of Mexico* (Lincoln, NE: University of Nebraska Press, 2017), 95–129. Transcripts of the correspondence from Spanish archives in Paz Cabello Carro, *Política investigadora de la época de Carlos III en el área maya. Según documentación de Calderón, Bernasconi, Del Río y otros* (Madrid: Ediciones de la Torre, 1992); further primary sources published in Manuel Ballesteros Gabrois, *Nuevas noticias sobre Palenque en un manuscrito del siglo XVIII* (Mexico: UNAM, 1960), 23–40; Ricard Castañeda Paganini, *Las ruinas de Palenque: su descubrimiento y primeras exploraciones en el siglo XVIII* (Guatemala: Tipografía nacional, 1946); Dolores Aramoni Calderón, 'Los indios constructores de Palenque y Toniná en un documento del siglo XVIII', *Estudios de Cultura Maya* 18 (1991): 417–38, at 425–33; Carlos Navarrete, *Palenque, 1784: el inicio de la aventura arqueológica maya* (Mexico City: UNAM, 2000).

rural priests, administrators, and Indian informants, the latter often remaining nameless in Spanish reports. Indigenous people maintained traditions and histories about abandoned cities such as Palenque, and Spanish officials had also been to some extent aware of the existence of ruins scattered across the territory of the Kingdom of Guatemala and the Yucatán peninsula. There had been at least two previous local explorations by Spanish officials of Palenque, but no further actions had been taken at the time.[7] In 1784, the local priest's breathless reports of a marvellous ancient city were received with enthusiasm by the highest-ranking official in Guatemala's colonial hierarchy, Audiencia president José Estachería, who commissioned a series of reports. Estachería first asked the Spanish magistrate of the village of Palenque near the ruins to confirm the existence of the city and send some more information. The president then sent an architect working in Guatemala City, Antonio Bernasconi, to compile a more reliable report in 1785. Soon after he had completed this, Bernasconi died suddenly. President Estachería, prompted by orders from the Madrid government and a request for more information from the royal cosmographer, Juan Bautista Muñoz, had to commission a follow-up report from a new team, consisting this time of a military officer, the artillery captain Antonio del Río, and a professional draughtsman, Ricardo Almendáriz. The engineer and the draughtsman produced a new report by 1787, which was eventually published in London in 1822 and made the ruins world-famous.[8]

The reports obeyed a particular bureaucratic logic with regards to exploring landscape and followed established patterns of communication. They were part of a system in which documents were accorded certain power, and strict procedure governed the handling of important correspondence. Ángel Rama's notion of an inextricably close relationship between power and writing in the Spanish empire has recently been developed further by historians such as Sylvia Sellers-García and Kathryn Burns. They have shown how the very act of creating reports, copying them, and sending them to the correct recipient in the right geographical location was a vital part of the act of governing.[9] There was power in recording knowledge, and an understanding that hierarchy and status

[7] 1784 report: Roberto Romero Sandóval 'Viajeros en Palenque, siglos XVIII y XIX: un estudio histórico a través de su bibliografía', *Boletín del Instituto de Investigaciones Bibliográficas* 2, no. 1 (1997): 9–40, at 11. Previous explorations: Aramoni Calderón, 'Los indios contructores', 418; Ballesteros Gabrois, *Nuevas noticias,* 23; Alcina Franch, *Arqueólogos o anticuarios*, 84.

[8] The 1822 publication was orchestrated by Félix Cabrera, an Italian living in Guatemala in the 1780s who published Del Río's report alongside his own work, the *Teatro Crítico Americano*, a fanciful speculation about the Mesoamerican past: Antonio del Rio, *Description of the ruins of an ancient city: discovered near Palenque, in the kingdom of Guatemala, in Spanish America* (London: H. Berthoud, and Suttaby, Evance and Fox, 1822).

[9] Ángel Rama, *The Lettered City*, edited and translated by John Chasteen (Durham, NC and London: Duke University Press, 1996 [1984]), esp. 29–35; Kathryn Burns, *Into the Archive: Writing and Power in Colonial Peru* (Durham, NC and London: Duke University Press, 2010); Sellers-García, *Distance and Documents*, 147–54. Vicente Cortés Alonso, 'La antropologia de

controlled the flow of this information. It is therefore unsurprising that the documents compiled by these explorations travelled along prescribed pathways within the bureaucracy, finally arriving at the Court of the Indies (*Consejo de Indias*) or the king in Madrid. Having been copied at numerous stages along the way, they would eventually be deposited in an archive.[10] Officials both in Central America and Madrid sometimes seemed to privilege adherence to formalities and careful documentation over the dissemination of content. For instance, once the drawings from the Del Río expedition had arrived in Madrid, the Minister for Justice of the Indies, Antonio Porlier, ordered that they be forwarded to Juan Bautista Muñoz, the royal cosmographer, who might use them to illustrate his grand *History of the Indies*. The official who received the order objected, since this would mean that the file (*expediente administrativo*) would be disrupted. Porlier acknowledged this objection as valid. The originals were sent to Muñoz a whole year later, after copies of the drawings had been made for the archives and official use.[11] Such attention to procedure and formality might leave us with the impression of a large, well-co-ordinated enterprise, but it also had the effect of delaying the dissemination of knowledge, concealing the work of participants in this larger scheme from each other, and most likely of ignoring local knowledge in favour of the methodologies that the governor's explorers were more familiar with.[12]

The men who were sent to study the ruins in the 1780s were well aware that their work took place in a relatively rigid bureaucratic context, which was both long-established and undergoing changes to reflect the priorities of Bourbon governments. One of the governor's emissaries, Del Río, followed established traditions when he renounced any personal interest that he may have had in the matter of the origin of the ruins, despite occasionally digressing into wild speculation. He was tempted to explore other, more recently built 'ruins of Indian kings' which he had heard were nearby, but quickly noted somewhat ruefully: 'but it seems to me that I have strayed a little from the object of my commission, to which I should limit myself, I have sacrificed its literal observance'. His tone was remarkably similar to the protestations of other officers of the Crown who wanted to make sure that they were not overstepping the boundaries of their office. For instance, a few years later a regional governor

América y los archivos', *Revista Española de Antropología Americana* 6 (1971): 149–78, at 161, provides a useful chart of bureaucratic and archival pathways.

[10] On the creation of a centralised Archive of the Indies, see Francisco de Solano, 'El Archivo General de Indias y la promoción del americanismo científico', in Sellés et al, *Carlos III y la ciencia de la Ilustración*, 278–9; Paquette, *Enlightenment, Governance, and Reform*, 47–55.

[11] Cabello Carro, *Política investigadora*, 41–2.

[12] For instance, it is unclear whether Captain Del Río was aware that the instructions he received from Estachería originally came from Juan Bautista Muñoz, the royal cosmographer: 'Real Orden al Presidente de Guatemala (Estachería)', El Pardo, 15 March 1786, in Cabello Carro, *Política investigadora*, 126.

who was shown a potentially valuable mineral felt unable to investigate the place where it had been found, since it was in the neighbouring jurisdiction rather than his own. This prevented him from going 'to survey it all in a clear manner'.[13] As historians have repeatedly pointed out, Spanish administrators in the Americas had leeway in the interpretation and implementation of orders – often contradictory – sent from Spain. Yet documentation of a conspicuous observance of rules and directives was a common theme in official correspondence.[14] Del Río had a series of detailed questions to answer, which included instructions to collect specimens. The governor had also asked for 'pieces of cement, stucco, etc' from the site, and Del Río promptly hacked off 'the head of … figure number 8 and the foot and leg of … number 11' to send alongside his report – a tragedy from the point of view of the site's cultural conservation, for Del Río an opportunity to demonstrate the quality of the stonework.[15] These structures of reporting, both bureaucratic and empirical, textual and material, formed the basis of recording information about landscapes across Spanish America, and therefore for constructing their meaning. As Paula de Vos has pointed out, traditional empirical bureaucratic frameworks alongside a novel emphasis on collection of physical specimens marked a new Spanish approach to nature in the eighteenth century, one which was tied not just to curiosity but to political economy.[16] The artwork which Del Río sent to Spain was perhaps not considered 'useful' in the sense of an exploitation of resources, but it obeyed the same framework and would have its place in the imperial cabinet of natural history.

'A Man of Dust and Mud'

Although antiquities had crept into the list of governors' priorities all over the empire by the late eighteenth century, explorers of Palenque were not gentlemen-scholars. Instead, they were architects, soldiers, and professional draughtsmen who went about this antiquarian mission much like any other assignment. The initial reactions to the ruined city therefore have the potential to tell us as much about Bourbon bureaucrats' geographical practices as their historical ones. In practical terms, those who explored ruins in 1780s Guatemala at the behest of the government made little distinction between

---

[13] 'Informe rendido por el Alc.e May.r de Suchitepeques, don J Rossi y Rubí', 8 July 1796. AGCA, A3.9, Leg. 158, Exp. 3084, fol. 23.

[14] John Leddy Phelan, 'Authority and Flexibility in the Spanish Imperial Bureaucracy', *Administrative Science Quarterly* 5, no. 1 (1960), 47–65; Peter Christopher Albi, '*Derecho Indiano* vs. the Bourbon Reforms: The Legal Philosophy of Francisco Xavier de Gamboa', in Paquette (ed.), *Enlightened Reform in Southern Europe*, 233–7.

[15] 'Informe de Antonio del Río', Palenque, 24 June 1787, in Cabello Carro, *Política investigadora*, 135–8.

[16] De Vos, 'Natural History', 210.

this and other enterprises related to reconnaissance of the land. At every turn, they reflected the idea that even if they had no specific conceptual or personnel solutions to the mysterious ruined buildings, they were prepared to deal with this landscape much as they had with other physical environments in Central America. Rather than curiosity about the site's potential indigenous histories, they employed models drawn from established interactions between government and landscape, that is road-building, military mapping, and dealing with the aftermath of natural disasters, to make sense of Palenque. These field-based activities stood in contrast to the work of Mexican intellectuals José de Alzate y Ramírez and Antonio León y Gama, who in the 1790s theorised about the antiquities of New Spain, but focused on the interpretation of particular artefacts and their meaning for the contemporary Creole elite rather than charting and excavating whole sites themselves.[17] The men commissioned by the president of Guatemala with the exploration of the Palenque ruins had little in common with these scholars.

The men whom Estachería initially singled out as suitable for further explorations of the ruins after the architect Bernasconi's death were military engineers. As Irina Podgorny has observed, 'how to dig, how to record, draw up plans, how to take measures – these problems were left to the engineers and surveyors'.[18] However, the two engineers whom Estachería thought most qualified for the task were otherwise occupied with work on harbour fortifications.[19] Estachería now settled on the artillery captain Antonio del Río for the third exploration of the ruins in 1786. Del Río at least had a military background, in a clear attempt by the president to incorporate the specific professional skills of a military man (if not a military engineer) into the exploration report. In addition, the draughtsman Ricardo Almendáriz accompanied Del Río to create maps and architectural drawings. While the images and maps sent to Spain along with the final expedition report preserved the simplicity of the original drawings, they were 'clean copies' which an engineer working in Guatemala by 1788, Josef de la Sierra, created on the basis of the draughtsman's sketches.[20] Estachería's insistence that an engineer play a role in the 'formatting' of Del Río's report before it was sent to Spain shows the

---

[17] Cañizares-Esguerra, *How to Write the History*, chapter 5.

[18] Irina Podgorny, 'The Reliability of the Ruins', *Journal of Spanish Cultural Studies*, 8, no. 2 (2007): 213–33, at 224.

[19] José Estachería to José de Gálvez, Guatemala, 12 August 1786, in Cabello Carro, *Política investigadora*, 128–9. On engineers see Horacio Capel, *Los ingenieros militares en España, siglo XVIII. Repertorio biográfico e inventario de su labor científica y espacial* (Barcelona: Universitat Barcelona, 1983); Janet Firestone, *The Spanish Royal Corps of Engineers in the Western Borderlands: Instrument of Bourbon Reform, 1764 to 1815* (Glendale, CA: A.H. Clark Co., 1977).

[20] José Estachería to Antonio Valdés, Guatemala, 9 July 1788, in Cabello Carro, *Política investigadora*, 150. Images *ibid.*, Lámina 37–44.

importance of the technical skills which these men were perceived to have, and the sort of visual representation that formed the 'gold standard' of recording landscapes at a distance in the colonial period. It is then hardly surprising that the documentation of the ruins which was sent to Spain resembled military maps of fortifications in their labelling, style, and technical sophistication. Although historical geographies had become a branch of cartography in eighteenth-century France and were being applied to sites of ancient civilisations, this was very much a geography of the here and now.[21] The architect Bernasconi's plans, too, included cross-sections and floorplans of buildings, as well as a topographic map of the entire terrain surrounding Palenque (Figure 1.1).[22]

The magistrate of Palenque, Calderón, cared less about these men's technical skills than their practical skills of exploration and physical resilience. In an initial report about the ruins of Palenque to his superior, Estachería, he suggested that if his own document were to be deemed unsatisfactory, the president should send somebody to Palenque to revise it. However, he cautioned, this man, 'apart from being knowledgable [*ynteligente*]', also had to be 'a man of dust, and mud, with the character of a soldier, who will be able to walk on foot, and alongside me: often using the hands in order not to fall to the floor'. Only such a man, who combined some scholarly expertise with the capacity to endure muddy and uncomfortable conditions during fieldwork, would be one who could 'do justice [*dé la perfeccion correspondiente*] to these sculptures' and other features of the ruins.[23] Scholars who only sat at their desks would not be suitable for this kind of work. The idea that only a 'man of dust and mud' could conduct reconnaissance was clearly born out of the same experience of Central American landscapes which led other officials to complain about roads destroyed by rains. Of course, the magistrate or explorer was not alone in this – Captain del Río was accompanied by a team of labourers, whose identity is not certain but who were likely such local villagers. Tropes of learning about landscapes by physically experiencing them in the process of administrative work such as road-building (but delegating physical work to Indian labourers) were well established.

The identification of ruined cities as part of the landscape as much as historical artefacts at Palenque was a function of bureaucratic traditions and the preoccupations of eighteenth-century governments of Spanish America. It was marked by longevity and remained even into independence. Engineers continued to provide interpretations for ancient sites as late as the 1830s, when new states increasingly

[21] Anne Godlewska, *Geography Unbound: French Geographic Science from Cassini to Humboldt* (University of Chicago Press, 1999), 267–80.

[22] Bernasconi's plans of Palenque: AGI-MP, Guatemala 257, 260; Military map e.g. Luis Diez Navarro, 'Plano del Puerto y sitio de San Fernando de Omoa, situado en la costa de Honduras' (1757), AGI, MP-Guatemala, 45BIS (digitised at pares.mcu.es).

[23] José Antonio Calderón to José Estachería, Palenque, 15 December 1784, in Cabello Carro, *Política investigadora*, 82.

used indigenous histories to transform them into national histories. When the engineer Miguel Rivera Maestre completed the first atlas of independent Guatemala in the early 1830s, he was sent to visit and map pre-Columbian sites including Quiché (Q'umarkaj), the former capital of the K'iché people in the Guatemalan highlands. Armed with Domingo Juarros's 1808 *History of the City of Guatemala* as his guide, he expected to encounter a great number of the 'ruined buildings of the Quiché'. Yet he found the site confusing, as he noted in the margins of his copy of Juarros. Juarros, citing the seventeenth-century chronicler Fuentes y Guzmán, had described how 'a deep ravine' surrounding the city 'functioned as a moat, and only left two very narrow entrances to the city which in turn were defended by the *Resguardo* castle which made it unassailable'.[24] The engineer believed that he had found the correct site based on this historical topographic description, but was disappointed by the small number of ruined buildings there. After observing the site that matched Juarros's description most closely topographically, he concluded that 'it was not possible that the small space that the site presented should contain the number of buildings to which the chronicler Fuentes refers, cited in this history by Fr. D. Juarros'. He did encounter a second site further north, but kept referring back to the jarring difference between Fuentes and Juarros's descriptions of the city and the initial site. The deep ravine and single fortification remained Rivera Maestre's main criterion for identifying the 'true' site of the capital of the Quiché, to the extent that he now seemed ready to ignore the collection of buildings to the north because, topographically speaking, they were in the 'wrong place'.

It cannot be another site because it is the only one in the vicinity which is endowed with the security of a natural fortification, with a single entrance that is defended with a single fort. If *Señor* Fuentes had said that all these grand buildings which he describes were to be found to the north of the place in which he wanted to pin them down, there would be more grounds for believing him, because here there are vestiges of a great population and the fortresses which defended them in various places, although they are missing the deep surrounding ravines which constituted the principal defence.[25]

---

[24] Domingo Juarros, *Compendio de la historia de la ciudad de Guatemala* (Guatemala: Ignacio Beteta, 1808), Vol. I, 66. For these archaeological investigations, see also Oswaldo Chinchilla Mazariegos, 'Archaeology and Nationalism in Guatemala at the Time of Independence' *Antiquity* 72 (1998): 376–86; Rebecca Earle, '"Padres de la Patria" and the Ancestral Past: Commemorations of Independence in Nineteenth-Century Spanish America', *Journal of Latin American Studies* 34, no. 44 (2002): 775–805.

[25] Handwritten note glued to back of map entitled 'Istapa. Vista desde Mar' in the copy of Miguel Rivera Maestre's *Atlas guatemalteco*, held at the New York Public Library, Map Division, New York. The note reproduces text that (according to the owner of the Atlas) Miguel Rivera Maestre left in his copy of Juarros' *Historia de Guatemala*. An ExLibris note on the front page as well as notes throughout the edition suggest that the owner of the book who transferred Rivera Maestre's words into this atlas was Miguel Larreynaga (1772–1848), Nicaraguan nation-builder, scholar, and writer. His involvement in the Guatemalan *Sociedad Económica* with which Rivera Maestre was also affiliated affirms the connection between these two men.

Rather than believe that Fuentes and Juarros had misunderstood the precise location of the city and investigating the larger group of buildings, the engineer stubbornly placed more emphasis on the detail that the city was on a defensible hill and stuck with the first site. In the end, he mapped the smaller of the two possible sites because it corresponded to the topographical description, not the buildings to the north which seemed the likely candidates according to the historical account. He marked the ravine, the dramatic plateau landscape, and the 'fort which defends the entrance to the buildings' clearly on the map (Figure 1.2).[26]

The works of both Fuentes and Juarros had described the city's narrow streets, specific edifices such as the royal palace and a school. However, it was the topographical and military information that the engineer drew on when reading their accounts.[27] While it is unsurprising that an engineer should look out for these to him so recognisable elements, it is also worth noting that he referred to the first volume of Juarros's *Compendio*, which incorporated a 'topographical description' of the kingdom, rather than the second volume, which gave more historical detail and indeed described the territory of the Quiché as having a series of defences.[28] Given a choice between his observations of the landscape and reports of a more historical nature provided by two eminent chroniclers, Rivera Maestre firmly sided with the evidence provided by the landscape. For the engineer, the 'true' location and centre of the capital was determined by what was known about its defence possibilities provided by its topography. This episode confirms the extent to which landscape as much as historical record played a part in understanding archaeological sites and shows that the Palenque explorations were just one example of a tradition that placed authority in the landscape itself. Although in this case, the histories on offer for Maestre were Spanish and not indigenous, it is also worth highlighting that putting one's faith in the landscape was a detached and 'scientific' way to ensure that questions of indigenous authorship could be discreetly swept under the metaphorical rug. The successive reports of Calderón, Del Río, and Rivera Maestre fused topography and history to create a practical assessment of archaeological sites that addressed their 'innate' economic or military potential alongside, or even before, their historical significance. It is therefore not surprising that in addition to using landscape to think about ruins, colonial explorers and their superiors also used ruins to help them understand landscape.

---

[26] Miguel Rivera Maestre, 'Plano del terreno en que se hallan situados los vestigios de los edificios antiguos del Kiché. Se levantó de orden del gefe del estado C. Dr. Mariano Galvez. Año de 1834', in *Atlas guatemalteco en ocho cartas formadas y grabadas en Guatemala: de orden del gefe del estado C. Doctor Mariano Gálvez* (Guatemala City, 1834).
[27] Juarros, *Compendio*, Vol. II, 66–7. [28] For example Juarros, *Compendio*, Vol. II: 11, 33.

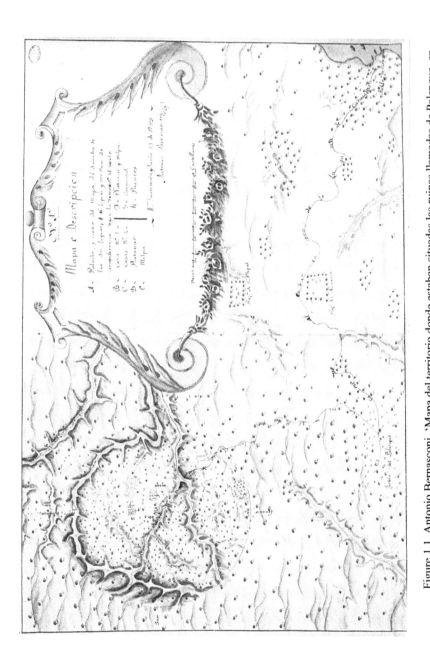

Figure 1.1 Antonio Bernasconi, 'Mapa del territorio donde estaban situadas las ruinas llamadas de Palenque, en la provincia de Ciudad Real de Chiapa', 1785. Ministerio de Cultura y Deporte. Archivo General de Indias. MP-Guatemala, 257.

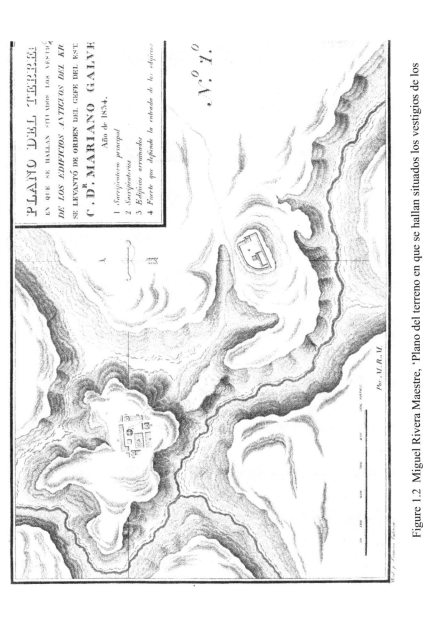

Figure 1.2 Miguel Rivera Maestre, 'Plano del terreno en que se hallan situados los vestigios de los edificios antiguos del Kiché. Se levantó de orden del gefe del estado C. Dr. Mariano Galvez. Año de 1834'. In *Atlas guatemalteco en ocho cartas formadas y grabadas en Guatemala: de orden del gefe del estado C. Doctor Mariano Galvez.* Guatemala City, 1834. The Lionel Pincus and Princess Firyal Map Division, The New York Public Library.

Ruins in a Destructive Landscape

The Chiapas explorations also touched upon a much larger question: the relationship between the landscape that Central Americans past and present inhabited, and their settlements. Several of the seventeen points that Estachería asked the architect Bernasconi to pay close attention to treated the ruins like any living city rather than a historical monument. In addition to gaining more information about the origin of the city and the people who built them, the governor prepared detailed instructions which showed his interest in Palenque's architecture, building materials, but also the city's natural surroundings. The questions seem influenced by contemporary interest in the landscape and its products, and use the flattening abilities of bureaucratic accounts to show concern with the site as a whole rather than the ruins in isolation. In addition, a preoccupying mystery loomed: what had led to the downfall of this civilisation? The initial report by the magistrate Calderón had spoken of steep and tall hills within the city. Estachería wondered whether these had anything to do with the fall of the city. He assumed that it was unlikely that these hills 'formed of the ruins of the buildings ... merely by their collapse'.[29] He deduced that the hills were perhaps the result of the eruption of a volcano. The decayed city might be the result of a destructive landscape. José Alcina Franch has argued that this reflected the president's desire to find the American equivalent of Pompeii or Herculaneum, since those cities, too, were destroyed by a volcano.[30] This is plausible, since Pompeii and Herculaneum were much talked about in Bourbon Spain.[31] Estachería also asked about paved roads and worked iron, since the original reports by Ordóñez had suggested their presence. This concern with solid, paved roads might once again be a sign that Estachería took Pompeii as a reference point, since Roman roads were considered an ideal template for modern road-building in eighteenth-century Spain.[32] Yet there are also explicit references to various aspects of the landscape, and a preoccupation with

---

[29] Estachería, 'Instrucciones para el reconocimiento de las ruinas de Palenque', Guatemala, 29 January 1785, in Cabello Carro, *Política investigadora*, 93–5.

[30] Alcina Franch, *Arqueólogos o anticuarios*, 91; Joanne Pillsbury and Lisa Trever, 'The King, the Bishop, and the Creation of an American Antiquity', Ñawpa Pacha 29, no. 1 (2008), 191–219, at 206–7, also make this link for Peru.

[31] Martín Almagro Gorbea and Jorge Maier Allende (eds.), *Corona y arqueología en el siglo de las luces* (Madrid: Patrimonio Nacional, 2010), chapters by Miguel Morán Turina and Jaime Alvar. Antonio Porlier, Minister for Justice of the Indies, opined that the king himself would be personally interested in the Palenque documents, since he had also taken an interest in recent excavations at Pompeii and Herculaneum: Cabello Carro, *Política investigadora*, 41–2.

[32] Michael Crozier Shaw, 'El siglo de hazer caminos. Spanish Road Reforms during the Eighteenth Century. A Survey And Assessment', *Dieciocho* 32, no. 2 (2009): 413–34, at 416–17. Cañizares-Esguerra, *How to Write the History*, 327, also suggests that Estachería was using roads as one of the markers of a 'great civilisation' that the reports of the 1780s did not confirm.

natural disasters, that suggest that the investigations of Palenque were more specifically rooted in familiarity with contemporary Central America.

The way in which Estachería expressed his mandate to look for evidence of stable roads, for instance, speaks of an acquaintance with the difficult terrain of tropical Chiapas and the problems of maintaining communications through the rainy season in this and other Central American regions. The president acknowledged that roads which were not built with special care in this region 'would be impassable during at least a good part of the year', so any surviving roads would certainly have to be of solid construction.[33] Parallels to Roman roads and Vesuvius might have been implied, but references based on experience of the Central American landscape, its washed-away roads, and the threat of natural disasters were explicit. Estachería's instruction required Bernasconi to 'make an exact and thorough reconnaissance of the material of which the hills are composed'. Demonstrating an acquaintance with local circumstances, he homed in on the possibility of finding a specific volcanic stone at the site, asking the architect to report back on whether 'said hills contain the materials emitted by volcano eruptions which normally consist of stones called, in this kingdom, *malpais*, similar to slag from the forge, and more properly similar to that which the working or manufacture of iron produces'.[34] Bernasconi replied that 'in none of the hills, or other parts of the ancient city' did he find any sign 'of the eruption of a volcano, nor any other sign denoting violent destruction'. The response to the question of whether a volcano had led to the abandonment of a once-great Central American city was in this case negative. Bernasconi deduced that the collapse of the city was simply due to being abandoned by its inhabitants.[35]

The detailed description on Estachería's part of the type of stone he expected to find, and Bernasconi's assurance that there were no volcanic materials to be seen, implies that both men shared a frame of reference in which volcanic destruction was an expected part of the environmental dangers that threatened society. In fact, when Estachería referred to the volcanic material known as '*malpais*, similar to slag from the forge', his words were remarkably similar to the language used by engineers to describe other Central American landscapes. In a report, chief engineer Luis Diez Navarro and his young colleague Josef María Alejandre described the products of a volcanic eruption as being of a particular type: 'vulgarly known in this country as *mal país*', similar to volcanic

---

[33] Estachería, 'Instrucciones', in Cabello Carro, *Política investigadora*, 93.

[34] *Ibid.* Contemporary geological publications assume that the word *malpais*, still used for specific volcanic formations in Mexico and the southern US, is simply a derivation of local descriptions of 'badlands' or barren land. However, this description suggests that it referred specifically to volcanic stone at least since the eighteenth century.

[35] 'Informe de Antonio Bernasconi', Guatemala, 13 June 1785, in Cabello Carro, *Política investigadora*, 114.

material that Alejandre had observed near San Salvador six years earlier. In its hardened appearance, it reminded him of 'slag from a metal forge'.[36] It is uncertain whether Estachería was directly quoting from this report, since it was a decade old by the time of the Palenque explorations and had been compiled for his predecessor. However, he was certainly not far removed from the kind of knowledge represented: when he was looking for a qualified engineer to lead the Palenque exploration, the now-established Alejandre was his first choice, although duties in Honduras prevented the engineer from leading the 1786 expedition. Ideas of volcanoes, earthquakes, and all manner of natural risks were thoroughly embedded in the politics and society of eighteenth-century Central America, so it is not surprising that Estachería and Bernasconi also tried to understand the ruined city of Palenque by looking for such destruction. Put in the context of preoccupations about the natural world, the president's interpretation of Palenque's landscape indicates more Central American than European reference points. Indeed, it is quite likely that, aside from the convenience of not having to deal with uncertain questions of history, the reason that landscape-based ways of deciphering the ruins became so important was the sense that the destroyed ruins of Palenque might allow explorers to learn more about the landscape's properties, including its destructive potential.

News of a ruined city in Chiapas, while received with some excitement in Guatemala City, must have also evoked some more glum connotations of a ruined city of a different kind. If the Palenque explorations reflected contemporary concerns about what made a city suitable for prosperity, or liable for destruction, several recent natural disasters and their resulting geographical reports were likely to be on Estachería's mind. Only ten years before Estachería became president of the Audiencia of Guatemala, a massive earthquake had destroyed the old capital of the kingdom, and the government took the extraordinary decision to move the capital Guatemala City to a new location in the Hermita Valley in 1773. The move of the capital was a bold resolution, though not a completely isolated event. The capital had already been moved once, in 1541, and the government in Peru also considered moving its capital city to a less disaster-prone site after an earthquake destroyed Lima in 1746.[37] For Estachería's predecessor Martín de Mayorga, who orchestrated the capital's move, the threat of natural disasters that might befall the old and new cities was a matter of basic safety, but also of deft political manoeuvering. The decision to

---

[36] Joseph Maria Alejandre, 6 July 1775, San Juan Amatitan; J. M. Alejandre and Luis Diez Navarro, Hermita, 15 July 1775. AGI, Guatemala, 462, fols. 15v, 52. It is worth noting that engineer Diez Navarro also linked volcanoes to a productive force. Perhaps taking the metaphor of 'slag from the forge' literally, and possibly influenced by the writings of Athanasius Kircher, he thought that the 'subterranean fires' that produced volcanoes also produced metals, and were therefore to be expected in America. *Ibid.*, fol. 52.

[37] Charles Walker, 'The Upper Classes and their Upper Stories: Architecture and the Aftermath of the Lima Earthquake of 1746', *Hispanic American Historical Review* 83 (2003): 53–82, at 67.

move the capital was fiercely contested by the governor's opponents, and power struggles between different social and political factions erupted in its wake.[38] Nevertheless, it was geographical reports that were at the forefront of the arguments that Mayorga and his allies put forward.

Reports that compared the site of the new city with the old made reference to the suitability of the terrain for habitation when it came to the potential for agriculture, the presence of wood for construction material, but also possible natural disasters. Alongside finding a plain with good lands and a good water supply, initial prospecting parties looking for a new site for the city should 'carefully' ascertain whether there 'were any volcanoes in the vicinity of the site, or whether it has experienced any quakes'.[39] In the aftermath of 1773, city officials also turned to the archives to ascertain the relationship between the old capital, nearby volcanoes, and past natural disasters in a bid to assess the risk that threatened the city. One report painstakingly listed fourteen major earthquakes that had befallen the capital between 1531 and 1773. Another unearthed documents from the city archives from the last major eruption of the volcano Fuego in 1717, and listed frightful descriptions of its eruptions going back to 1581.[40] Guatemala's volcanoes and earthquakes became part of a historicised landscape, demonstrating the deep link between geography, governance, and history that is also visible in the Palenque explorations. In this way, the governor framed the move of the capital not as political, but as resulting from an understanding of the landscape, including its inherent threats. The very earth was full of dangers, but careful research into the landscape would prevent future disasters. The eruption had led to a wider re-assessment of the city's geographical situation within a landscape that had originally been deemed secure.

It is quite likely that Estachería was familiar with the details of these arguments when he started his tenure as governor in a city still under construction. The architect Bernasconi, who compiled the 1785 Palenque report, was also familiar with these events, since he had arrived in Guatemala as one of the chief architects of the new capital city. Particular knowledge of the history of

[38] Political opposition, see Christophe Belaubre, 'El traslado de la capital del Reino de Guatemala (1773–1779). Conflicto de poder y juegos sociales', *Revista De Historia*, 57–8 (2008): 23–61, and Julio Martín Blasco and Jesús María García Añoveros, *El arzobispo de Guatemala don Pedro Cortés y Larraz* (Belchite: Ayuntamiento, 1992), 99–105, 188–96. For a general summary of the move, see Zilbermann de Luján, 'Destrucción y traslado'.

[39] 'Testimonio de los Autos formados ... de la inspeccion por maior del paraje nombrado el Valle de Jumay, y el de Xalapa. Oficio de Peñalver.' 1774. AGI, Guatemala, 661, 13v–25.

[40] Felipe Cadena, *Breve descripcion de la noble ciudad de Santiago de los Caballeros de Guatemala* (Mixco, Guatemala: Oficina de don Antonio Sanchez Cubillas, 1774), 7; Gonzáles Bustillo, *Extracto, ó Relacion methodica, y puntual de los autos de reconocimiento* (Mixco, Guatemala, 1774). See also Hutchison et al., 'The 1717 Eruption of Volcán de Fuego, Guatemala: Cascading Hazards and Societal Response', *Quaternary International* 394 (2016): 69–78.

the destruction of Pompeii and Herculaneum was, then, hardly necessary in order to speculate whether Palenque, too, had been destroyed by a natural disaster. Examples were to be found much closer to home, and these considerations about the way that natural threats or opportunities influenced society continued to be a part of political discourse. The idea that places influenced history had appeared in the French seventeenth-century geographer Jean François's work, but the idea of men's history and culture being shaped by geography and nature, rather than the other way around, was more formally expressed in nineteenth-century Europe by German geographers such as Carl Ritter and Alexander von Humboldt.[41] Guatemalans also left the door open to such ideas in the eighteenth century when they discussed the effects of earthquakes and volcanoes as forces which society and politics must heed. Humboldt's geographical theory itself had of course also been heavily influenced by South American and Mexican nature. Moreover, there was a comparable preoccupation with man's relationship with nature in Stephen Hales and Voltaire's work after the London and Lisbon earthquakes, suggesting that experience of seismic events was also linked to a theoretical reassessment of environmental influences on society.[42] Living in a seismically active region, Central Americans were part of this trend and perhaps even at its vanguard. Although they generally embraced enlightened optimism when it came to expressing the possibility of man's control over nature, their geographical thinking allowed for incontrovertible forces that showed societies the limits of such visions.

The records of administrators of this era are full of examples of an approach to governance that seamlessly connected considerations of the natural environment and natural disasters with government, a tendency which Estachería and the explorers he sent also followed. While governor Mayorga commissioned his file on the Pacaya volcano in the midst of an acute natural disaster, other documents, particularly bishops' *visitas*, also show an ongoing preoccupation with geography, earthquakes, and volcanoes. Archbishop Pedro Cortés y Larraz had produced an exceptionally thorough and famous *visita* in 1768, expressing his belief that knowledge of physical and human geography formed the basis of good governance and a moral society.[43] His successor, archbishop Cayetano Francos y Monroy, also tied spiritual and earthly governance to understanding natural risk in his 1784 *visita* of the archbishopric. The

---

[41] Godlewska, *Geography Unbound*, 118; Chenxi Tang, *The Geographic Imagination of Modernity: Geography, Literature, and Philosophy in German Romanticism* (Stanford University Press, 2008), esp. 220–3.

[42] Richard Grove, *Green Imperialism: Colonial Expansion, Tropical Island Edens and the Origins of Environmentalism, 1600–1860* (Cambridge University Press, 1995), 165–6.

[43] Pedro Cortés y Larraz, *Descripción geográfico-moral*, edited by Julio Martín Blasco and Jesús María García Añoveros (Madrid: CSIC, 2001). For a thorough discussion, see also Sellers-García, *Distance and Documents*, 55–75.

conclusions he drew from his travels echoed the concerns that his predecessors had expressed about the spiritual well-being of parishioners, and in particular the abuses that regional judicial officials perpetrated against the Indians supposedly in their care. As the archbishop himself admitted, these failings of the state when it came to supporting its most disadvantaged subjects were not new. However, he was able to provide at least one concrete idea to ease the burden imposed on rural parishioners: a change to church architecture. He claimed that thanks to repeated earthquakes, there was 'in this whole archbishopric scarcely one church' that had not been destroyed to the point of being unusable. This meant that instead of attending mass in suitable buildings, parishioners were instead merely constantly reconstructing the churches.

His solution to this was to ask for the king to impose a law to regulate church architecture that would take into account the natural forces that acted on buildings in Guatemala. Churches should be only of moderate height and vaulted ceilings should be replaced with coffered *artesonado* ceilings, or be based on earthquake-resistant wooden pillars. Although these buildings would 'seem provisional at first sight, in reality, they would be more permanent, and more useful'.[44] If not for the spiritual benefits of better church buildings, Francos y Monroy slyly suggested, a physical way of co-existing with the threat of earthquakes would also mean freeing up tax money for other purposes, instead of having to contribute to the constant re-edification of churches. The archbishop's report makes clear that natural risk and the congregation's physical surroundings should be considered as factors in the quest for effective governance and taxation of the kingdom. There are clear parallels to his contemporary Estachería's understanding of Palenque as a city conditioned by its natural surroundings. Archbishop Francos y Monroy's *visita*, conducted in the same year that the Palenque explorations started, highlights that both Spanish secular and religious administrators were preoccupied with similar questions about environments. The archbishop may have been pandering to one the king's most obvious priorities when he suggested that understanding nature would lead to more effective uses of tax money, but he was also speaking the common language of the Bourbon reform period when he tied natural environment to earthly government and its economic consequences.

Ruins in a Productive Landscape

President Estachería's instructions about Palenque also connected natural resources with the commercial possibilities of the city's mysterious architects more explicitly. The questionnaire as a whole can be explained by a thoroughly

---

[44] Cayetano Francos y Monroy, 'Visita General', Copy of 15 August 1784. AGCA, A1 Leg. 1532 Exp. 10097, fols. 28v–29.

practical, gubernatorial interest in the land. Jorge Cañizares-Esguerra has alluded to this in describing Estachería's 'research programme' as rooted in the way eighteenth-century 'social science' imagined a great civilisation.[45] His questions echoed the sort of questions asked by Bourbon governors about Spanish American cities and populations: what sort of trade, road networks, and industry sustained them? Economic productivity through knowledge of the land was after all the guiding thread of Charles III's minister, the Count of Campomanes's reform agenda, and an idea that thoroughly affected the language of governance and civic reform attempts in Spanish America. The 'industry, commerce, or means of subsistence' of the city's inhabitants was one of the main points that interested Estachería. One of his questions for Bernasconi was about metals. He should look for signs of 'worked metal, or minted coins', since such a magnificent city, in the president's eyes, could not have been built by a culture that merely practised subsistence agriculture. He suspected that they had another, richer source of income, 'such as mines', or another resource that 'would have provided the settlers with a lucrative and active trade with which to attract happiness to themselves, and the [sort of] abundance at which the magnificence of the city now hints'.[46] This rhetoric of 'happiness' brought about by trade, and the interest in the possibility of there being mines or other economically useful features, was steeped in the thought of the Bourbon Reforms. Estachería's instructions themselves probably leaned on the influential questionnaire that Antonio de Ulloa, scientist and officer of the Spanish Crown, had drawn up to gather information about the viceroyalty of New Spain in 1777. Ulloa's questionnaire urged respondents to identify knowledge about the geography, mineralogy, and metallurgy of New Spain, but also antiquities. In addition to asking questions about the vestiges of these civilisations, which would help identify clues about their origin, what they 'accomplished, and the way in which they behaved', the questionnaire would also reveal information on their 'government and economy'.[47] For Ulloa, as for Estachería, the material remains of a city might give clues to the economic underpinnings of that society. Places could always encompass within them the promise of improvement.

That Estachería looked for evidence of a past society's prosperity at these ruined sites presupposed adherence to a particular Enlightenment epistemology that privileged material evidence and observation over other sources, as Jorge Cañizares-Esguerra has pointed out. The governor approached the exploration of Palenque within the framework of *relaciones* and *informes* with which he was familiar, and privileged the trained eye and formally recognised expertise

---

[45] Cañizares-Esguerra, *How to Write the History*, 326.
[46] Estachería, 'Instrucciones', in Cabello Carro, *Política investigadora*, 91–2.
[47] Francisco de Solano, *Antonio de Ulloa y la Nueva España* (Mexico City: UNAM, 1987), LVII; Alcina Franch, *Arqueólogos o anticuarios*, 79.

of Bernasconi. In particular, Cañizares-Esguerra contrasts Estachería's episte-mology with the allegorical, Baroque 'literary speculation' of a group of Creole scholars led by Ramón Ordóñez y Aguiar.[48] The historiography of these explorations of the Palenque ruins provides diverging explanations of the extent to which a bureaucratic way of looking at the world was at odds with the interpretations that these scholars put forward.[49] The status accorded to indigenous documents in the interpretation of the ruins is especially contested. Ordóñez, who had never visited the site, developed his own interpretation based on his reading of Maya manuscripts, and was dissatisfied with the Crown bureaucrats' approaches.[50] However, despite generally privileging geo-graphical and natural-historical sources, Estachería did not seem opposed on principle to the idea of using indigenous documents. The Enlightenment epistemology represented by government officials certainly represented one particular perspective, but it was not an entirely rigid one. After all, Estachería did consult an indigenous K'iché manuscript which reached him after the architect Bernasconi's 1785 report. He even asked for it to be sent 'with the greatest speed'. In the end, he did not regard it as an illuminating addition to his knowledge of Palenque because it turned out to refer to another town by the same name in the Guatemalan highlands, but he did appear to take it seriously.[51] While he had received Bernasconi's report in late June 1785, he did not forward it to Madrid until August 1785. This timing may indicate that he specifically waited for the K'iché manuscript from the highlands to reach him before he finalised his own report. Regardless of this manuscript, the Creole literary scholars and Estachería would never see eye to eye when it came to the epistemologies of indigenous sources. And yet, their positions were

[48] Cañizares-Esguerra, *How to Write the History*, 329, and Note 136, 343, 398–9; Rosa Casanova, 'Imaginando el pasado: el mito de las ruinas de Palenque, 1784–1813', Cuadernos de la Asociación de Historiadores Latinoamericanistas Europeos 2 (1994): 33–90.

[49] Scholars have disagreed on just about every aspect of these explorations. In particular, Cañizares-Esguerra (*How to Write the History*, 327), Cabello Carro (*Política investigadora*, 38), and Rosa Casanova ('Imaginando el pasado', 52–4) ascribe different motivations to Estachería, questioning whether he wanted to explore the ruins further after Bernasconi's initial report, or whether questions from Madrid dictated the process. It is perhaps a reflection of the difficulty of discerning an individual's opinion while operating in a bureaucratic system that it is unclear from the documents cited which description is most accurate. While it took Estachería several months to order Del Río to explore the ruins (he left in May 1787 instead of October or November of 1786), this could simply denote a lack of faith in Del Río's suitability for such exploration on Estachería's part as much as a judgement about the whole enterprise's validity.

[50] Cañizares-Esguerra, *How to Write the History*, 329–38 and 205–20, develops this as a case study for a historiographical tradition he terms 'literary Creole patriotism', championed by some Creole scholars in the 1780s and 1790s; Roca to San Juan, 27 November 1792, in Ballesteros Gabrois, *Nuevas noticias,* 25.

[51] José Estachería to José de Gálvez (1786), in Cabello Carro, *Política investigadora,* 110; 'José Estachería al alcalde mayor de Totonicapán y Huehuetenango', 22 June 1785 and 'Francisco Geraldino a Jose Estachería', 2 July 1785, both in Navarrete, *Palenque*, 80, n. 11.

not necessarily so different from each other. Both leaned on the rhetoric of the Bourbon reforms and Enlightenment ideas about improvement and progress.

Both interpretations were suffused with the rhetoric of 'improvement' and the idea of studying nature in order to benefit the *patria*. Such patriotism was relevant to Ordóñez's localistic epistemology, but was also a hallmark of the 'enlightened improvement' that shaped public life in Bourbon-era Spanish America. The Dominican Tomás Luis de Roca, who as Ordóñez's associate had criticised the crown's methods, was convinced that 'this old, until now unknown, and wonderful [*famosa*] city can bestow much lustre on our nation', and that Ordóñez's studies would be 'useful to the State'. This sort of optimism, and the belief in the useful nature of new discoveries, was a critical part of the language of the Bourbon reforms and the Spanish American Enlightenment. In this vein, Roca mused that, according to a 'French author' he remembered reading, 'the provinces of Chiapa would flourish more than any other in America if the Spanish would appreciate that which is worth more than gold, or silver' – that is, the productions of the land itself. Estachería and Roca also employed similar rhetoric when it came to discussing the land that surrounded Palenque, again mirroring the priorities of reformers regarding the improvement of agriculture and industry. Such plans went hand in hand with an optimism about the land and its potential to provide agricultural productions. Roca stressed the 'most fertile' qualities of the Chiapas terrain, which 'does not envy Oaxaca its cochineal, or Guatemala its indigo, nor Tabasco its cacao', and went on to list the many plants of nutritional, industrial, and medicinal value which in Chiapas formed 'the voluntary products of this uncultivated terrain'.[52] Such a rhetoric of exceptionalism may lead us to view Roca as a 'Creole patriot' intent on singing the praises of his American homeland. Indeed, members of the clergy, whether born in Europe or the Americas, often embraced the ideology of 'Creole patriotism'.[53] However, such patriotism was also compatible with the ideologies of Bourbon reformers. For Roca, recognising the past greatness of Chiapas was one part of a wider benefit which might be gained from a better understanding and recognition of the region. In his subsequent call for a liberalisation of trade to facilitate the commercial exploitation of the region, and his characterisation of the land as the source of a nation's riches, he subscribed to the same language of enlightened reform that members of the bureaucracy also employed. That Roca cited a 'French author' in his recognition of the potential of the province of Chiapas is a significant detail, since Spanish American Enlightenment reformers often cited French *philosophes* to show awareness of wider debates within enlightened scholarship. That Roca and Ordóñez on the one hand and Estachería and

---

[52] Roca to San Juan, 27 November 1792, in Ballesteros Gabrois, *Nuevas noticias,* 23, 25.
[53] Cañizares-Esguerra, *How to Write the History,* 209.

Del Río on the other, for all their epistemological disagreements, all employed arguments that were suffused with the rhetoric of improvement shows how widespread these ideas were by the 1780s among lettered men in Guatemala at least. Ideologies of improvement, in much more explicit, practical, and self-consciously 'reform-minded' ways, would inform government and civic projects for decades to come.

The connection between ruins, the land around them, and its economic potential was evident in the writings of both Roca and Estachería. It was not only typical of eighteenth-century approaches to improvement, but continued to resonate with enlightened reformers at least until the early nineteenth century. A report on the province of Chiapas, compiled by an anonymous author around 1805, saw in the 'luxurious' buildings of the ruined city of Palenque evidence for the past economic wealth of the region. The author observed that:

It is likely that the Indians before the conquest made use of the fruits and possibilities that nature itself offered to them. About twenty years ago now, a city was found a mile from Palenque, whose magnificence was not matched even by the Greeks or Romans. ... all this proves the extraordinary luxury of a wealthy population, the product of a flourishing commerce.[54]

The anonymous author then continued with a description of architectural details and views, suggesting that he had personally observed the buildings and the nature around it before drawing these conclusions. He also turned to historical evidence, but rather than showing an interest in indigenous history, he was drawn to stories about its economic status. He mentioned that in the early days after the Spanish conquest, the city of Chiapas hosted a market that was so grand it attracted people from across New Spain and Guatemala. This past status mattered because it provided the possibility of it becoming so again. Changing prices, a locust plague, and subsequent emigration had led to Chiapas's current poverty. However, the evidence from Palenque and the historical memory of the market led him to the conclusion that 'this province was in other times wealthy', just as Estachería had seen stable roads as a potential sign of a prosperous trading city. The ruined city contributed to an understanding of the Chiapas countryside as fertile and full of potential. These comments about Chiapas fell under the headlines both of a 'topographic description' and of political economy, further bolstering the link between ruins, the land, and economic potential. The knowledge about past glories was for him incontrovertible evidence of future economic promise. The author

---

[54] Anonymous, 'Noticia topografica de la Yntendencia de Chiapa', 86–9. BL, Add. MS 17,573. Reference to the first excavations of Palenque happened 'around twenty years ago' and to the *Real Expedición de la Vacuna* dates the document to around 1805.

expected not just a 'regeneration', but even hoped for 'greater wealth' than before in Chiapas.

The vague conclusions that Estachería, Bernasconi, and Del Río had started to draw from the landscape in the 1780s were now being made much more explicit on the basis of personal observations of the site. A later Bourbon intendant of Chiapas, Agustín de las Quentas Zayas, made similar calculations when he established a new village near Palenque in 1798. He claimed that his town would certainly be a success, firstly because of its healthy climate and abundant terrain which denoted agricultural potential, but secondly also because 'pieces of millstones, of pots, vessels and candlesticks, flutes, oars, and very disintegrated human bones' had been found nearby. These were 'all signs that in this place in a different time there was a town, maybe of people of the same kind as those of Palenque'.[55] The presence of the ancient town made the success of the new settlement certain for him. At around the same time, bishop Martínez Compañón in Peru also used evidence from the pre-conquest ruins at Chan to prove that there was a basis for the future success of a settlement. However, in the Peruvian case, Emily Berquist Soule has argued that the material evidence of ancient cities for the bishop was proof that Indians had the capacity for civilisation.[56] In 1791, Hipólito Unanue, the Peruvian Creole philosopher, also looked to studying the ancient 'monuments of the Incas' to establish connections between the past and present, and tied their significance to the 'soil' on which they stood as a physical link between the past and present. As Mark Thurner has argued, for Unanue, this soil quickly became 'mythopoetic', a timeless symbol of the emergent nation.[57] Countering the insistence of colonial magistrates on splitting off past indigenous histories from contemporary Indians, Unanue and to some extent also Martínez Compañón defended the continuity of Indian nations. Not so the Central American administrators, who 'read' the material remains of the cities for evidence of the past and of future potential of the land and its geographical location in a more literal way.[58] The land might pose dangers, but bureaucrats also took ruins and abandoned settlements as signs of the future.

While bureaucrats might base their belief in the potential of the land on physical vestiges of ancient cultures, it did not extend to an appreciation of the indigenous population whose ancestors had built the ruins. Explorers and bureaucrats rarely equated pride in local nature with a positive valuation of the region's native inhabitants. Many commentators, including Antonio Del Río, believed that the great ruined cities of the jungle must have been built by

---

[55] 'De Chiapa', *Gazeta*, Vol. 2, no. 91 (10 December 1798), 327.
[56] Berquist Soule, *Bishop's Utopia*, 65–70.   [57] Thurner, *History's Peru*, 89–91.
[58] Compare the idea of land as a geological archive in Nancy Applebaum, 'Reading the Past on the Mountainsides of Colombia: Mid-Nineteenth-Century Patriotic Geology, Archaeology, and Historiography', *Hispanic American Historical Review* 93, no. 3 (2013), 347–76, at 352.

'*ultramarinos*', that is, Romans, Phoenicians, or a similar 'civilised people'. In this dismissal of indigenous people as 'uncivilised' and incapable of building such edifices, they followed earlier Central American and Spanish authors. An anonymous Dominican friar acknowledged in the early eighteenth century that some of the ancient buildings found in Guatemala were made by 'Barbarous Indians', but that the more outstanding examples showed a 'splendour and skill' that meant that 'the Barbarian Indians did not make them'. For the friar, the splendid pre-contact cities of Copán and Ocosingo must therefore have been 'works of other civilised people who lived or were present in these lands', perhaps one of the 'lost tribes of Israel' or 'Phoenicians'.[59] A 1576 report shows that this attitude went back a long time, its author claiming that 'such a barbarous spirit [*ingenio*] as the natives of this province possess' could not have created buildings of 'such awe and sumptuosity' as the ruins of Copán.[60] The author of the early nineteenth-century 'topographical description of Chiapas', too, described the intricate architectural details of Palenque as 'asiatic'. Some Creole scholars did put much more emphasis on the importance of indigenous sources than most members of the Spanish government, but even they connected the ruined city and the people who built it to biblical stories of the lost tribes of Israel and other European traditions.[61] Many Spaniards and Creoles continued to separate their arguments about the potential of the land from those about the Indian population for decades when they mentioned ruined cities, and peddled the related trope that separated contemporary Indians from any past civilisations for decades. In 1830, the influential states-man José del Valle was convinced that 'the educated class' of Indians disap-peared with the Spanish conquest, and 'only the ignorant and unfortunate Indians remained'.[62]

As a result of these long-held stereotypes about the crude and savage nature of Indians, Creoles, and Spaniards who did accept the ruins' Indian authorship did so grudgingly. Just as European archaeologists would do in the nineteenth century, eighteenth-century Spaniards tended to produce 'archaeological sub-jects by splitting contemporary non-European peoples off from their precolo-nial … pasts'.[63] Estachería recognised the possibility that the present native population was related to the builders of Palenque, but struggled to reconcile his prejudices with the architectural finds. Having concluded from

---

[59] Anonymous, *Isagoge historico apologetico general de todas las Indias*, edited by José María Reina Barrios (Madrid, 1892), 100, 109.

[60] Licenciado Palacios to Philipp II, Guatemala, 8 March 1576, in Cabello Carro, *Política investigadora*, 123.

[61] Cañizares-Esguerra, *How to Write the History*, 334–6.

[62] 'Historia', *Mensual de la Sociedad Económica de Amigos del Estado de Guatemala*, no. 3 (Guatemala City: Imprenta de la Union, June 1830), 64.

[63] Marie Louise Pratt, *Imperial Eyes. Travel Writing and Transculturation* (New York: Routledge, 2008 [1992]), 133.

Bernasconi's report that the buildings were apparently to be attributed to the 'uncivilised' Indians of pre-conquest times, Estachería could not find a satisfactory theory of the relationship between past and present 'Indians'. He reported that 'it appeared to be true' that the Lacandón Indians who lived in the vicinity of Palenque 'descended from the ancient population [of that city]'.[64] 'We have to wonder', he wrote in his summary of the explorations, 'at the solidity and multitude of buildings in a people today so little inclined to habitations of any ... permanence'.[65] In this, he repeated a popular trope that assumed Indians of the colonial period to be 'barbarous', having lost the 'civilisation' of their pre-conquest ancestors, an assumption prevalent until well into the independent period.[66] Juan Bautista Muñoz, the historian and cosmographer, showed himself incredulous at the descriptions that were coming from Guatemala, in particular at references to a tower with a spiral staircase by the magistrate of Palenque, Calderón: 'I was struck with admiration upon reading such pronouncements, never having believed Indian architects to be capable of constructing a spiral staircase.' Spiral staircases, Muñoz explained, were something that not even the Inca or Aztecs had known, although they were regarded as the highest of American civilisations.[67] This scepticism as to the capabilities of the native population may have also contributed to the emphasis on studying ruins in the context of landscapes instead of their human builders. If Spanish administrators lacked the respect for the Indian population to consider them the architects of these edifices, then their emphasis on landscape as a phenomenon that could be studied within the pre-established categories of American governance could add logic to sites that were otherwise deeply puzzling. Even the anonymous author who, in the economic description of Chiapas cited above, evaluated Palenque based on the potential of the land around it, stressed that the 'Indians before the conquest made use of the ... possibilities nature offered to them' – with the clear implication that they no longer did so.[68] While they looked for continuities in the landscape and topography in their dealings with the buildings, even Spanish and Creole officials who accepted the indigenous authorship of the site drew a stark line between current and past indigenous societies.

The convoluted investigations into the authorship of the ruins on the one hand showed the elasticity of the boundaries of traditional administrative reports, but on the other also exposed their limits when applied to a complex

---

[64] José Estachería to Antonio Valdés, 9 July 1788, in Cabello Carro, *Política investigadora*, 150.
[65] José Estachería to José de Gálvez, 26 August 1785, in Cabello Carro, *Política investigadora*, 109–10.
[66] For instance 'Historia', *Mensual de la Sociedad Económica*, no. 3, 64.
[67] Juan Bautista Muñoz to José de Gálvez, 7 March 1786, in Cabello Carro, *Política investigadora*, 120.
[68] Anonymous, 'Noticia topografica de la Yntendencia de Chiapa', 86. BL, Add. MS 17,573.

and ideologically charged question. A genre which was best suited to a plain description of buildings and topography could not fully resolve such conceptually difficult topics. We can contrast the bureaucracy's stubborn refusals to lend credence to the notion of Indian architects with the opinion of Vicente José Solórzano, an otherwise unknown friar and village priest who had by chance been able to examine the findings of Del Río's expedition. He recorded that Antonio del Río, 'as he was returning from this commission, [earlier] this month, ... granted me a view of some fragments of carved [*amoldados*] figures from the ruined houses'. He called the ruins by their local name 'stone houses' and rejected Del Río's interpretation of their authors as 'Phoenicians, Goths, Carthaginians, or Romans', since he recognised some of Del Río's finds as being similar to the objects he understood as part of Indian culture. He named these objects as '*chalchiguites*, that is, small green common stones' and '*chayas*, which in the province of Soconusco is the name for those arrowpoints from glass-like material'. From this he deduced that 'Indian gentiles' must have once inhabited these ruins. Solórzano's knowledge of Indian languages and customs gained from living in Soconusco evidently served to interpret the evidence. This knowledge further allowed him to state that the 'gentile Indians, especially Mexicans, had talent and skill in various arts of painting and sculpture ... they were famous architects'.[69] This suggests not just the rather obvious conclusion that, locally, the existence of these ruins was well known. It also shows that there were men within the lower rungs of the Spanish bureaucracy who were quite ready to acknowledge the Indian authorship of the ruins on the basis of general anthropological traits of their culture with which they were familiar, and did not subscribe to the idea of a great divide between past and present indigenous culture. The priest Solórzano's opinion stands apart from the bulk of the debates that make up the paper trail regarding the land, geography, and buildings of Palenque. He was a relative outsider in this debate who possessed neither the official clout of Estachería to contribute to the report that would eventually land on the king's desk, nor the social connections that Ordóñez was able to exploit to put his argument forward. His opinions vanished save for this document, which appears not to have been sent to Spain, demonstrating that beneath the surface of this much-debated episode and outside the authoritative and regulated pathways of the transmission of information, an even wider range of epistemological approaches existed that were unlikely to find wide circulation before the existence of the sort of public forum for discussion that would emerge in the following decade through civic Enlightenment efforts.

---

[69] Vicente José de Solórzano, 'Parecer sobre el origen de la construcción de las casas de piedra', Yajalón, 14 July 1787, in Aramoni Calderón, 'Los indios constructores', 425–28.

Conclusion

The Palenque explorations are notable for the way in which they make visible the actions of a local government, as well as its opponents, trying to make sense of an unfamiliar set of monuments. As Jorge Cañizares-Esguerra and Miruna Achim have highlighted, it is absolutely a story of administrators who would not entertain the possibility of indigenous people as authors of these architectural wonders because of enduring prejudice and their marginalised position in colonial society. The bureaucracy's approaches to the investigation were certainly one-sided, but they also clearly laid out the logic of turning a question about the past or even future of a place into one connected to geographical approaches. The role that landscapes played in the interpretation of the ruins is striking. In an age preoccupied with systematic classification of objects, archaeological monuments defied easy categorisation, but the elastic boundaries of administrative reports were able to accommodate them within surveys that blurred the distinctions between roads, buildings and landscapes, the man-made and the natural environment. They expanded these categories to allow an analysis of a 'flattened' landscape that contained any and all of these elements, and the ability to analyse them in relation to each other. Bourbon officials and reformers extrapolated information about the economic potential of the region from the landscape they observed, and applied their own experience of inhospitable landscapes and natural disasters to analyse the surroundings of the ruined city and draw conclusions about its fate. Later approaches to ruins, including Miguel Rivera Maestre's 1834 map of the city of Quiché, maintained these attitudes of prioritising geographical facts over historical ones, or indeed finding history in the landscape as much as in books, suggesting that such landscape-based interpretations of place were well established by the nineteenth century.

If the Palenque explorations demonstrate the logic and practices of the Spanish bureaucracy's knowledge-gathering processes, they also highlight their limited reach. Although the genre of the bureaucratic report could incorporate a broad range of information including economic and botanical knowledge, it transferred the ultimate responsibility for interpreting these excavations to the cosmographer Muñoz in Madrid, reducing their immediate local relevance. The circulation of the government's and its opponents' reports was limited and excluded indigenous voices as well as dissenting opinions even within the broader apparatus of Spanish governance, such as the priest Solórzano's. The priorities of Bourbon reformers, centred around concepts such as 'public happiness' and economic improvement through ascertaining a landscape's potential, informed these reports but did not generate broader discussion. At the time, no newspaper existed in Guatemala which could have reported the findings. The ruins of Palenque dwindled into relative obscurity within Guatemalan discourses after the 1780s expeditions, confirming the limitations of the pathways along which information was able to travel at the time of the ruins' 'discovery' in 1785.

More than ten years later, reports on the city were still scarce. Two articles in the newly founded *Gazeta de Guatemala* mentioned the city several years apart. The first one spoke of 'palaces and houses of hewn stone, and other notable antiquities', but despite the editor's promises, no second article with more details followed.[70] Another on 'ruins in Chiapas' was not the extensive notice that the editors had promised, either. Most of the article was taken up by excerpts relating to Chiapas from Gregorio García's 1607 *Origen de los indios de el Nuevo Mundo e Indias Occidentales*, rather than any more recent scholars, suggesting that no detailed recent reports were available.[71] A further expedition to the ruins under Colonel Dupaix (1805–1808) seems to have created little resonance at the time in Guatemala, perhaps because the Napoleonic invasion of Spain overshadowed such news, or because it was directed from Mexico City and therefore did not directly involve scholars from Guatemala.[72] One Honduran official, for instance, reported hearing about the ruins of Palenque for the very first time sometime between 1806 and 1809 from the Italian Félix Cabrera (an associate of the Creole circle) through a chance encounter while the latter was on his way to Europe with the intention of publishing his treatise on the ruins.[73] Only after the 1822 publication of Del Río's report in London were the ruins the subject of renewed attention, local as well as international.[74] The bureaucratic correspondence networks of the 1780s had therefore been most successful in generating discussion within their prescribed frameworks, but not so much in broader society. They would soon be rivalled by a novel way of communicating knowledge about Central American landscapes for the purpose of enlightened progress with foundation of a patriotic society and its newspaper.

---

[70] 'De Chiapa', *Gazeta,* Vol. 2, no. 91 (10 December 1798), 327.
[71] 'Antiguedades americanas', *Gazeta,* Vol. 5, no. 240 (14 November 1801), 632–3.
[72] Alcina Franch, *Arqueólogos o anticuarios,* 140–2.
[73] José Cacho, 'Extracto del resumen estadístico, corográfico histórico del Departamento de Gracias, escrito por el señor Don José Maria Cacho en el año 1834', *Revista del Archivo y Biblioteca Nacional de Honduras* 4 (1908): 690, note 4a.
[74] Irina Podgorny, "Silent and Alone": How the Ruins of Palenque Were Taught to Speak the Language of Archaeology', in Ludomir Lozny (ed.), *Comparative Archaeologies* (New York: Springer, 2011), 527–54.

2   Networks of Knowledge and Action

New possibilities for communication and knowledge creation opened up in the Kingdom of Guatemala in the 1790s. When officials of the Spanish colonial state had gathered information about the ruins of Palenque and the nature surrounding it in the 1780s, their correspondence networks rested on the set pathways of documents associated with governance and read by a relatively small circle of men. Other circuits of communication existed within which administrators, priests, and scholars created and passed on information, but they were largely confined to direct exchanges between individuals and are less visible in the archival record. This changed through the work of an association of government officials, engineers, merchants, priests, and members of Guatemala City's elite, the *Sociedad Económica de Amantes de la Patria de Guatemala*. The Society's regional and global networks enabled new patterns of communication about matters that preoccupied its members: knowledge about nature, scientific knowledge, but above all the practical challenge of converting these into economic prosperity.

Although the Economic Society's first informal meetings took place in 1794 and the king officially approved it in 1795, the first recorded mention of the possibility of establishing such a group in Guatemala appears in archbishop Caetano Francos y Monroy's 1784 report on his *visita general* of the archbishopric. Not coincidentally, this was the same report in which he suggested that churches should be built with earthquake-resistance architectural features. Here, too, he focused on a narrative of destruction: the archbishop was convinced that the tragic effects of earthquakes would always be an obstacle to Guatemala's development. Aside from this unfortunate situation, however, he found that the Central American terrain was 'one of the best and richest'. Few other places could boast such uncommonly promising potential for becoming 'magnificent and brilliant', especially if an Economic Society were to be established to channel such improvement. No other place, he concluded, needed these '*sociedades de amigos del pays*' more than the archbishopric of Guatemala.[1] It is fitting that the idea for this association should have come from

---

[1] Cayetano Francos y Monroy, 'Visita General', Spring 1784 – Copy of 15 August 1784. AGCA, A1 Leg. 1532 Exp. 10097, 27v.

someone who had just completed an extensive survey of the state of the province, since a *visita*'s emphasis on the interlinking natural, cultural, and economic elements of the landscape was precisely also the type of knowledge that the Economic Society would investigate.

At the core of the Society's mission was a belief, much like Francos y Monroy's, that Guatemala was a land with incredible potential that could be made prosperous through the actions of determined men. Just as the archbishop had envisioned, Society members saw the association as a catalyst for improvement in the name of Enlightenment. The novelty of the Society within the context of Central America was to take the established language of concern about the land, and the interest in using geographical and natural-historical knowledge to improve the governance and economy of the region, and convert it into the basis of practical projects for the improvement of the country. For instance, different sub-commissions turned their attention to exploring the possibility of cultivating cotton around the capital city, to the school where girls would learn to spin the cotton so harvested, and the state of the Guatemalan weaving industry.[2] A school of mathematics (dedicated to principles of engineering) and one of drawing (established after a more ambitious plan to open a school dedicated to all fine art including sculpture did not materialise), were also among the Society's projects.[3] Here the founding members shared the archbishop's concern with managing Central America's sometimes hostile geological environment, declaring that their project of mathematical education for young Guatemalans would produce architects capable of building houses that would not collapse in earthquakes, putting mathematical knowledge to good use for managing landscapes.[4] By planning specific interventions in society and the economy, and an optimistic belief in the productions of the land itself, they displayed some of the hallmarks of *proyectismo*, as Spanish historians have termed the project-based improvement strategies of eighteenth-century reformers. However, their aims were more urgent, and more prone to challenging accepted conventions than would fit into historian José Muñoz Pérez's original definition of *proyectismo* as characterised by 'moderation' and a total lack of even a 'hint of subversion'.[5] Furthermore, top-down reform projects directed from its base in Guatemala City would not be sufficient to fulfil the Economic Society's ambitious plans for gathering and spreading using knowledge throughout the region.

---

[2] 'Concurrencia General 22 Sept 1795', HSA, HC 418 / 563; Real Sociedad Económica, *Noticia de la publica distribucion*.

[3] Luque Alcaide, *Sociedad Económica*, 135–8.

[4] Quoted in Paquette, 'State-Civil Society Cooperation', 287.

[5] José Muñoz Pérez, 'Los proyectos sobre España e Indias en el siglo XVIII: El proyectismo como género', *Revista de estudios políticos* 54 (1955), 169–95.

The Economic Society's members were convinced that if they were to achieve any of its reform aims, it would be with the help of their regional and transnational networks. The Society createned new communication networks that opened the possibility of contributing to enlightened progress beyond bureaucratic networks, through informal and formal meetings, correspondence, and print culture. These were networks both of knowledge and of practical action – they shared information, but also served to organise agricultural interventions and shared botanical specimens. The chief medium of this collaborative Enlightenment would be the *Gazeta de Guatemala*, a weekly newspaper published in Guatemala City between 1797 and 1807. The surviving traces of the Economic Society's correspondence archive as well as a close reading of the *Gazeta* demonstrate that the members of the Society and the editors of the newspaper sought out a socially and geographically remarkably wide range of contacts in order to gather source material for articles, and co-ordinate initiatives for the progress of the country. Guatemala City, as the major city in the region, was the most obvious place at which knowledge networks constructed by private correspondence and contributors to newspapers converged, but smaller towns and rural settings were also important. Beyond the Audiencia, their network included correspondence with their counterparts in Havana, while some high-ranking officials in New Spain also became supporters of the Society or accepted honorary membership.[6] The *Gazeta* facilitated and amplified many of these exchanges through its public platform, and helped to tie together different types of expertise that might not otherwise have shared the same intellectual space. In addition, it proved a catalyst for placing regional practical improvement projects in the context of global influences from places as far away as Philadelphia, Jamaica, and Sumatra. The expansion and effective utilisation of networks of correspondence and learning would create the deep understanding of the natural world that was necessary for the desired practical management of landscapes.

The Economic Society's Networks

The desire to establish communication networks was built into the foundation of the Economic Society. Within months of its establishment, and even before the Society received official approval from the king in December 1795, its membership expanded to include men who lived outside the capital city. Domingo Espina, a resident of the town of Mineral de los Encuentros in El Salvador, was incorporated as a member in October 1795. He had solicited admission to the Society through his correspondent Ignacio Guerra, who was

---

[6] 'Inventario del archivo de la Sociedad Economica', 1847. AGCA, B. Leg.1390, Exp. 32099. 'Havana, 17 diciembre de 1797', *Gazeta*, Vol. 2, no. 61 (4 March 1798), 108–9. Shafer, *Economic Societies*, 216.

a Guatemala City resident, Society member, and Audiencia scribe. In November of the same year, the Society enlisted another 'corresponding member' based outside of Guatemala City, one Antonio Norverto Serrano Polo, a regional official in Ciudad Real, Chiapas. He had asked the Society member nearest to him to facilitate his joining. That member was the *provisor* (vicar-general) José de León y Goicoechea, himself a relative of the learned friar José Liendo y Goicoechea in Guatemala City. Goicoechea duly wrote to Society members, who unanimously voted to admit Serrano Polo to their ranks. Serrano Polo was a highly educated man, and at some point in the 1790s even maintained a correspondence with the Spanish naturalist Mariano Mociño.[7] His introduction by Society member Goicoechea may be considered a mere formality, since Serrano Polo was already well connected and presumably familiar to many of the members.

The process of recruiting new members allows us to see a convergence of previous correspondence patterns and membership, with the formalisation of familial and social ties and the transfer of private into the association's correspondence. Another well-connected clergyman, Juan Francisco Vilches of León in Nicaragua, was similarly introduced to the Society through a correspondent in Guatemala City. He reported that he had received word of the Society's work from a printed 'booklet', probably the record of the Economic Society's general assembly, from his correspondent in late 1797. In direct response to this, he wrote to the Society's secretary, and promptly joined the association in 1798.[8] Existing correspondence and social networks were clearly at the foundation of the Society's membership, but it expanded and formalised them. A shared interest in enlightened projects and the ability to self-declare as a 'patriot' were among the motivations for new members.[9] President Gonzáles did not doubt the genuine 'desire' of these men to contribute to the 'public good', but argued that the Economic Society was able to pursue its useful projects partly because it could motivate its members with honours (*incentivos honoríficos*).[10] Even simple membership of the Economic

---

[7] 'Testimonio sobre establecimiento de la Sociedad Económica'; 'Concurrencia General del Sabado 3 de Octubre de 1795'; 'Concurrencia General, 27 Noviembre 1795'. HSA, HC 418/563. In the 1790s, Serrano Polo held the rank of *teniente letrado* and *teniente asesor*, and by 1796 worked as an attorney for prisoners for the Inquisition: Christophe Belaubre, 'Serrano Polo, Antonio Norberto', *Diccionario AFEHC*; Mark Burkholder, *Biographical Dictionary of Audiencia Ministers in the Americas, 1687–1821* (Westport, CT and London: Greenwood Press, 1982), 322.

[8] Juan Francisco Vilches to Jacobo de Villaurrutia. Leon, 23 December 1797. AGCA, A1.1, Leg. 2007, Exp. 4347, fol. 39. 'Socios correspondientes', AGCA, A1.6, Leg. 2008, Exp. 13842.

[9] For members within the capital who could attend weekly meetings, sociability must have also been important: Dena Goodman, *The Republic of Letters: A Cultural History of the French Enlightenment* (Ithaca, NY and London: Cornell University Press, 1994), 12–15.

[10] Gonzáles Mollinedo, 'Sobre restablecimiento de la Sociedad Económica', AGCA, A1, Leg. 4035, Exp. 3118.

Society appeared to bestow a certain status, with the certificates that members received to confirm their participation becoming sought-after credentials. When the outgoing vice-secretary of the Economic Society Tomás de Moreda relinquished his post, he specifically requested a 'patent or title' confirming his membership and service as vice-secretary, perhaps to include it in a future petition for promotion within the colonial bureaucracy.[11] However, the Society was never content to be a mere discussion club within the capital.

Documents dating from the earliest days of the Society's establishment confirm that a regional reach beyond the capital city, and beyond their own existing correspondence partners, was fundamental to its members' ambitions of agricultural reform. Effective correspondence networks would enable the Economic Society to build up a presence in the countryside by creating proxies and representatives. In this way, they would connect Guatemala City's intellectual elite with the rural population, whom they expected to carry out their visions for rendering nature useful. Within the first few sessions of the Economic Society's governing body in 1795, the founding members resolved to write to the priests of seven parishes across southern Guatemala (Mazatenango, Cuyotenango, Samayeque, Retalhueu, Mita, Jocotan, and Chiquimula de la Sierra), informing them of the establishment of the Society and its plans for the encouragement of harvesting and spinning cotton. They may have targeted these specific villages for the existence of a nascent cotton industry, and Society members believed that parish priests were key figures in transforming it. They were offered membership of the Society and were expected to become its agents within their villages. They would be responsible for putting into practice on a local scale the projects of enlightened productivity and progress which the Society envisioned.

Even before any of the priests replied to accept or decline the offer of membership, the Society members decided to send them the willowing and scutching implements that the Society had imported for the easier processing of cotton anyway, along with instructions on how to use them. The Society's letters would exhort the priests to make these tools and their advantages known to their parishioners. Even more specifically, priests should 'persuade the growers by whatever methods seem prudent to them' to plant and harvest more cotton.[12] Some priests certainly saw themselves as such agents of change. Tomás Calderón, the priest of San Agustín Acasaguastlan and a supporter of the Economic Society's projects, proudly wrote to his Society contact in the city that 'the first ten bushels that I ordered to be sown in June and July were

---

[11] AGCA, A1.1, Leg. 2007, Exp. 4347, fols. 2–7.
[12] 'Concurrencia General del sabado 8 de Agosto de 1795', 'Testimonio sobre establecimiento de la Sociedad Económica'. HSA, HC 418/563.

harvested in September'.[13] Whether the parishioners welcomed such paternalistic interventions into their agricultural practices in the name of enlightened progress is of course doubtful, but priests were able to direct their parishioners' harvests to some extent. The Society recognised this when it promised a gold medal to the priest whose parishioners (whether Indian, Creole, or 'poor Spaniards') cultivated the most flax.[14] The importance of parish priests here paralleled the role accorded to them in the Spanish Count of Campomanes's plans for popular reform set out in his 1774 *Discurso sobre el fomento de la Industria popular* and among other Spanish reformers of the 1790s, but also took into account the important role of priests as representatives of colonial authority in the Americas.[15]

The Society continued to actively recruit correspondents and representatives in the countryside. In addition to cotton, the Society had set its sights on cultivating more cacao in the region. In 1797, they sent twenty-two letters to the parish priests and secular officials (*corregidores* and *subdelegados*) of Guatemala's lowland provinces with the aim of recruiting them to their project of planting cacao.[16] This was merely one of several batches of letters, with the first one sent on 1 August of that year, and a third on 7 October. One reply came from a treasury official from the parish of Santa Ana near Guatemala City. Although his speedy reply, within two weeks, suggested an engaged correspondent, the content of the reply was perhaps a little disappointing. He merely confirmed that he had 'filed' or 'posted' (*fichado*) the Society's printed leaflets in 'the usual places'.[17] Others replies were similarly prompt, and more committed. One Matías Ferrandiz of Mazatenango, perhaps a judicial official, responded to secretary Sebastián Melón within a month, acknowledging receipt of two booklets about cacao cultivation. He assured the secretary that 'it has been explained to the Indians and the [booklets] have been put up in public places so that notice of this may be given to all', although he also cautioned that the lessons could not be applied until the following year, because cacao would be planted in May.[18] The Economic Society had found at least one effective correspondent, although his own reply made it clear that when officials and priests became agents of reform in their towns, this often meant delegating the

---

[13] Tomás Calderón to Antonio Porta y Costas, San Agustín, 9 November 1798. HSA, HC 418/563. In the letter, Calderón also hoped that Porta would show samples of the 'not very fine, but regular' socks that the Indians in his village school had produced.

[14] 'Premio', *Gazeta*, Vol. 2, no. 58 (21 April 1798), 88.

[15] Elena Serrano, 'Making Oeconomic People: The Spanish Magazine of Agriculture and Arts for Parish Rectors (1797–1808)', *History and Technology* 30, no. 3 (2014), 149–76, at 152.

[16] 'Alexandro Espeña, Escribano de la Sociedad Económica', 27 August 1797. AGCA, A 1.1, Leg. 2007, Exp. 4347, fol. 10

[17] Salvador de Coubtiño to José Victoria de Reyes, Santa Ana, 19 October 1797, replying to letter of 7 October, AGCA, A 1.1, Leg. 2007, Exp. 4347, fol. 12.

[18] Matías Ferrandiz to Sebastián Melon, Mazatenango, 9 September 1797. AGCA, A 1.1, Leg. 2007, Exp. 4347, fol. 13, replying to letter of 1 August.

practical aspect to Indians. The town of Mazatenango, and the surrounding province of Suchitepéquez were agricultural areas with a history of cacao cultivation and had, like much of Guatemala, undergone an economic depression in the 1790s. This may explain the enthusiasm of its residents and officials for the Society's project. A second reply from Mazatenango also arrived, this time from a university-educated land owner (*hacendado licenciado*) by the name of Manuel Garrote, who promised to contribute ideas to the Society that would help to 'revive this province'. The next year, he followed this up with his opinion on how to promote the crop in Mazatenango.[19] These detailed responses suggest that the Society had quickly found like-minded correspondents who were willing to act on their behalf in the countryside.

This included using their secular offices to further the cause of enlightened agriculture, much as priests used their influence within parishes. An administrative document of 1802 even implied that there was an official directive for Indians to plant at least twenty *cuerdas* of cacao in the western Guatemalan province of Suchitepéquez. Unsurprisingly, this particular initiative of Society member and governor of Suchitepéquez José Rossi y Rubí met with resistance. Rossi had shown concern with this crop before: in 1799, he had drawn up a 'census' of cacao trees in the province.[20] Landowners of Mazatenango now complained that Indians neglected their *haciendas* in favour of cacao. Indians in nearby Samayeque protested that the governor seemed to want to replace their sugarcane harvests with cacao, and accused him of not understanding their economic realities.[21] We can hear the exasperation in the Indian complainants' voices when they argued that 'we do not have any cacao trees, and we cannot pay the [tribute to king and Church] if we cannot plant sugarcane'. The farmers understood quite well that cacao would be the version of agriculture politically and socially acceptable to the governor, since, unlike sugarcane, it would not be used for distilling spirits (*aguardiente*) illegally. However, the farmers refused to take blame for what others chose to do with their crop. Turning the old Spanish colonial stereotype of Indians as lazy drunks on its head, they stated categorically: 'We sell the sugarcane, but we do not consume *aguardiente*. It is the *ladinos* who use it, and because of these *ladino* vices, our sugarcane harvests are being hindered.'[22] They would not switch crops to satisfy an enlightened project, moral or economic.

---

[19] Manuel Garrote Bueno to Josef Vitoria de Reyes, Mazatenango, 8 October 1797, AGCA, A 1.1, Leg. 2007, Exp. 4347, fol. 6; 'Agricultura', *Tercera Junta Pública de la Real Sociedad Economica de Amantes de la Patria de Guatemala: celebrada el dia 9 de diciembre de 1797* (Guatemala, 1798), 4; Antonio García Redondo, *Memoria sobre el fomento de las cosechas del cacaos* (Ignacio Beteta: Guatemala, 1799), 3.

[20] Real Sociedad, *Quinta junta pública*, 8.

[21] 'Varios vecinos de Mazatenango', 23 April 1802, AGCA, A1, Leg. 207, Exp. 4170. A *cuerda* was a unit of distance or area that could have different local definitions.

[22] 'Los Yndios de Samayaque sobre que el Alcalde maior les deje libre el Comercio de Panelas', 1796, AGCA, A1, Leg. 207, Exp. 4169.

Such agricultural projects took place within established systems of land ownership and labour. Most Society members were personally more likely to grow herbs in a kitchen or monastery garden than several acres on a farm. Director of the Society, Villaurrutia, apparently grew flax 'in his house', while one of the Society's plans for a reform of cacao cultivation was spearheaded by a canon who 'rarely went to the trouble of leaving Guatemala City'.[23] The ability of the Economic Society to make landscape interventions therefore depended on a member's ability to direct and control Indian labourers, conceived as the actions of benevolent landowners and priests something akin to the 'patrician version of improvement' that Daniels, Seymor, and Watkins have observed in rural England.[24] The Society strongly criticised the 'violence' that characterised forced tributary labour in the form of *repartimientos*, even while recognising that Indian labour was fundamental to the kingdom.[25] *Repartimientos* or *mandamientos* had officially been outlawed by the Crown, but different forms of tributary and forced labour continued throughout the colonial period.[26] Aside from Rossi's particularly blunt instrument of trying to enforce a decree that would legislate an enlightened crop into existence, most examples of the Society encouraging particular types of agriculture seemed to leave responsibility for the planting to individual landowners or priests. Perhaps it was because of this abdication of responsibility that a merchant accused the Society's agricultural projects of amounting to no more than 'papers and words'. In the case of cacao plantations, however, they felt able to defend their practical effects. They provided a round-up of cacao plantations, citing the names of individual landowners and the number of cacao trees they had planted, or the amount of cacao they had harvested. They stressed that 196,000 cacao trees had to their knowledge been planted in the last eight years,

[23] Christophe Belaubre, 'Lectura crítica de la "Memoria sobre el fomento de las cosechas del cacao" del canónigo Antonio García Redondo', *Boletín de la AFEHC n°39* (2008): 11; Antonio Lafuente, 'Institucionalización metropolitana de la ciencia española en el siglo XVIII', in Antonio Lafuente and José Sala Catalá (eds.), *Ciencia colonial en América* (Madrid: Alianza, 1992), 103.

[24] Stephen Daniels, Susanne Seymore, and Charles Watkins, 'Enlightenment, Improvement, and the Geographies of Horticulture in Later Georgian England', in David Livingstone and Charles Withers (eds.), *Geography and Enlightenment* (Chicago, IL and London: University of Chicago Press), 354–71, at 346. See also further discussion of labour practices in Chapter 5.

[25] *Gazeta,* Vol. 2, no. 58 (21 April 1798), 88.

[26] Lovell, *Conquest and Survival*, 119–27; Belzunegui Ormázabal, *Pensamiento económico y reforma agraria*, 59–92; David McCreery, *Rural Guatemala, 1760–1940* (Stanford University Press, 1994), 89–112; Wortman, *Government and Society*, 180, for Guatemalan labour. For a recent revisionist history that explains *repartimiento* in terms of the mutual and interdependent interest of hacienda owners, *Consulado* and *alcaldes mayores*, and argues that it was Indian peasants' dependence on credit rather than state coercion that sustained the institution, see Jeremy Baskes, *Indians, Merchants, and Markets* (Stanford University Press, 2000). AGCA Leg. 207 Exp. 4169 shows that mandamiento was not practised in Rossi's Suchitepéquez. An example of *mandamiento* labour continuing in Jocotán in 1804 is AGCA, A1. leg 177 Exp. 3630.

since 1792. This number included 20,000 trees owned by Manuel Garrote Bueno, one of the Society's early correspondents. 'Is that nothing?' they asked mockingly.[27] Ultimately, cacao never had the intended consequence of transforming Central America's economic fortunes, but another summary also took credit for the 'progress' symbolised by new plantations in Soconusco, Escuintla, Chiquimula, and Suchitepéquez, in Guatemala.[28]

Beyond correspondence networks that would eventually affect the country-side through various intermediaries, the Society further tied in members from other parts of the country through *juntas de correspondencia* (associations of 'corresponding members') in different parts of the Kingdom of Guatemala, established with the express aim of contributing chorographic and economic knowledge concerning their respective provinces.[29] The first and most prominent of these associations, the *junta* in Trujillo, met every eight days according to its own report. It was led by the military officer Tadeo Muniesa and among its fifteen members counted Juan Ortiz de Letona, the Crown's chief treasury officer in that town.[30] Individual members also contributed to the Society's reach beyond the capital. Agustín Rodríguez de Zea, a member of the Economic Society (and a relative of the influential reformer Blas Rodríguez de Zea) offered to act on behalf of the Society in the small town of Amatitlán, planting and distributing cotton if the Society could send him 'a considerable number of seeds'.[31] The Economic Society's insistence on priests, governors, and landowners as their agents, and, by extension, their plan to effect change in the agriculture of the countryside through Indian labour, shows the practical ambitions associated with the Society's communication networks. Even less 'elite' members of society could be drawn into these networks, like the two 'farmers [*cosecheros*]' who were asked by the government to provide correspondence between the Society and interested individuals, played a vital part in their aims for agricultural reform and to bring Enlightenment to rural Central America.

The *Gazeta de Guatemala*'s Regional Network

Beyond putting individual connections between men who considered themselves '*ilustrados*' in the spotlight and bringing them together as members of the Economic Society, the new communication network also facilitated the

[27] 'Junta de Gazeta', *Gazeta*, Vol. 4, no. 145 (24 February 1800), 171.
[28] Belaubre, 'Lectura crítica de la "Memoria sobre el fomento de las cosechas del cacao"', 3; García Redondo, *Memoria*, 5.
[29] Real Sociedad, *Junta Pública de la Real Sociedad Económica de Amantes de la Patria de Guatemala: celebrada en 12. de diciembre de 1796* (Guatemala, 1797), 20; Luque Alcaide, *Sociedad Económica*, 56.
[30] 'Junta de Correspondencia de Truxillo', *Gazeta*, Vol. 2, no. 76 (27 August 1798), 225.
[31] Agustín Rodriguez de Zea, Amatitán, 2 April 1797, AGCA, A 1.1, Leg. 2007, Exp. 4347, fol. 5.

Society's most ambitious project: the region's first modern newspaper, the *Gazeta de Guatemala* (1797–1807). It was published as a weekly newspaper comprising four to eight octavo pages (sixteen by 1803), and from the first year onwards was also envisaged and available as a bound volume with all the year's issues. The succession of well-connected editors, the support of the highest government officials including the captain-general, and an engaged readership all contributed to making it a paper of significant geographical and social reach and intellectual impact, as Jordana Dym and Catherine Poupeney Hart's analyses have also shown.[32] On the most basic level, the newspaper was an excellent way of spreading news within Central America, news which otherwise might only be transmitted to narrow circles, to the government through official letters, through the contacts of merchant guilds or rumours.[33] For instance, the editors ensured that the latest news from Europe reached its readers even when the official post from Madrid failed to arrive, in one case reproducing material from papers that had arrived in an alternate way, on a merchant ship coming from Veracruz.[34] It covered some day-to-day news that one might expect from an official gazette, such as royal appointments and regulations published by the Audiencia government, but also more locally relevant news, such as the arrival and departure of commercial ships in the region's ports, prices of goods, and the odd announcement of property for sale in the capital. However, the paper's ambitions went much further. It became to a great extent synonymous with the Economic Society's reform ambitions, its circle of action, and grew to be the region's most important forum for public debate.

A commission of Economic Society members led its publication. It included Guatemala's archbishop Antonio Carbonel, several friars and high-ranking government officials, and Alejandro Ramírez, a Spaniard who was only 20 years old when he took charge of the paper, and acted as editor for at least two years. Later editors included Simón Bergaño y Villegas and Antonio García Redondo, although they always remained anonymous on the printed page.[35]

---

[32] Dym, 'Conceiving Central America'; Poupeney Hart, 'Entre gaceta y "espectador"'. The years 1797–1807 form the core and most distinctive years of this publication. Although a paper under the same name followed it (1808–1816), this was quite a different publication, more akin to an official gazette, containing articles almost exclusively related to, and influenced by, government policy. See also Catalina Barrios y Barrios, *Estudio histórico del periodismo guatemalteco: época colonial y siglo XIX* (Guatemala City: Editorial Universitaria, 2003). An earlier version of the analysis of the *Gazeta* in this chapter has also previously been published as Sophie Brockmann, 'Sumatran Rice and "Miracle" Herbs: Local and International Natural Knowledge in Late-Colonial Guatemala', *Colonial Latin American Review* 24 (2015).

[33] Compare information circuits in New Granada, which in the absence of a newspaper were more heavily based on correspondence, travel, and rumours: Rebecca Earle, 'Information and Disinformation in Late Colonial New Granada', *The Americas* 54, no. 2 (1997): 167–84.

[34] 'Nueva Guatemala 19 de Junio', *Gazeta*, Vol. 1, no. 20 (19 June 1797), 158.

[35] Belzunegui Ormázabal, *Pensamiento económico*, 238. Poupeney Hart, 'Parcours journalistiques en régime colonial'.

The director of the Economic Society, *oidor* Jacobo de Villaurrutia, also happened to be the country's new officer in charge of censoring printed works (*juez de imprenta*). He put himself in charge of establishing a new newspaper to follow the previous official gazettes. Ignacio Beteta, whose printing press would publish the newspaper, had already gained royal permission for such an endeavour in 1794, after he hatched a plan to publish a monthly gazette, although this never got off the ground. Villaurrutia proposed that Society member Ramírez, who was also the tutor of his children, should help Beteta. In his dealings with the Economic Society, Beteta sometimes appears to be little more than a contractor, for instance sending a bill in 1798 for publishing the Society's communications.[36] Although it is difficult to distinguish the authors of many of the articles or notes written by an editorial voice (always in a plural and anonymous 'we'), we can identify Beteta as a central figure in the first years of the paper. He was the author of a few crucial articles, showing the close relationship between the newspaper and the Society's agenda of knowledge production and recruiting correspondents. For instance, Beteta sent and signed a letter containing the draft of an important article on geographical knowledge production three months before it was published in the paper.[37] Finally, the learned friars and Society members Antonio de Goicoechea and Antonio García Redondo were put in charge of revising the paper as censors. As a result, Economic Society members almost wholly controlled the newspaper.[38]

Modelled on the Peruvian newspaper *Mercurio Peruano* (Beteta's original 1793 petition referred to it, and Economic Society member José Rossi y Rubí was one of the co-founders of the Lima paper before arriving in Guatemala) and influenced by José Alzate y Ramírez's *Gacetas de literatura*, the *Gazeta* set out to provide its audience with a wide range of practical and learned knowledge on commerce, political economy, industry, and 'all the sciences'.[39] One of the core ideas of the newspaper was the diffusion of scientific knowledge that would lead to practical improvements for economy and society. Sebastián Melón,

---

[36] 'Cuenta de lo que se ha trabajado para el uso de la Real Sociedad', Guatemala, 18 May 1798, AGCA, Leg. 2008, Exp. 13844, fol. 37.

[37] Ignacio Beteta to Regidor Syndico Joseph Manuel Laparte, 17 February 1797, AGCA, A1 Leg. 5361 Exp. 45314; and 'Memorias para hacer una descripcion puntual del Reyno de Guatemala', *Gazeta*, Vol. 1, no. 14 (15 May 1797), 107.

[38] Medina, *La imprenta en Guatemala*, L–LI, 294–310; Luque Alcaide, *Sociedad Económica*, 121.

[39] 'Prospecto de ampliacion', *Gazeta*, Vol. 1 (1797), unpag.; Dym, 'Conceiving Central America', is a thorough overview and interpretation of the *Gazeta*, its authors, and its reading public. For other newspapers see Mariselle Meléndez, 'Spanish American Enlightenments: Local Epistemologies and Transnational Exchanges in Eighteenth-Century Newspapers', *Dieciocho, Anejo* 4 (2009): 115–33; Rosa Zeta Quinde, *El pensamiento ilustrado en el Mercurio Peruano 1791–1794* (Piura: Universidad de Piura, 2000); Francisco Puerto Sarmiento, 'José Antonio de Alzate y Ramírez ante la ciencia española ilustrada' in Patricia Aceves Pastrana (eds.), *Periodismo científico en el siglo xviii* (Mexico City: UNAM, 2001), 79–107.

secretary of the Economic Society, was convinced that the agricultural and industrial advances of the 'northern nations' could be in some part attributed to having an educated population. After all, the academies and scientific societies of those countries regularly published their proceedings, and the public bene-fited from having access to these 'useful papers and notices'. For Melón, this was a clear argument for publishing summaries of the Guatemalan Economic Society's activities, and the *Gazeta* would be the tool of this public Enlightenment.[40]

The editors set out their hope that the publication would be a challenge to 'inactive minds', stimulating them to 'think, discuss, and invent useful things'.[41] Most Latin American newspapers of the time that considered themselves 'enlight-ened' subscribed to a similar mission of educating their readers. According to the editors, Guatemalan society was in need of the education that this newspaper could provide. The general public in their opinion lacked 'learned correspon-dents', and had 'little affinity for reading', perhaps in a continuation of concep-tions of the 'public' as an uninformed 'multitude' more common in the early eighteenth century.[42] Slightly paradoxically, the *Gazeta* nevertheless envisaged the introduction of '*las luzes*' to Central America as a process that rested as much on the willingness of its readers to act patriotically to improve the region or 'country' (*patria*) as on erudite correspondence and a one-directional dissemina-tion of ideas. The editors vehemently rejected an opinion that they had received by post, which stated that anyone who had 'lived among the Guatemalans for a long time, he would not imagine ... that there could be men of ideas in these provinces'.[43] Despite initial scepticism about the Central American population's capacity for learning and reading among editors and readers alike, the *Gazeta* would, in this ambitious plan, not just diffuse enlightened knowledge from acknowledged authorities, but also become a catalyst for the creation of such knowledge. At least one letter by a reader demonstrated an appetite for scientific topics, complaining that not enough articles on the subjects of 'physics, medi-cine, economy, and agriculture' were being published, even though they had been promised in the *Prospectus* that set out the editors' intentions to potential subscribers.[44]

---

[40] Real Sociedad Económica de Amantes de la Patria, *Quarta Junta Pública de la Real Sociedad Economica de Amantes de la Patria de Guatemala: celebrada el dia 15. de julio de 1798* (Guatemala City, 1798), 3–6.

[41] 'Prospecto de ampliacion', *Gazeta*, Vol. 1 (1797), unpag.

[42] 'Memorias para hacer una descripcion puntual del Reyno de Guatemala', *Gazeta*, Vol. 1, no. 14 (15 May 1797), 107; Roger Chartier, *The Cultural Origins of the French Revolution* (Durham, NC and London: Duke University Press, 2010), 27–8.

[43] The wording and timing of the 'letter', supposedly by one Juan Huron, suggests that it may have been written by the editors as a straw man argument to stir up debate. 'Carta', *Gazeta*, Vol. 1, no. 2 (20 February 1797), 12–15 and no. 3 (27 February 1797), 19.

[44] 'Carta', *Gazeta*, Vol. 2, no. 60 (7 May 1798), 100–1.

The *Gazeta* did not have to build communities of interested men from scratch. From the beginning, the editors stressed the potential of the *Gazeta*'s collaborative model, asking for the help of readers in fulfilling the plan of publication set out in the Prospectus. They mentioned that some 'curious' and 'enlightened' men, 'lovers of the *patria*', had already promised to contribute their time and expertise to the success of the paper.[45] This seems to be a reference to the relatively predictable network of educated men who also composed the Economic Society. But over a short time, the *Gazeta* also expanded this network. The editors promised in the Prospectus to 'print any paper which is sent to the editor, as long as it is considered to have a useful aspect' (with the usual caveats not to criticise Church or government), and also welcomed any criticism of their own work. Objections were welcome 'as long as they treat useful materials, and are written with decorum and courtesy'.[46] Readers' letters constituted about a fifth of all articles, while the number of articles based on information contributed by readers across Central America was certainly even higher.[47] Readers were actively involved in the creation of this periodical, and this involvement was publicly displayed in its pages. Aside from letters to the editor, it also included letters from readers responding to each other. Surveying the state of the newspaper after its first year of publication, the editors noted with pleasure that they had already partly succeeded in 'invigorating' the minds of their readers. They pointed to the lively debates that their correspondents had engaged in throughout the first year of publication, and were satisfied that the paper was playing a useful role.[48] The *Gazeta* and its reading public were therefore central to the formation of a Central American 'public sphere', although it coalesced around government, the clergy, and elite institutions more than in the capitalist and bourgeois model developed by Habermas.[49] Perhaps, then, the intellectual situation in Guatemala was not as dire to start with as the editors had dismissively suggested. Following Dena

[45] 'Prospecto de ampliacion', *Gazeta*, Vol. 1 (1797), unpag.

[46] 'Contravenenos', *Gazeta*, Vol. 7, no. 293 (28 February 1803), 36.

[47] 'Prospecto de amplicacion', *Gazeta*, Vol. 1 (1797), unpag. The percentage of readers' letters is 19 per cent when counting only those articles which are marked in the index or their title as expressing a reader's opinion.

[48] 'Introducion [sic] al tomo segundo', *Gazeta*, Vol. 2, no. 49 (19 February 1798), 1–2.

[49] Jürgen Habermas, *The Structural Transformation of the Public Sphere* (Cambridge, MA: MIT Press, 1989); The 'Guatemalan reading public' is also discussed in Dym, 'Conceiving Central America'; and Belzunegui Ormázabal, *Pensamiento económico*, 139, also stresses the importance of critical thought being taken out of the archives and to the public. On the 'public sphere' in Latin America, see for instance François-Xavier Guerra and Annick Lempérière (eds.), *Los espacios públicos en Iberoamérica: ambigüedades y problemas, siglos xviii–xix* (México: Fondo de Cultura Económica, 1998) – the editors' Introduction, 9–10, discusses the inapplicability of the 'bourgeois' public sphere to colonial Spanish America; Francisco Ortega Martínez and Alexander Chaparro Silva (eds.), *Disfraz y pluma de todos. Opinión pública y cultura política, siglos XVIII y XIX* (Bogotá: Universidad Nacional de Colombia, 2012); Elias José Palti, 'Recent Studies on the Emergence of a Public Sphere in Latin America', *Latin American*

Goodman's argument, to subscribe to a periodical and have one's name published at the end of the volume represented, much like Society membership, an 'ideological commitment' to, and participation in, a new public.[50] Subscribers were part of a new Republic of Letters within Central America.

Because the *Gazeta* published letters by a variety of contributors who otherwise do not feature in the records of the Economic Society or the government bureaucracy, any definition of the newspaper's 'reading public' also helps us to see intellectual connections otherwise hidden from the archival record. The correspondence base made visible in the *Gazeta* shows the range of people who participated in the new information networks that connected rural and urban Central America, even beyond the Society's regular correspondence. Matías Ferrandiz, the official from Mazatenango who had enthusiastically replied to the Society's circular about cacao cultivation, immediately became a subscriber to the *Gazeta* in its first year of publication, in 1797. Perhaps surprisingly, so did the treasury official from Santa Ana whose reply to the Society seemed reticent. Manuel Garrote Bueno from Mazatenango, the landowner who had replied with interest to the Society's letter, does not appear in the subscriber lists, but a priest and subscriber by the name of Ponciano Garrote Bueno, possibly his younger brother, sent contributions to the newspaper by 1798 from the same town.[51] Like the Society itself, the newspaper tapped into pre-existing communication networks and provides an additional printed record of these already existent correspondence patterns.

The newspaper created a new platform for communication that in its ideals at least aspired to a democratic pool of participants. In practice, like much of Latin American public life, it still excluded more people than it included. The range of subscribers and correspondents was formally only constrained by the requirement that a public postal service be available in the subscriber's location.[52] Indeed, the *Gazeta*'s distribution would not have been possible without the postal network that had undergone major reforms and improvements in the previous decades.[53] The postal service also carried readers' letters back to the capital, enabling the collaborative authorship which made the *Gazeta* so distinctive. In practice, of course, many of the newspaper's

Research Review 36 (2001): 255–66; Renán Silva, *Prensa y revolución a finales del s. xviii: contribución a un análisis de la formación de la ideología de independencia nacional* (Bogotá: Banco de la República, 1988); Victor Uribe Urán, 'The Birth of a Public Sphere in Latin America During the Age of Revolution', *Comparative Studies in Society and History* 42 (2000): 425–57.

50 Goodman, *Republic of Letters*, 175–82.
51 'Subscriptores a este tomo', *Gazeta*, Vol. 1 (1797).
52 'Prospecto de amplicacion', *Gazeta*, Vol. 1 (1797), unpag. As the number of letters from readers grew in the first year of the *Gazeta*'s publication, the editors reprimanded their correspondents that they could no longer expect the *Gazeta* to pay for the cost of letters that arrived without the correct postage: 'Aviso', *Gazeta*, Vol. 1, no. 40 (6 November 1797), 320.
53 Sellers-García, *Distance and Documents*, 103–37.

subscribers were part of the educated or ruling elite of Spaniards and Creoles.[54] Their subscriptions to both the *Gazeta* and the Society helped to fund the paper, although the Society also obtained additional funding from a short-lived lottery in its first years.[55] Two hundred subscribers are known for the *Gazeta*'s very first year – a mostly elite minority, yet as a proportion of the kingdom's around 40,000 Spaniards and Creoles, comparable in its circulation to other papers such as the *Mercurio Peruano*.[56] Subscribers included governors, militia captains, treasury officials, parish priests, friars, cathedral canons, the archbishops of Guatemala and Chiapas, as well as a number of individuals without an official title, who may have been merchants or landowners. Where provincial village priests or rural landowners contributed to the newspaper, they represented a sector of society that would not normally have been included in elite discussions on political economy or natural history. An example of a contributor who was not a part of the Guatemala City elite was Antonio Muro, a poor, self-taught Bethlehemite friar residing in Mexico. Between 1797 and 1805, he was a regular contributor to the *Gazeta* and a distinguished Society member (*socio de mérito*).[57] The Society's correspondence networks through the *Gazeta* and through individual corresponding members established a possibility for transmitting information across different spheres of society. Items of knowledge from diverse sources of expertise and different parts of Central American society came to be tied together in one publication.

While the Indian and *mestizo* population – the majority of the population – were not excluded from the Society and its paper in theory, in practice there is no evidence that their participation went much beyond being the subject of various schemes to paternalistically integrate Indians more closely into the Audiencia's economy.[58] Announcing it in the *Gazeta*, the Economic Society might, for instance, offer a prize for agricultural activities such as 'the Indian who harvests the most wheat'.[59] The exclusion of people from a supposedly equal and democratic 'public sphere' based on gender, race, and class was, of

---

[54] 'Lista de subscriptores', *Gazeta*, Vols. 1, 3, 5.

[55] Economic Society correspondence 1797, AGCA, A 1.1, Leg. 2007, Exp. 4347, fols. 10, 29–34; 'Contra Josef Gregorio Torres', 1796, AGCA, A1.6, Leg. 4035, Exp. 31115.

[56] Dym, 'Conceiving Central America', 105–6, 115–18, for an extensive discussion of the *Gazeta* and its readership.

[57] Belzunegui Ormazábal, *Pensamiento económico*, 305; Rojas Lima, 'Muro, Antonio', in *Diccionario histórico-biográfico*, 658–9; Maldonado Polo, *Huellas de la razón*, 232–3. See also announcement of prize in Matías de Córdova, *Utilidades de que los Indios se visten y calzen a la española* (Guatemala City: Ignacio Beteta, 1798).

[58] Belzunegui Ormazábal, *Pensamiento económico*, 229–30.

[59] *Noticia de la publica distribucion de los premios*, 1796. 'Junta de Correspondencia en San Salvador', 1812, AGCA, A.1, Leg. 2008, Exp. 13856, fol. 3. Prizes and competitions were also important in Economic Societies not just in Spain but the rest of continental Europe: see Lowood, *Patriotism*, 103–9; Jeremy L. Caradonna, *The Enlightenment in Practice: Academic Prize Contests and Intellectual Culture in France, 1670–1794* (Ithaca, NY and London: Cornell University Press, 2012).

course, entirely consistent with manifestations of a 'public sphere' in a number of different eighteenth-century contexts.[60] Members nevertheless understood that agricultural workers of rural areas would be the ones to make their projects a success. 'Indian' was the category reformers normally used to refer to all rural labourers, without a nuanced view of different population groups, or taking into account city-dwelling Indians. As one article argued over several issues, Indian labour was the 'most useful' one of all, equated as it was with agricultural work. Other *castas* may have occupations, such as trades, which could be exercised indoors, in the shade, but it was the 'sweat of the brow' of the Indians, their exposure to the elements and cruel overseers, that put food on the table for everyone else.[61] One article filled half a page of the newspaper with a list of all the vegetables and fruits which Indians cultivated and sold, from *achiote* to *zanahoria*.[62] While readers did not want a radical change in the social system, many did recognise the inequalities that existed between Spaniards and Indians, and the disdain in which the former often held the latter.[63] A series of articles conceptualised Indians and other *gente de color* as poor, but not fundamentally different from Spaniards – much like a peasant from Livonia might differ from a peasant from England, in this author's comparison. The author looked for a way to improve the Indians' lot without wanting to break down the barriers between these 'classes' even while he recognised that in America, 'a white man's skin gives him privileges of opinion that in Europe are the reserve of certain classes'.[64] Useful people and their useful actions built stories of inclusion and exclusion, since, as one woman in late-eighteenth-century Madrid argued, it followed at least in theory that nobody who could be 'useful' could be excluded from this community.[65] As Jordana Dym points out, the *Gazeta*'s editors even coined the term *guatemaltecos* or *guatemalenses* in an effort to find an inclusive word for the entire population.[66]

Women's participation in these public exchanges of opinion was similarly possible in theory, but the evidence for it is ambiguous. The common practice

---

[60] Mary Terrall, 'The Uses of Anonymity in the Age of Reason', in Mario Biagioli and Peter Galison (eds.), *Scientific Authorship: Credit and Intellectual Property in Science* (New York and London: Routledge, 2003), 91–112, at 94–5, 106–7; Nancy Fraser, 'Rethinking the Public Sphere: a Contribution to the Critique of Actually Existing Democracy', in Craig Calhoun (ed.), *Habermas and the Public Sphere* (Cambridge, MA and London: MIT Press, 1992), 109–42.

[61] 'Trabajo de los Indios', *Gazeta*, Vol. 5, no. 236 (29 October 1801), 615.

[62] 'Modo de conservar los granos', *Gazeta*, Vol. 5, no. 231 (12 October 1801), 600.

[63] 'Cartas del Cura de N', *Gazeta*, Vol. 1, no. 31 (4 September 1797), 243.

[64] 'Subscripcion', *Gazeta*, Vol. 1, no. 33 (18 September 1797) and 34 (25 September 1797), quote 266.

[65] Theresa Ann Smith, *The Emerging Female Citizen: Gender and Enlightenment in Spain* (Berkeley and Los Angeles, CA: University of California Press, 2006), 105.

[66] Dym, 'Conceiving Central America', 112.

of submitting comments to the newspaper under a pseudonym may have helped women or other 'outsiders' to enter into public debates. Some of the letters also arrived under female pseudonyms, although it is possible that editors or male readers sometimes 'usurped' a female or Indian identity through pseudonyms in order to stir up debate.[67] Editors and Society leaders were happy for women to contribute to society by learning to spin yarn in the Society's school for girls, and considered the 'initially a little disgusting' task of dealing with silkworms women's work. The perception among the Society that students came from the capital's 'deserving' poor, and petitions from mothers about their daughters' admission to the school show the position of these female students as halfway between useful citizens and 'beggars' in the Society's eyes.[68] The Society and *Gazeta*, by contrast, rarely acknowledged women's participation in public life unless it was connected to stereotypically 'female' topics. There was, for instance, the irate '*criolla* from Guatemala' who defended Guatemalan women's fashion against attacks by a Yucatecan woman.[69] This particular letter-writer announced that she was moved to respond after a male relative mockingly read aloud an article in the *Gazeta*. Women were certainly present at intellectual events in the capital. A leaflet printed by the Society about a public examination on Linnaean botany recorded that the daughter of President Domás y Valle asked a question about the application of natural history to 'whimsy and fashion', supposedly in order to steer the discussion towards a topic of relevance to the ladies. Maria Josefa Domás was satisfied with an answer about the extraction of dyes from butterflies and flowers.[70] Despite this patronising commentary, the account makes clear that several women attended this presentation. Society members also recognised the 'patriotic' work of Micaela Carvajal, who studied in the Economic Society's School of Drawing, learnt the art of taxidermy from the Royal Botanical Expedition member José Longinos Martínez, and worked alongside her father to present a collection of minerals to the Society in 1812.[71] In the 1810s, the young intellectual María Josefa García Granados contributed to the intellectual life of the capital (sometimes using a male pen-name).[72] The evidence of female intellectuals in the ambit of the Society presents a tension with the dismissal of most women as

[67] 'Aviso', *Gazeta*, Vol. 1, no. 42 (20 November 1797), 338; Poupeney Hart, 'Entre gaceta y "espectador"', 16; See also Smith, *Emerging Female Citizen*, chapter 3.

[68] Real Sociedad, *Quinta junta pública*, 7–9; Letters in 'Solicitudes hechas en 1796, 97, 98 y 99', HSA, HC 418 / 563.

[69] 'Una criolla incognita, á los Señores Editores', *Gazeta*, Vol. 7, no. 330 (21 November 1803), 433–4.

[70] Real Sociedad, *Noticia del establecimiento del museo de esta capital de la Nueva Guatemala* (Guatemala City, 1796), 17.

[71] Real Sociedad, *Quinta Junta Pública*, 14; Real Sociedad, *Novena Junta Pública de la Sociedad Economica de Amantes de la Patria de Guatemala* (1812), 30–1; Maldonado Polo, *Huellas de la razón*, 418–19.

[72] 'García Granados, María Josefa', in Rojas Lima, *Diccionario histórico-biográfico*, 437.

only concerned with fashion and 'female things', and highlights both the democratising and obfuscating possibilities inherent in the format of a collaborative and anonymous newspaper.

The Economic Society's membership and *Gazeta*'s subscribers never represented a true cross-section of Guatemalan society. However, scattered clues for the participation of women and non-elite men suggest that within the narrow confines of a network that was fundamentally shored up by male Spanish elites, the *Gazeta de Guatemala* demonstrated reasonable success in drawing on the participation of a range of different members of society. In addition, it did succeed in covering a broad geographical area through its membership and participants. Given that the seat of the Economic Society, with many of its most active members, was in Guatemala City, it is not surprising that many of the readers who submitted opinions to the *Gazeta* were from that city. However, subscribers to the *Gazeta* also came from Guatemala's other provinces, El Salvador, Nicaragua, and Chiapas, and included some men from New Spain and the editors of the *Correo mercantil* newspaper in Madrid. Contributions and news from places outside of the capital were not just welcome, but actively encouraged, although the *Gazeta* was not able to paper over long-standing resentments which some inhabitants of 'the provinces' felt against capital-dwellers. A female reader from Danlí in Honduras complained that she had been called a *guanaca*, or country bumpkin, for not having been born in Guatemala.[73] If anything, the Economic Society reinforced these divisions when its statutes declared that it would work for the kingdom, but 'especially this capital and its [surrounding] province'.[74] However, this was less visible in the newspaper, since participation in this community did not presuppose physically attending meetings. Neither in the Society nor the *Gazeta* was Nueva Guatemala given a special status independent of its agricultural base – one's knowledge could not exist without the other.[75] Moreover, despite the Society's 1795 statutes giving 'Guatemala and its provinces' this slightly preferential status, the director of the Society went out of his way by 1798 to stress that he expected the fruits of enlightened transformation to be shared equally, perhaps as a result of the *Gazeta*'s reach. The residents of Chiapas and León should not enjoy Central America's 'natural advantages' any less than the capital.[76] Compared to the *Mercurio Peruano*, the Peruvian newspaper that the *Gazeta* was modelled on, which frequently highlighted Lima as the

---

[73] 'Defensa parcial', *Gazeta*, Vol. 8, no. 341 (9 April 1804), 35.
[74] Real Sociedad Económica de Amantes de la Patria, *Estatutos de la Real Sociedad Económica de Amantes de la Patria de Guatemala, aprobada por S.M. en real cedula fecha en S. Lorenzo á 21. de Octubre de 1795* (Guatemala City: Ignacio Beteta, 1796), 1.
[75] For the importance of agricultural hinterland, see Dym, *From Sovereign Villages*, 20–4.
[76] Real Sociedad, *Quinta Junta Pública*, 3. There are of course echoes of Anderson's *Imagined Communities*, chapter 2, in this community function of the newspaper, although since it was weekly rather than daily it did not have the same temporal implications as Anderson's examples.

embodiment of the Peruvian *patria* and locus of knowledge, the Central American newspaper often defined the Audiencia as a whole as the geographical space in which the enlightened knowledge they sought to gather and disseminate lived.[77]

Using the Networks

The most effective new communication patterns came from connecting the geographically dispersed readership of the *Gazeta* itself. In addition to reproducing the Society's correspondence with key members, the *Gazeta* also provided a platform for members of the 'reading public' to communicate information to a wider audience. This public exchange of information made it possible for individuals outside the Society's leadership to communicate information to the rest of the readership. By melding the patterns of the Economic Society's original correspondence base of priests with the information-gathering possibilities that stemmed from a collaborative newspaper, they ensured that different members of Central American society came to be connected with each other in novel ways. A particularly significant case study of these patterns of interaction and debate was the story of the *algalia*, a plant with antivenomous properties that appeared in *Gazeta* articles between 1799 and 1802 and became a success story of the newspaper's collaborative ideal.[78] The commentary about this plant in the newspaper allows us to follow interactions between Society members, the newspaper's publishers, and more marginal participants in the Society's Enlightenment project. The newspaper offered a new platform through which different modes of communication could intersect. In particular, it offered an avenue for bringing the content of some of the traditional routes of information between priests, bishops, and secular officials into the public sphere.

The *Gazeta* published its first article on the topic of this useful medicinal plant, the *algalia*, in 1799. The event that set off discussions was the visit of the archbishop of León, Lorenzo Tristán, to Guatemala City. Archbishop Tristán was on his way to New Spain, and it seems that he made sure to include a visit to the Society's leadership in his travels. There, the archbishop 'left the seeds of a bush which he called *algalia*' in the care of the Economic Society.[79] This was a geographical transfer on the one hand, and one of knowledge and plant

---

Anderson's own arguments about the Latin American elite as fundamentally un-lettered in chapter 4 are more problematic.
[77] Mariselle Meléndez, '*Patria, Criollos* and Blacks: Imagining the Nation in the *Mercurio peruano*, 1791–1795', *Colonial Latin American Review* 15 (2006): 207–27, at 208–12.
[78] Brockmann, 'Sumatran Rice', note 22 for the identification of the *algalia* as *abelmoschus moschus*.
[79] 'Aviso', *Gazeta*, Vol. 3, no. 133 (29 November 1799), 150.

material from indigenous people to a Spanish cleric, since Tristán had received these seeds from 'Caribbean Indians', probably in his diocese of León de Nicaragua. The Society's members were impressed by the potential that the seeds of this medicinal plant offered, and announced the news in the *Gazeta*. In response to this article, Eugenio Merino, the parish priest of the village of Texacuangos (El Salvador), requested the *algalia* plant. Merino was also a *Gazeta* subscriber and long-term participant in the Society's projects.[80] Two years after he requested the seeds, the *Gazeta* reported that not only had the priest been able to receive the plant from one Ignacio Somosa, he had also distributed it throughout his parish and beyond: in El Salvador, a seven-year-old girl in Texacuangos and five people in Zacatecoluca had been healed after being bitten by snakes.[81] Rather than reporting on this plant passively, the *Gazeta* played an active part in spreading the medicinal plant across Central America. The newspaper had provided the platform and network that made this transfer of the plant between two individuals and further into El Salvador possible.

In this series of articles, it became clear that the plant had travelled from Nicaragua across Central America over the previous decade. For one reader, the *algalia* article had not been news, and he made sure that the editors knew it. The *corregidor* of Quetzaltenango, Prudencio Cozar, had obtained *algalia* seeds from Masaya in Nicaragua almost twenty years previously, in 1781, and he had himself attempted to distribute this useful plant around Central America. Cozar emphasised his own physical contribution to the seed's propagation, having 'taken it with me on expeditions and commissions, as it is known, I have travelled all over this kingdom [in their pursuit]'. It was on one of these official trips that Cozar had visited the town of Mazatenango, in the Guatemalan province of Suchitepéquez in 1793, where he gave some of the seeds to the priest Ponciano Garrote. The *Gazeta* had mentioned Garrote in its original 1799 article, but did not seem to know this back story.[82] Further letters also started pouring in. The priest Garrote himself also wrote to the *Gazeta* to confirm the *algalia*'s curative powers for the bites of 'all manner of snakes'.[83] The Society was also able to add further examples of the plant's propagation, which esteemed members of the Economic Society had witnessed with their own eyes: in Esquipulas in Guatemala, a man suffering from rabies had been cured, and Society members saw him walking around the town the day after receiving the *algalia* treatment. In Quiriguá, the landowner and Society founding member Juan Payés y Font had healed a man with it who was thought to

---

[80] 'Mencion honrosa del cura de Texaquangos', *Gazeta*, Vol. 3, no. 131 (18 November 1799), 132; 'Cuenta y cargo de data', AGCA, A.1, Leg. 2006, Exp. 13799.
[81] 'Algalia', *Gazeta*, Vol. 5, no. 216 (3 August 1801), 531–2.
[82] *Ibid.*, no. 222 (10 September 1801), 560–1.   [83] *Ibid.*, no. 235 (26 October 1801), 612.

have been dying. With these examples, it seems that the Society also wanted to claim responsibility for spreading this useful plant.[84]

More than tracing the origins of this plant, though, the *Gazeta* was now becoming a hub for new transfers of knowledge and plant material. Cozar's concern was not so much with claiming credit for the distribution of the plant (although there was an element of that), it was that he disagreed with a botanical detail that Merino of Texacuangos had provided. Merino had claimed that he knew when the seeds had gone off by their lack of distinctive smell when compressed. However, Cozar was convinced that the seeds never expired – he had 20-year-old seeds in his possession, which he assured the editors had not lost their smell. He had promptly sent in the seeds alongside his letter. Having examined the seeds, the editor now promised to plant them and report to the readers whether they grew. The prominent Society member José Rossi y Rubí, writing from Mazatenango (also the priest Ponciano Garrote's parish), now reported on a series of experiments that he had carried out with the plant. In order for the Society to be able to 'verify' them, he sent 'a small box of the same seeds, freshly picked', with the hope that hospitals and professors of medicine would use them. In the next issue, the *Gazeta* clarified that this box was at the house of one of Rossi's friends in the capital, and that the surgeon-general Narciso Esparragosa had already performed successful experiments with it.[85]

The initial success in promoting the *algalia* was a testament to the strength of the Society's routine networks of travel and correspondence as they intersected with administrative journeys, but its further spread was more specifically due to the *Gazeta*'s ability to forge new connections. Many early contributors to the discussion and propagation of the *algalia* had been in the ambit of the Society since the beginning. The Garrote family from Mazatenango had been part of the Economic Society since the early days of their project for cacao cultivation. Another Society member by the name of Francisco de Campo, who now contributed his observations on the *algalia* in relation to healing animals as well as humans, had previously sent in a report on opportunities to improve the kingdom's pottery industry shortly after its foundation.[86] These exchanges were significant by themselves, since they revealed that the Society functioned as its founders had envisioned, with members and correspondents contributing and gathering useful knowledge on a variety of topics including botany, agriculture, and industry. However, these exchanges would most likely not have been quite so effective were it not for the additional networks of the

---

[84] *Ibid.*, no. 216 (3 August 1801), 531–2.
[85] *Ibid.*, no. 234 and 235 (22 and 26 October 1801), 609, 612.
[86] 'Concurrencia General del sabado 8 de Agosto de 1795', HAS HC 418/563; 'Cuenta y cargo de data', AGCA, A.1, Leg. 2006, Exp. 13799; 'Aviso', *Gazeta,* Vol. 3, no. 133 (29 November 1799), 150.

Gazeta. These were residents of rural Guatemala and of El Salvador who were less well connected with each other than members in the capital who were able to attend regular meetings. The *Gazeta* allowed them to engage with each other directly. The discussion of the *algalia* as it played out in the pages of the newspaper makes visible a range of contacts, maintained in person and through correspondence, through administrative duties and private interest, which connected Central Americans. Be it Cozar distributing it on his travels or Merino in El Salvador, these journeys and contacts would not be visible without the *Gazeta*'s reports. After Cozar had given the seeds to the priest Garrote, for instance, it took six years and the arrival of the *Gazeta*'s communication networks for this transfer to become visible to the public, and therefore the historical record. Rossi y Rubí also reported to have been in its possession for six years, and again did not communicate his empirical observations on the topic until the *Gazeta* made it an issue.[87] Because the newspaper reproduced the exact reports sent in by readers, we can follow their interactions and motivations in vivid anecdotal detail. In discussions of the *algalia*, the newspaper contains rich clues about the practical and social context from which plants came. Newspaper articles therefore deepen our understanding of the social interactions which formed the basis of the *Gazeta*'s local and regional networks of knowledge exchanges.

Knowledge networks could create and disseminate useful information, but ultimately, this knowledge was at its most valuable when it led to enlightened interventions in the landscape. While the discussion of the *algalia* had played out largely among readers, the Economic Society also used its networks to solicit practical action from the *Gazeta*'s readers, just as the Economic Society had asked parish priests to intervene in the agriculture of their towns and villages. Articles exhorted readers to collect seeds of exotic plants from a particular citizen's house and try to grow them, or to collect any useful plants they encountered on their travels.[88] Medicinal plants were not the only context in which the reach of these networks proved their efficacy. The *Gazeta* even advertised an offer to collect the larvae of silkworms from the house of Society member and student of natural history Pascasio Letona in Guatemala City. The Society had arranged for the worms to be sent from Oaxaca, and any interested reader could now pick them up for free, along with instructions on how to raise and use them.[89] Readers who participated in such experiments would be fulfilling a vital role in the Society's broader plans. In the case of the silkworms, there was

---

[87] 'Algalia', *Gazeta,* Vol. 5, no. 234, (22 October 1801), 610.
[88] For instance, 'Yerva de Guinea', *Gazeta*, Vol. 8, no. 384 (11 March 1805), 398; 'Semilla', *Gazeta*, Vol. 5, no. 196 (23 March 1801), 426.
[89] *Gazeta*, Vol. 1, no. 43 (27 November 1797), 346. Pascasio Ortiz de Letona had studied botany under Mociño and Longinos Martínez in 1796: Real Sociedad, *Noticia del establecimiento del museo*, 6–12.

an Economic Society sub-committee whose purpose it was to encourage the production of silk in Guatemala as part of one of their projects to diversify the economy and reduce its reliance on indigo production.[90] The *Gazeta* helped to advertise this project and tried to recruit its readers' active participation.

The Economic Society's membership, sometimes independently from the *Gazeta*, also worked together to gather useful materials in the realm of dyes with potential industrial applications. These were clear examples of exchanges that promised economic utility, and it was clear that only practical experience and physical examples of the plant material mattered to the Society. Its networks could succeed in locating plants where even the professional botanists of the botanical expedition had failed. The Society had asked the expedition's botanists to look for a dye-producing plant by the name of *rubia* in Guatemala, but they were apparently unable to locate it. The Society then attempted to import the plant's seeds from Spain twice, but they arrived spoilt. However, when the Society widened its search in 1816 and was about to announce a prize for whomever might find 'this precious plant', the provincial of the Santo Domingo convent presented them with three ounces of the precious seeds, and a commission of three members of the Society then distributed them across Central America.[91] The friar presumably searched for the seeds in response to the Society's renewed efforts. He was a distinguished member (*socio de mérito*) and had previously sent the association plant samples in the hope they might be considered useful.[92] The success in finding the *rubia* plant represents an example of the members' network providing positive results where the botanical expedition could not.

Networks of members and *Gazeta* readers continued to perform such services useful to industry. In 1816, the Society sorely lamented the loss of a small colony of dye-producing cochineal insects that they had been sent from Mexico in 1811. However, the eminent Dominican friar Antonio López from the Baja Verapaz town of Cubulco came to the rescue where the Society's own project had failed. As the editors of the newspaper approvingly noted, the friar had kept the 'little animals' alive and had even found a way to raise them on common cactuses (*nopal del país*) rather than the Mexican plants (*nopal fino*). These actions, the Society's leadership concluded, made the friar a true patriot. Finally, the cochineal experiments started paying off, and the first harvest of 70 pounds of cochineal made its way towards Spain in 1818. In these cases, the network of correspondents and contributors fulfilled

---

[90] 'Concurrencia General 22 de Julio de 1795', HSA, HC 418 / 563.
[91] 'Continua el resumen de las actas de la Real Sociedad, leido por el Secretario en la junta publica general de 27 de Diciembre ultimo', *Periódico de la Sociedad Económica*, No. 19 (1 February 1816), 298–9.
[92] 'Carta del MRP Provincial de Santo Domingo al Secretario de la Sociedad', *Periódico de la Sociedad Económica*, No. 16 (15 July 1815), 90–1.

its early promise of supplementing and even surpassing the knowledge and capacities of the Economic Society's central branch. Moreover, they show the importance of the membership's extended contacts that were able to draw additional experts into the ambit of the Society: while the Society blamed the lack of 'practical knowledge' (*inteligencia practica*) for the initial loss of cochineal, the new insect colony succeeded because Antonio López sent it in the care of an expert (*inteligente*). This anonymous 'expert', possibly an indigenous cultivator from Oaxaca, was employed by the Society for a year and a half and demonstrated the correct planting, maintenance, and harvest of the insect dyes.[93] Jeremy Caradonna has observed that prizes and other invitations to communicate with learned and economic societies in Enlightenment France were marks of a 'participatory Enlightenment', an opportunity to 'mine expertise' from within a wider population.[94] These examples of the Guatemalan Economic Society as a hub for material exchange show that beyond merely participating in an urban elite's competitions and prizes, a range of contributors were encouraged to produce new enlightened knowledge and practical results for the Society's day-to-day operations..

Global Networks

While many of the letters submitted by the readers were intended to communicate knowledge or material between different parts of Central America, other contributions were the result of members of the Economic Society reporting back from their travels or drawing on correspondence from outside of Central America. Two of these travellers were Alejandro Ramírez y Blanco (editor of the *Gazeta* from 1797) and the merchant Francisco Sosa, who returned from an extraordinary Caribbean voyage in 1801 with a valuable botanical shipment: eighteen boxes of live 'exotic' plants, and a collection of seeds from Sumatra. They had departed two years previously: Sosa as the commercial representative of the influential merchant Juan Bautista Irisarri; Ramírez, with his knowledge of French and English, as an interpreter on the same ship. Although he was officially in charge of dealing with 'correspondence and contracts', Ramírez later stressed that he did not receive a share of the trading voyage's profits. For him, this was chiefly an opportunity to travel to new countries, and acquire new knowledge.[95] This was indeed a rare opportunity for a young Spanish traveller.

---

[93] 'Grana', *Periódico de la Sociedad Económica*, No. 23 (1 April 1816), 329; Antonio López, *Instrucción para cultivar los nopales y beneficiar la grana fina* (Guatemala City: Ignacio Beteta, 1818). For the role of indigenous expert cultivators of cochineal who travelled from Oaxaca to Guatemala with the precious cargo, see the work of Deirdre Moore (PhD dissertation, Harvard University, forthcoming).

[94] Caradonna, *Enlightenment in Practice*, 90–117.

[95] 'Certificación de méritos y servicios' [Alejandro Ramírez], transcribed in Héctor Humberto Samayoa Guevara, *Ensayos sobre la independencia de Centroamérica* (Guatemala:

Ramírez and Sosa had received an official permit to visit not just Spain's Caribbean possessions, but also 'foreign colonies'. While there were well-established trading channels for the contraband route between Honduras and Jamaica, it was only during a brief moment after 1797 that, thanks to a royal decree allowing trade with 'neutral countries' and several exceptions granted by Guatemala's Audiencia court, a Central American trading journey to the British as well as French Caribbean could legally take place.[96] Perhaps to deflect attention from the controversial permits on which their journey depended, the travellers did not describe their route in detail, but it seems certain that they visited Jamaica, since they were unlikely to have found the collection of plants and seeds anywhere else. In early 1801, they arrived back in Guatemala via the harbours of Trujillo and then Santo Tomás. The *Gazeta de Guatemala* published a letter with information about the 'exotic' botanical bounty that they had brought with them over several issues.[97]

The journey of the plants and seeds to Guatemala had been a long one, but it was shaped throughout by designs of enlightened improvement. The origin of the collection of Asian seeds that now appeared in Guatemala, as Alejandro Ramírez explained in the *Gazeta*, was an assortment of seeds and their description made by one Dr Campbell, 'a botanist employed in the service of the [East India Company] resident in Sumatra'.[98] Charles Campbell, as it happened, was engaged in a project similar to Ramírez's own. As a botanist in the employ of the East India Company, he was very familiar with the aims of economic botany. Between 1798 and 1803, for instance, he was involved in attempts to acclimatise and propagate nutmeg and cloves around the East India Company's Fort Marlborough in Sumatra. The collection of Asian seeds that the *Gazeta* ended up reporting on was a product of this context. Campbell must have compiled the collection by late 1800, and initially sent it to Britain. There, it was apparently designed to contribute to the work of a society for agrarian

Ibarra, 1972), 36–7; Poupeney Hart, 'Parcours journalistiques en régime colonial, 7. *Gazeta*, Vol. 5, no. 221 (7 September 1801), 554, refers to the paper's first editor having been absent for two years.

[96] Carolyn Hall and Héctor Pérez Brignoli, *Historical Atlas of Central America* (paperback ed., Norman, OK: University of Oklahoma Press, 2003), 132–3. Ramírez spoke about voyaging to '*las islas*' (*Islas de Barlovento*, that is the Spanish possessions of Cuba, Puerto Rico, and Santo Domingo within the Lesser Antilles), rather than giving details of their journey to Jamaica. The short-lived permission for trade with 'neutral countries' happened in 1797 on the occasion of war with Britain blocking other Atlantic traffic. For the controversy of Irisarri's permits, see Dewitt Chandler, 'Jacobo de Villaurrutia and the Audiencia of Guatemala, 1794–1804', *The Americas* 32, no. 3 (1976): 402–17. Ramírez's *relación de méritos*, see previous note, specifically refers to a permit to visit '*colonias extranjeras*'.

[97] 'Catálogo de plantas traídas á Trugillo por D Alexandro Ramirez y D Francisco Sosa', *Gazeta*, Vol. 5, no. 194 (9 March 1801), 413–15; 'De semillas de otras plantas asiaticas', *Gazeta*, Vol. 5, no. 195 (16 March 1801), 421–2 and no. 196 (23 March 1801), 425–6.

[98] 'Carta', *Gazeta*, Vol. 5, no. 195 (16 March 1801), 421–2, 'Semillas', *Gazeta*, Vol. 5, no. 196 (23 March 1801), 425–6.

improvement not unlike Guatemala's Economic Society in its goals: the British Board of Agriculture. Campbell's description of the Sumatran seeds appeared in the Board of Agriculture's own periodical publication, the *Annals of Agriculture and Other Useful Arts* of that year.[99] In fact, the three full pages of the description in the *Gazeta* are a faithful translation into Spanish of Campbell's description as it appears in the *Annals*. Both descriptions presumed that the reader had access to the seed collection, since they include detailed references to the accompanying seed packets. For instance, the 'packet marked A' contained an 'upland, or dry rice', while packet B contained 'a shorter and light-coloured grain, esteemed a stronger and more nutritive food': information reproduced in both the Spanish- and English-language articles. This suggests that the Board had sent out many examples of the Asian seeds to its correspondents alongside copies of Campbell's letter, encouraging its members to cultivate them.

Ramírez's source must have been a copy of the original letter circulating in Jamaica or a personal acquaintance with a person in charge of the plants, since he was in possession not only of the description, but also the highly prized seeds. Moreover, the publication dates of the two articles make it clear that Ramírez was drawing on a copy of the original letter rather than the British publication, since the article in the *Gazeta de Guatemala* appeared in the same year as its equivalent in the *Annals of Agriculture*, in 1801. With a publication date of March 1801, the *Gazeta* article may well even predate the British version by a few months. Ramírez probably also had some access to the Jamaican botanical gardens, since in addition to the Sumatran seeds, he also brought live plants with him from Jamaica, including breadfruit, cinnamon, jackfruit, ackee, and mango. These plants were not native to Jamaica, but had rather been acclimatised in one of the island's two botanical gardens. Their arrival in Guatemala therefore made them but the latest stop in a multitude of British and French imperial journeys that had started in places as diverse as Tahiti, West Africa, and Mauritius.[100] The travel of plants was not out of the ordinary, in that botanical materials regularly moved around the British, Spanish, and French empires for projects of economic botany and acclimatisation, and ostensibly peripheral places like the Jamaican gardens could also operate an autonomous centre of botanical collection and distribution.[101]

---

[99] On Campbell, see William Marsden, *The History of Sumatra: Containing an Account of the Government, Laws, Customs, and Manners of the Native Inhabitants* (London: McCreery, 1811), 147–8. On the Board, see C. A. Bayly, *Imperial Meridian: The British Empire and the World 1780–1830* (London and New York: Longman, 1989), 121–4; Jonsson, 'Scottish Tobacco and Rhubarb', 129–47. Publication of Campbell's description: 'Oriental Plants Cultivated in the East Indies', *Annals of Agriculture* 37, no. 214 (1801): 557–61.

[100] 'Catálogo de las plantas traídas', *Gazeta*, Vol. 5, no. 194 (9 March 1801), 414–15; for more detail on the origin of individual plants, see Brockmann, 'Sumatran Rice', 96.

[101] McAleer, 'A Young Slip of Botany'.

However, these were usually carefully guarded national projects and it is less usual to see a collection of these imperial products move from one empire to the other. While we do not know the precise story of how Ramírez received the seeds, plants, and plant catalogue, it seems likely that it was based on scholarly or commercial contacts with the director of the Jamaican botanical garden or other scholars on the island. It shows that there was a possibility of local agents sharing botanical samples, probably based on commercial or intellectual rather than diplomatic connections between the Spanish and the British. The movement of these botanical materials and information around the globe was therefore the culmination of an extraordinary series of individual journeys and informal contacts between representatives and subjects of different empires. Ramírez's journey to Jamaica gave him an inroad into a world of imperial journeys, establishing a new lateral contact that was more connected to illicit Caribbean trading networks and scholarly exchange than the high-level politics of the British and Spanish empires.

The subsequent acclimatisation of the plants and seeds in Central America, too, created a parallel world to the British acclimatisation project that they were originally intended for. The *Gazeta* made the 'exotic plants' its new project, just as the newspaper had helped to distribute the *algalia* plant across the region. Editors informed readers about the plants' progress and took an active role in their distribution and cultivation. The Sumatran seeds which Ramírez and Sosa had sent to the capital were distributed to various persons who planted them in the capital and on the southern coast. A notice in the *Gazeta*, for instance, mentioned that Ignacio Guerra Marchán, government scribe and an early member of the Economic Society, had successfully grown a type of plum tree that appeared on Ramírez's list as *ciruelo del mar del Sur* on his hacienda in Escuintla on the southern coast.[102] Another article on the 'Method of cultivating *yerba de guinea* [Guinea grass]', again a plant on the list, sought to encourage even more readers to introduce these plants to Guatemala by giving guidance on how to grow them.[103] The parallels between the British Board of Agriculture, who commissioned the collection, and the Economic Society of Guatemala, who now took up the mantle of this originally foreign acclimatisation project, are striking. Just as the aim of the *Gazeta* in publishing the account of these plants was to demonstrate the effectiveness of its information networks and encourage Guatemalan patriots to grow these plants, the aim of the *Annals of Agriculture* in publishing the description was to show the travel of these plants within the context of the British empire's acclimatisation projects and to ask every correspondent to 'transmit . . . an account of his success to the President' of the Board.

---

[102] 'Plantas exóticas', *Gazeta*, Vol. 6, no. 283 (1 November 1802), 282–3. The 'plum tree' in question was most likely *spondias dulcis*.
[103] *Gazeta*, Vol. 7, no. 294 (25 February 1803), 45.

As the *Annals*'s editor wrote, 'it is presumed that [they] will succeed as well in the inter-tropical colonies of Great Britain, as in their native climate'.[104] Through the *Gazeta*'s work, the Economic Society members' gardens and *haciendas* across Central America became sites of experimentation analogous to those of the British empire, although perhaps unwittingly.

The newspaper and Society's propaganda efforts appeared to pay off. Some of the initial experimentation in plant acclimatisation and seed planting had taken place in the botanical garden which Society member (and Crown treasury official) Juan Ortiz de Letona had established. For instance, he had cultivated the fruit-bearing *bilimbi* tree there. The *Gazeta* happily reported that Letona had enjoyed eating the fruits ('similar to a salad cucumber') and that the tree had a fast sprouting time. The first examples planted were now growing so nicely that they already appeared as if this were 'their soil and climate [*suelo y clima*]', and it had been worth planting them for a second time.[105] However, it was the propagation beyond the botanical garden that would truly prove the plants' incorporation into the local flora. Beyond scattered examples of Society members experimenting with specific seeds, at least one of Ramírez's Sumatran seeds, a type of rice, seemed to find its way into Central American agricultural production. In the area around Trujillo in Honduras, where the 'exotic' plants had first arrived, there is evidence of the plants and seeds from this shipment thriving years later. In late 1803, two settlers in Chapagua, a new settlement just outside Trujillo, reported in the *Gazeta* on the 'abundant' quantity of rice they had harvested, and sent a sample of it to the capital city to prove its quality. The settlers specified that it was 'not an indigenous product of this part of America', which led the newspaper editors to believe that this rice was part of the seeds shipped 'from Sumatra which in the year 1801 was brought to this kingdom'. It is evident that the newspaper editors related the rice sample they were sent directly to Campbell's original description of Sumatran rice, matching its taste and qualities to the botanical information that the newspaper itself had published two years earlier. After all, the British botanist had spoken of 'Packet B' of his rice seeds as a 'stronger and more nutritive food' than the first packet, and the *Gazeta* editors now described the Honduran rice as 'stronger and more nutritious than common rice'.[106] The circuitous journey of the Sumatran rice had seemingly led to one instance of enlightened improvement to local cultivators, although not in quite the way that Dr Campbell had intended.

---

[104] 'Oriental Plants Cultivated in the East Indies', *Annals of Agriculture*, Vol. 37, no. 214 (1801), 557.

[105] 'Plantas exóticas', *Gazeta*, Vol. 6, no. 279 (2 October 1802), 252 and no. 283 (1 November 1802), 282–3.

[106] 'Progresos de la agricultura en Truxillo', *Gazeta*, Vol. 7, no. 334 (19 December 1803), 469–70.

The 'exotic plants' led a long 'afterlife' once they arrived in Guatemala, not just in their cultivation in Trujillo and across the isthmus, but also in the publications and conversations they generated. The successful introduction of the plants to Guatemala was no mean feat. It was celebrated accordingly in the *Gazeta*, and was even picked up by newspapers beyond Central American shores. The Spanish economic and agricultural journal *Correo mercantil* reprinted several articles from the *Gazeta* about the exotic plants being introduced in Guatemala, as well as their further cultivation within the Audiencia.[107] The *Correo*'s mission was the dissemination of economic information between Spain and its colonies. Due to its editors' economic philosophies supporting the liberalisation of Atlantic trade and frequent complaints about the lack of accurate information, especially from the colonies, the newspaper was not universally popular.[108] The *Gazeta*'s editors, however, saw an ally in the Spanish paper and embraced its potential to amplify their message. Thanks to the *Correo*'s wider circulation, the reprints of *Gazeta* articles provoked additional responses. The *Gazeta* in turn commented on the articles that were appearing and reprinted advice on rice cultivation offered by the *Correo*, proudly noting that correspondents in Cuba, New Spain, and elsewhere were 'badgering' the editors for more information on the Sumatran plants.[109] While information about local plants such as the *algalia* signified a success story because of their local usefulness, the Sumatran seeds and Jamaican plants formed an internationally acclaimed coup for the Economic Society and its newspaper. They created a meaningful transatlantic conversation that unfolded in the pages of newspapers about acclimatisation and the practicalities of rice growing.[110] It had been proven that the networks of knowledge that tied together locations from Masaya to Guatemala City and Trujillo could also be projected outwards, to the borders of the Spanish empire and beyond.

This transatlantic exchange was significant because Guatemalan reformers placed their ability to connect with the world beyond on a pedestal. Being part of transnational networks of knowledge could be seen as an asset in itself for what it meant for Guatemala's place in the world. That at least was the reason the editors gave when they defended their practice of reprinting articles from elsewhere that were perhaps no longer 'news'. The editors here were referring to material that might have originally appeared in Philadelphia or Britain, but had probably

---

[107] For instance, *Correo mercantil de España y sus Indias*, no. 38 (10 May 1804), 298–9. See also Brockmann, 'Sumatran Rice', 98.

[108] Barbara and Stanley Stein, *Edge of Crisis: War and Trade in the Spanish Atlantic, 1789–1808* (Baltimore, MD: The Johns Hopkins University Press, 2009), 14–19.

[109] 'Plantas exóticas', *Gazeta*, Vol. 6, no. 273 (23 August 1802), 197–9.

[110] Brockmann, 'Sumatran Rice', 98–9.

arrived in Guatemala already translated and abstracted into other newspapers or journals.[111] If periodicals in Spain were in the habit of reprinting articles, then the *Gazeta* should do the same, the editors argued. It could even be an act of 'pure patriotism' to re-print the articles of Spanish newspapers, because it signified that 'Guatemala should not be less than Jerez de la Frontera, Gerona, or Salamanca', that is, a place of reading and science equal to Spain.[112] Given that a host of anecdotal evidence shows a lively intellectual culture and information exchange, the editors sometimes over-emphasised the alleged imbalance between the knowledge available in Guatemala and the rest of the world. Nevertheless, they maintained the ideology that the newspaper and the Economic Society's members worked to overcome intellectual distance conditioned by geographical distance.[113] Being able to bring exotic plants and seeds to Guatemala and effecting a change in agricultural practices was merely one easily traceable example of a process that encouraged travellers and correspondents to use their networks for the purpose of improving the region, whether through writings on botany, medicine, geography, or political economy.

The *Gazeta* frequently sang the praises of José Felipe Flores, the country's leading physician and supporter of the Society's projects, who undertook a voyage to many of the scientific centres and universities of Europe and North America in 1797 and sent back information about the 'current state of the sciences' in Philadelphia and other locations, which the *Gazeta* duly published.[114] José Mariano Mociño, the only American-born member of the Royal Botanical Expedition, also maintained regular correspondence with several of the Guatemalan reformers after the expedition's departure. As both an acclaimed botanist in the service of the Crown and a Creole from neighbouring New Spain, his ability to use this unique position to connect Guatemala to the wider world was not lost on the Economic Society's members. Alejandro Ramírez, as editor of the *Gazeta*, repeatedly highlighted the close connection between Mociño and the Economic Society, as well as his own friendship with the Mexican naturalist. The newspaper effectively combined private communications with its public ambitions when it published several treatises by Mociño that were not otherwise available in Guatemala. In 1797, Mociño had become interested in the question of indigo production after conversations with both Mexican and Guatemalan indigo farmers who all suffered from the same problem – the devastatingly unhealthy, cumbersome, and unreliable process of

---

[111] See also Sophie Brockmann, 'Retórica patriótica y redes de información científica en Centroamérica, 1790–1814'. *Cuadernos de Historia Moderna*, Anejo XI (2012): 179–81; Brockmann, 'Sumatran Rice', 94–5.

[112] *Gazeta*, Vol. 5, no. 221 (7 September 1801), 554.

[113] Compare Bettina Dietz, 'Making Natural History: Doing the Enlightenment', *Central European History* 43 (2010): 25–46, at 39, on availability of knowledge in Europe.

[114] 'Carta del Dr Flores. Filadelfia. Mayo 17 de 97', *Gazeta*, Vol. 1, no. 45 (11 December 1797), 356–66.

extracting dye from the plant.[115] Encouraged by the director of the Society, Jacobo Villaurrutia, he continued to work on this question even after leaving Guatemala.[116] A year later, Mociño had finally completed his short treatise on the matter and even developed an oven especially for this new method. However, as the editor of the *Gazeta* pointed out, the treatise existed for now as a private communication and had not been published. It had, however, been presented to the Economic Society, whose members were preparing to publish an annotated version. A contributor to the *Gazeta* who had seen the treatise was in a position to present excerpts to the newspaper, almost three years before the full publication was available for purchase in Guatemala.[117] This roundabout, but rapid way of disseminating at least some of the treatise's content shows how seriously the newspaper and its informants took the opportunity to circulate information more quickly than via established publication routes, and its ability to make available scientific news on topics which might in the past have remained within the closed information circuits of correspondence between naturalists or government officials. The newspaper also published Mociño's work *Noticias de Nutka* in instalments in 1803 and 1804. Mociño had written this account of the people and natural world of Nootka Sound in 1793, after journeying to what is now Vancouver Island, Canada, as the naturalist on the Spanish expedition led by Francisco de Bodega y Quadra. His observations were partially published in the account of another Spanish expedition in 1802, and Alexander von Humboldt consulted the manuscript of Mociño's natural history papers when he visited Mexico City in 1803.[118] However, this was the first publication of the text in the Americas, again signifying a successful deployment of the Society and *Gazeta's* impressive network.

When the *Gazeta* reported on Central America's connections to the rest of the world through correspondence networks and travel, their value was often measured in the knowledge or materials that they contributed to Guatemala, and the utility that could be achieved for the region through their application. While it is useful to establish an idea of the 'nodes' between which letters, persons, and botanical material traversed Central America, networks of knowledge were not disembodied entities, but tied to the travels of imperial officials, priests, and merchants. Eighteenth-century reformers seemed to acknowledge this when they defined utility not just through the travel of plant material or scientific treatises, but also the character of the traveller. If the cache of 'exotic plants' was one of

[115] Belzunegui Ormázabal, *Pensamiento económico*, 76.
[116] José Moziño, Chinandega, 6 May 1797, AGCA, A1.6, Leg. 2006, Exp. 13819.
[117] Real Sociedad, *Quinta junta publica*, 7; 'Economia rural', *Gazeta*, Vol. 2, No. 185 (17 December 1798), 331; *Gazeta*, Vol. 4, No. 185 (1 December 1800), 374.
[118] José Espinosa y Tello, *Relacion del viage hecho por las goletas Sutil y Mexicana en el año de 1792* (Madrid: Imprenta Real, 1802), chapters XVI–XIX. Humboldt quoted in Taracena Arriola, *Expedición científica*, 120.

the Economic Society's success stories, another was the introduction of particularly productive and sought-after beehives from Havana by a merchant, Ventura Batres. The physician José Felipe Flores had previously suggested that the introduction of such hives from Havana to Guatemala would be a great asset, and in 1798 Batres promptly brought two from Havana. Like Ramírez and Sosa, the merchant placed his travels in the pursuit of utility for the country. In an announcement entitled 'Useful Acquisitions', the editors characterised Batres as an involved Enlightenment traveller, with the 'active will of a man of reason'. He had taken care of the hives 'himself' on the journey, which was quite different from transporting such valuable cargo by 'mercenary hands' with no investment in the outcome. The article further lauded Batres for bringing an instruction 'regarding the method of utilising' the hives with him, a 'complete treatise on the matter'. The pairing of the beehives with an accompanying instruction, just as Ramírez's plants had been accompanied by a treatise, appeared to elevate the importation of the hives from a simple merchant's journey to a learned one. For the author of the article it was clear that 'this Guatemalan strives to distinguish himself, and if he leaves his country it is not to return to it with empty hands, and a hollow head, as so many other travellers of other places'. He held up Batres as the ideal contributor to the Economic Society's Enlightenment project, in contrast to other 'indolent and lukewarm correspondents', who it would be futile to solicit such valuable imports from.[119] It was not necessarily enough to extend the networks of which Central America was part – they had to be made useful by active and dedicated travellers like Ramírez and Batres. Knowledge networks mattered more when the 'right people' constituted and maintained them. Ventura Batres, José Felipe Flores, and Alejandro Ramírez had succeeded in expanding their personal geographical horizons, and, by extension, the range of information which was available to interested men in Guatemala. As intermediaries between Guatemala City and the wider world, local Central American networks of botany and the international world of science, they had brought useful knowledge to a new audience.

Conclusion

From the Economic Society's initial resolution to send instructions and tools for the cotton harvest to priests as rural agents of change in its very first year, to the elaborate networks of communication that formed out of the replies of *Gazeta* readers from across Central America to each other, the Economic Society and its newspaper generated new avenues for the exchange of information in the 1790s. The *Gazeta de Guatemala* was a participatory enterprise. It formalised

---

[119] 'Adquisiciones utiles', *Gazeta*, Vol. 6, no. 273 (23 August 1802), 199–200. 'Extracto de una carta del Dr D José Flores', *Gazeta*, Vol. 1, no. 16 (22 May 1797), 123.

and made visible established paths of correspondence and personal contact, but also tied them into a new public discourse about progress and Enlightenment by opening its pages as a discussion forum and mediator between correspondents. Gathering and distributing useful knowledge would be the basis of progress. The editors' emphasis on identifying the names and geographical locations of the contributors suggests that, in addition to the substance of the information provided, another factor was important here: the Society and newspaper's ability to reach into the countryside in order to understand it, and ultimately to transform it. The geographical location of members of their network itself stood for the Society's ability to transform physical landscape.

The Economic Society would not achieve its aim with just words: practical action deserved the highest praise. Encouraging the distribution of material units of Enlightenment like the shoots and seeds of useful plants rather than just information was one of the ways in which the editors of the *Gazeta* planned to achieve a more enlightened Central America. The distribution of Ramírez's Sumatran rice and the medicinal *algalia* herb were seen as particular success stories, examples of the *Gazeta*'s networks effectively reaching into rural Central America and beyond to produce tangible effects within the landscape. These long-distance journeys recorded in the *Gazeta* were themselves then extensions of the Economic Society's initial forays into recruiting priests and rural landowners as proxies (recruited through letters, not the periodical press) who would grow wheat, cocoa, flax, or cotton as part of a larger improvement project. The networks also provided dye-producing natural material: a plant as well as a colony of cochineal insects, and a person who knew how to handle them. These were concrete results that exemplified the Society's patriotic ambitions: to connect rural and urban, local and global knowledge and apply it to its target landscapes.

3   Making Enlightenment Local

In a speech of 1798, the outgoing vice-secretary of the Economic Society, Sebastián Melón, drew a clear link between Enlightenment (*las luces*), patriotic behaviour, and knowledge. Without 'true knowledge [*instruccion*]', there was not going to be any patriotism, and patriotism in turn was the foundation of *las luces*. Patriotism meant to 'come together for the public good', and Melón gave several examples, pointing to the specific projects that the Society's 'united dedication [*zelo unido*]' had produced, despite being up against very scarce resources. The Society had 'performed' the role of 'lovers of the *patria*' implied in their title admirably. The proof was in the detailed reports of the work of the sub-committees on agriculture and manufacturing (*industria*).[1] The patriotism that the Society had set itself as its aim was not a feeling, but an action, and it could not happen without possessing the required knowledge. However, the origins and interpretation of this true knowledge were not always straightforward. Patriotic naturalists all over Spanish America had emphasised practical and locally generated visions of scientific knowledge.[2] José de Alzate and Hipólito Unanue in particular argued vociferously that grand European systems of knowledge simply could not to Mexican and Peruvian nature respectively.[3] This chapter argues that Central American challenges to the translatability of knowledge were distinctive. Reformers delineated the boundaries of what knowledge could be good for, and how to gather it, in particularly detailed terms with reference to their own perceived peripherality. The identification of Central America as a remote place, whether justified or not, compounded other concerns about the verification of knowledge.

It was clear that the science of making nature useful depended in some way on the networks that could provide Central Americans with the knowledge that

[1] Real Sociedad, *Quinta junta pública*, 5.
[2] Cañizares-Esguerra, *How to Write the History*, 282–4; Thurner, *History's Peru*, 87; José Luis Peset, 'Ciencia e independencia en la América española', in Antonio Lafuente, Alberto Elena, and María Luisa Ortega (eds.), *Mundialización de la ciencia y cultura nacional* (Madrid: Ediciones Doce Calles, 1993), 195–217, at 205; Antonio Gonzáles Bueno and Raúl Rodríguez Nozal, *Plantas americanas para la españa ilustrada. Génesis, desarrollo y ocaso del proyecto español de expediciones botánicas* (Madrid: Editorial Complutense, 2000), 68.
[3] Cañizares-Esguerra, *Nature, Empire and Nation*, 63.

they needed to make interventions in land and society in the name of Enlightenment. The Economic Society had successfully demonstrated the reach of their networks and reported progress by connecting some farmers with plant material. However, these successful case studies, trumpeted in the *Gazeta*, belied the immense epistemological difficulties that editors and Society members were presented with every time they received new information. The key question was, as always, whether knowledge was not just 'true', but true for the context to which it was to be applied. In evaluating knowledge drawn from a wide range of social and geographical contexts that expanded traditional administrative and correspondence networks, reformers were faced with more profound methodological doubts than the colonial administrators of the 1780s who had worked within more established formats of knowledge-gathering at Palenque. Not all knowledge was considered infinitely transferable, especially when it arrived from far away in the form of text without accompanying material evidence. The rhetoric of universal knowledge now clashed with everyday assumptions about the Central American countryside, its agriculture, its people's sources of food, and their botanical remedies. The *Gazeta*'s participatory format produced quarrelsome contributors who could quickly find a fault in a source that assumed the applicability of knowledge without good justification. Perhaps in response, definitions of utility started to include more parochial aspects as the *Gazeta* became more established. In early 1798, the newspaper's editors dedicated their work to 'the utility of *this* public (*utilidad de éste publico*)', and went on to emphasise how important the contributions of Central American subscribers specifically were to the paper.[4]

An emphasis on the importance of local knowledge by no means implied opposition to the universal principles that reigned in eighteenth-century science in the European tradition. Different geographical scales could coexist in the same sentence. One resident of León in Nicaragua saw 'a city, a province, a kingdom' as the possible targets of Enlightenment.[5] Vice-secretary of the Economic Society, Sebastian Melón, was proud to say that 'there are in this Kingdom climes not just appropriate for the cultivation of [flax], but they also produce with an exuberance ... of which there is no example in the old continent', displaying pride in both the Kingdom of Guatemala and America in the same sentence.[6] It was therefore possible to regard knowledge from different global scales as an integral part of 'local' progress. The Prospectus that introduced the *Gazeta* to the reading public in 1797 even argued that 'there is no utility that is so local that it does not [also] extend to more than one place',

---

[4] 'Introducion al tomo segundo', *Gazeta*, Vol. 2 (1798), 3–5.
[5] 'León y 8 de Octubre', *Gazeta*, Vol. 2, no. 40 (6 November 1797), 320.
[6] Real Sociedad, *Quinta junta pública*, 8.

and the editors promised issues that were not transient but 'of interest to humanity', and beneficial to 'any country'.[7] In fact, across Europe, the eighteenth-century definition of 'patriot' usually denoted someone who 'worked to aid humanity in general as well as his more immediate fatherland'.[8] While eighteenth-century concepts of universal knowledge often served to entrench Eurocentrism in Enlightenment science, they therefore also contained the powerful promise of bringing 'utility to mankind'. The ideal of benefiting 'all of humanity' was certainly at the heart of the smallpox vaccination trials that the Audiencia's government and reformers (many of whom were Economic Society members) conducted.[9] Yet, there were branches and applications of knowledge outside these formal medical circles where questions of applicability were more controversial.

By paying attention to small challenges to the rhetoric of 'universal' knowledge, we can follow how the Economic Society and the newspaper understood empirical knowledge (local and from elsewhere) as a series of practical test cases for each piece of information that arrived in the readers' hands or on the editors' desk, whether from within the Audiencia, another Spanish American territory, or a non-Hispanic country. Medicinal plants, the environmental and health attributes of plantains, and the ability to transfer plants from one part of the globe to another all became questions of verifying knowledge that travelled, and opportunities to tie information to local places. Natural-historical discussions are especially valuable case studies because they tended to draw a large number of participants who explicitly referred to the geographical situatedness of both the source and applications of such knowledge. By investigating the specific ways in which the Economic Society and *Gazeta* deployed such affirmations of the importance of rooting knowledge in its appropriate geographical context, we can understand how their patriotic rhetoric framed Central America as distinctive against the world beyond.

Truth, Taxonomy, and Local Knowledge

As President Gonzáles Mollinedo explained, only with 'a true knowledge of the locality and the circumstances which are present in it' would Central America's fortunes be improved.[10] Knowledge gathered within 'the locality', that is presumably the Audiencia de Guatemala, was certainly to be part of the solution.

---

[7] 'Prospecto de amplicacion', *Gazeta*, Vol. 1 (1797), unpag. See also 'Carta', *Gazeta*, Vol. 2, no. 50 (26 February 1798), 13–14, for a description of José Celestino Mutis's work as bringing 'utility to mankind'.

[8] Denise Philipps, *Acolytes of Nature: Defining Natural Science in Germany, 1770–1850* (University of Chicago Press, 2012), 37.

[9] Few, *For All of Humanity*, 11, 165–96.

[10] 'Representacion que el Presidente de Guatemala dirije a S.M.' (February 1802), AGI, Guatemala, 481.

Medicinal plants were a prime example of the *Gazeta*'s potential to compile useful knowledge, since they were closely tied to contributors' local knowledge about nature, and of obvious potential benefit to both the local population (and, given the right circumstances, even 'all of mankind'). The basis of any such knowledge would be detailed information about the plants' appearance and habitat from readers who had seen them in the field. Such empirical knowledge drawn directly from Central American landscapes, to the editors, would facilitate 'progress'.[11] In this, the *Gazeta* followed the Mexican periodicals *Diario Literario de Mexico* and *Mercurio Volante*, which had also invited a collaborative approach to medicinal matters in the 1770s.[12] However, demonstrating the relevance and truth of such information was only marginally easier than applying knowledge from 'abroad'. Strategies for demonstrating the veracity of the readers' knowledge came to play an important part, since the contributor had to convince not just the reader, but also the editors in the first instance.

Within such local parameters of knowledge-gathering, the phrase 'persons worthy of trust' (*personas dignas de fé*) frequently appears in descriptions of the ideal provider of natural-historical information.[13] Credibility, for the Society and *Gazeta*, was acquired in one of several ways. A person respected for a role in the community, for instance an Economic Society member, was considered a reliable witness. The truth of a report in the *Gazeta* by a regular contributor, the friar Merino, was 'undoubtable, given that all who know him know his prudence, enlightenment [*luzes*], and forthrightness'.[14] When the Society requested submissions of leather, soap, and oil samples directly from the readers as part of a competition for artisans, they stressed the need for the samples to be authenticated by the 'society member closest to the workshop' or other 'reliable persons' – the elite gatekeepers of the Society.[15] There are obvious parallels to the measures credibility applied in the early Royal Society, as described by Steven Shapin and Simon Schaffer.[16] However, being a trusted member of society did not negate the importance of empirical knowledge and experiments drawn from a group of readers. Even when the surgeon-royal and *protomédico* Narciso Esparragosa personally sent a report

[11] 'Breve descripción', *Gazeta*, Vol. 8, no. 379 (4 February 1805), 555.
[12] Paul Ramírez, 'Enlightened Publics for Public Health: Assessing Disease in Colonial Mexico', *Endeavour* 37 no. 1 (2013): 3–12, at 3–7; Alberto Saladino García, *Ciencia y prensa durante la ilustración latinoamericana* (Mexico: Universidad Autónoma del Estado de México, 1996), 69–70; Fiona Clark, 'The *Gazeta de Literatura de Mexico* (1788–1795): The Formation of a Literary-Scientific Periodical in Late-Viceregal Mexico', *Dieciocho* 28 no. 1 (2005): 7–30, at 10.
[13] E.g. 'Contravenenos', *Gazeta*, Vol. 7, no. 293 (28 February 1803), 38–9.
[14] 'Algalia', *Gazeta*, Vol. 5, no. 216 (3 August 1801), 531–2.
[15] 'La Real Sociedad Economica de Guatemala ofrece los siguientes premios', *Gazeta*, Vol. 4, no. 157 (19 May 1800), 262.
[16] Steven Shapin and Simon Schaffer, *Leviathan and the Air-Pump* (Princeton University Press, 1989), 22–79.

about the healing qualities of a medicinal plant, the *Gazeta* encouraged its readers to conduct 'new experiments'.[17] Although Esparragosa's experience carried weight on account of his education and status, a single experience with a medical plant for the *Gazeta*'s editors was inferior to collaboratively acquired evidence from different sources. The 'valuable specific' would only be widely used if 'new observations should result as favourable as the experiment [*tentatiba*] of Dr. Esparragosa'.[18] Another attempt to incorporate the empirical information that the editors found so important was by drawing on those who could demonstrate direct experience of nature. Indeed, the *Gazeta* editors declared the experience of a herb gained by a 'simple labourer' more informative 'than any analogy which the most able naturalist may be able to deduce'.[19] One correspondent, for instance, called himself '*el guanaco observativo*', which may be translated as 'the observative simpleton'. While mocking his apparent inability to couch the story in more scientific terms, he nevertheless clearly considered himself an expert by virtue of his experience and observations. In this case, he was concerned with identifying the spider whose bite could be cured with the *camacarnata herb*, so the antivenom recipe might be applied to the right wounds and prevent unnecessary deaths. He described the characteristics of the spider in great detail, but also added further context, giving an account of the circumstances in which he last encountered it – when a spider of this sort bit 'an Indian of [the village of] Jocotenango' and was subsequently healed.[20] This detailed story seemed designed to corroborate his claims, becoming a strategy of verification through circumstantial detail that Stephen Shapin has called 'virtual witnessing'.[21] The 'observative simpleton' declared that he had 'repeated the experience of this often, and always successfully'. The *Gazeta* had established a clear model for verifying and testing the truthfulness of locally acquired knowledge based on both social reputation and empirical observations.

Experiential knowledge was important not just for the verification of a plant's effects, but also its very identity. The newspaper's articles on the *camacarnata* plant prioritised practical and local usage over learned definitions. The editors noted that their description of the plant was 'not in

---

[17] The words *experimento* (experiment) and *experiencia* (experience) were often employed interchangeably in this context, since *experimento* had traditionally been an 'all-purpose' word denoting a range of trials rather than the specific definition relating to an experiment in a controlled setting. See Lorraine Daston, 'The Empire of Observation, 1600–1800', in Lorraine Daston and Elizabeth Lunbeck (eds.), *Histories of Scientific Observation* (University of Chicago Press, 2011), 81–114, at 83–5.

[18] 'Breve descripción', *Gazeta*, Vol. 8, no. 379 (4 February 1805), 554–5. [19] *Ibid.*

[20] 'Receta para curar la picadura de una araña llamada Cazampulga, o Chiltuca', *Gazeta*, Vol. 3, no. 102 (29 April 1799), 24.

[21] Steven Shapin, 'Pump and Circumstance: Robert Boyle's Literary Technology', *Social Studies of Science* 14 (1984): 481–520, at 497.

accordance with botany', but it was 'enough to recognise it'. Accordingly, even though it was a learned friar from Guatemala City's Santa Clara convent who first described the plant in the paper, his description was one of precise local relevance, naming the villages around Lake Atitlán in Guatemala near which the plant grew and was cultivated. A note that its vine-like roots could not be extracted 'without a *machete*, or sabre' suggested that this was a description oriented towards practicalities. A description of its appearance and fruits (but not flowers) followed, as well as an explanation for the etymology of the indigenous name for the plant. The friar even gave a brief account of the traditional uses of the plant: Indians used its roots either boiled in water or as a paste to be applied to wounds.[22] In this description, the location of the plant within a particular local geography as well as its indigenous name formed an integral part of its authenticity. Because 'local' knowledge in a colonial context, as Kavita Philip has explained, paraphrasing Foucault, relied on the lived experience of the observer's body existing in a specific natural system, it was intrinsically more closely linked to indigenous knowledge.[23] This had the effect of separating 'the indigenous' from 'the scientific' in a pejorative way among nineteenth-century naturalists, but for the *Gazeta*'s authors actually represented an advantage. Naming a plant correctly and accurately was an important factor in producing useful knowledge, and the use of indigenous names might make it easier to identify plants in the future with the help of local informants. Indeed, the editors advised their readers to take into account 'popular traditions', specifically indigenous knowledge, when it came to medicinal plants.[24] According to the author of the *Gazeta* article, the *camacarnata* plant was also called *rox iyuin umùl*, a name derived from the Indians' word for 'rabbit's ear', noting that its leaves indeed resembled the ears of that animal.[25] In this case, the *Gazeta*'s editors disregarded the plant's Spanish-Mexican, Nahuatl, and Linnaean names which the author had also provided, but highlighted the Guatemalan Kakchiquel term in the heading of the article. Hispanic naturalists in both Europe and America had acknowledged the 'expressive nature' of Nahua and Quechua plant names. Beyond patriotic allegiances to

---

[22] 'Breve descripcion de la yerva ó planta llamada *Camácarnate* ó *Rox iyuin umùl*, que sin ser con arreglo à la botánica, basta para qué puéda conocerse. Y noticia de sus virtudes', *Gazeta*, Vol. 8, no. 379 (4 February 1805), 554–5; 'Plantas medicinales. La Camacarnata', *Gazeta*, Vol. 8, no. 360 (20 August 1804), 407. The Linnaean name for the *genus* appears to be *passiflora* rather than *pasionaria*. The plant is most likely *Passiflora Mexicana*.

[23] Philip, *Civilizing Natures*, 268; see also Michel Foucault, *The Order of Things: An Archaeology of the Human Sciences* (London and New York: Routledge Classics, 2002), 350.

[24] 'Diccionario de unas plantas', *Gazeta*, Vol. 9, no. 424 (17 February 1806), 753. José Antonio de Alzate, *Gacetas de literatura de Mexico*, Vol. 2 (28 August 1792), 438–9, proclaimed a similar reliance on 'the poor'.

[25] Kakchiquel: *ruxikin*, ear of, and *umùl*, rabbit. Filiberto Patal Majzul Lolmay's *Diccionario estándar bilígüe Kaqchiquel-Español* (Antigua Guatemala: OKMA, 2013) notes similar plant names – mushrooms named *ruxikin kuk*, squirrel's ear, and *ruxikin xar*, blue ear.

different systems, there were practical reasons for this. Unlike their Linnaean counterparts, indigenous taxonomies often referenced the 'appearance, location or uses' of plants, as Helen Cowie has pointed out.[26] It follows that knowledge rooted in local geography and practical approaches to finding the plant in the Guatemalan countryside were just as valuable as references to botanical authorities, if not more so.

Reformers were of course not always unqualified champions of indigenous knowledge. The *Gazeta*'s triumphant narrative around the movement of the *algalia* medicinal herb, for instance, discussed in Chapter 2, actively promoted the status of (mostly Creole and Spanish) Economic Society members and *Gazeta* readers as being in possession of useful, practical knowledge, even though the archbishop who had brought the plant from Nicaragua had declared that he had received it from indigenous people. A description of the priest Garrote, who planted the *algalia* in his parish, as generously providing the remedy to a local population who were 'running for it' to his house, subverted the traditional image of Spaniards prising the 'secrets' of nature from indigenous informers.[27] As Paul Carter notes for the case of place-names, the use of indigenous names by colonisers could also mean an 'assertion of possession'.[28] The use of indigenous names could therefore be a question of practical guidance rather than a deep appreciation of indigenous knowledge. Contributors to the *Gazeta*, however, did not usually see indigenous knowledge as a resource for Spanish or Creole naturalists, to be abstracted into a 'scientific' unit of information. Abstraction signified a loss of locally relevant descriptions, while the *Gazeta*'s own detailed accounts rested on specific knowledge that would guide readers who were connected by the shared experience of Central American nature. At least in theory, each reader had the potential to access this nature. Michel Foucault points out the obvious problem with ever-more detailed descriptions: they are 'inaccessible' and the opposite of the systematic abstractions that natural history requires; they are not 'natural history'.[29] The *Gazeta*'s authors, however, were clearly happy to 'do not-natural-history' here, to refrain from classifying in favour of practical advice and the place-specific descriptive indigenous names of plants. This was not necessarily for a lack of

---

[26] Helen Cowie, *Conquering Nature in Spain and Its Empire, 1750–1850* (Manchester University Press, 2011), 122–3.

[27] See also Brockmann, 'Sumatran Rice', 93–4. The original trope of Indians guarding natural secrets was of course also present in the *Gazeta*; see, for instance, 'Remedio para la mordedura de culebras venenosas', *Gazeta*, Vol. 6, no. 270 (13 September 1802), 223–5.

[28] Paul Carter, *The Road to Botany Bay: An Exploration of Landscape and History* (Minneapolis, MN and London: University of Minnesota Press, 2010 [1987]), 65. See also Raquel Álvarez Peláez, *La conquista de la naturaleza americana* (Madrid: Consejo Superior de Investigaciones Científicas, 1993).

[29] Foucault, *Order of Things*, 150–8.

ambition, but because reformers seemed to believe that this was more applicable in the local context than 'universal' botanical systems.

The *Gazeta*'s editors reminded the reader that theoretical botanical knowledge did not count for much. Book-learnedness was of little use in a science like botany, which was not 'sedentary', but required its followers to 'climb mountains'. Being interested in the theoretical cause of a plant's healing abilities was rather 'useless' compared to observing its healing effects. After all, the editors were convinced that practical experience in the form of 'fortuitous events' rather than theories had driven many 'advances in the sciences'.[30] They were therefore not too interested in one of the biggest theoretical debates in contemporary botany, the discussions around plant nomenclature and especially the Linnaean system. From Hipólito Unanue in Peru to Antonio de Alzate in Mexico and Antonio Liendo y Goicoechea and Mariano Mociño in Guatemala City, scholars across Spanish America had voiced doubts about the applicability of the Linnaean system to describing American nature, and often developed their own rival nomenclatures.[31] In Foucault's explanation, the universality of the Linnaean system was rooted in its insistence on being 'content with seeing – with seeing a few things systematically'.[32] Many Guatemalan reformers were dissatisfied with such taxonomies precisely because it hid the knowledge gathered from other senses and lived experience. The fact, drawn from experience, that one needed a *machete* or sabre to cut the thick stem of the *camacarnata* plant, was more important than the appearance of its flowers, less systematic though it was.

Even those who generally subscribed to 'universal' ideas about science were impatient with the confusing naming conventions of botany. The Guatemalan physician and naturalist José Felipe Flores thought they impeded practical utility. Like the *Gazeta* in its description of the *camacarnata*, he steered the discussion away from classification and towards practical concerns when he reported from Philadelphia on the latest advances of the sciences in a letter to the Economic Society (printed in the *Gazeta*) in 1797. He praised the deluge of new publications on new plant species, but warned:

This science [botany] is still in its infancy, and it does not easily stay in its box, because it is shaken up every now and then by all the botanists. They name [one plant] this way, the next one that way, so there is a countless number of names, to the point where each plant resembles the baptism of a prince.[33]

[30] 'Diccionario de unas plantas', *Gazeta*, Vol. 9, no. 424 (17 February 1806), 753; Vol. 10, no. 463 (1 December 1806), 907.

[31] Antonio Lafuente and Nuria Valverde, 'Linnaean Botany and Spanish Imperial Biopolitics', in Londa Schiebinger and Claudia Swan (eds.), *Colonial Botany* (Philadelphia, PA: University of Pennsylvania Press, 2014), 134–47, at 138–42; Maldonado Polo, *Huellas de la razón*, 289; Berquist Soule, *Bishop's Utopia*, 178–9; Bleichmar, *Visible Empire*, 130–1.

[32] Foucault, *Order of Things*, 146.   [33] 'Carta', *Gazeta*, Vol. 1, no. 45 (11 December 1797), 357.

Flores was exasperated with the debates about nomenclature. They were merely a step that had to be taken on the road to the true purpose of botany: the practical use of plants. Once botanists could agree on a uniform system of names, they would finally focus on the truly 'important' aspects of plant science: 'to learn about their qualities, and what they can be useful for'. That said, the Economic Society's members did not have one monolithic block of opinion. In 1805, an article in the *Gazeta* lamented that 'Guatemala still does not occupy a single page in the history of the natural sciences', wishing for Central America to be visible in European worlds of knowledge, with the implication that it would raise Central America's profile in Europe as a whole. To the author, the Spanish naturalist Mociño was 'the only scholar who has travelled through some of its provinces with intelligence' (that is, formal, probably Linnaean, taxonomies).[34] And yet, in the *Gazeta*, practicality generally prevailed. The Linnaean system within Guatemala had its place at institutions of learning, such as the University of San Carlos and at natural history demonstrations, but it was not suitable for the practical botany of the *Gazeta*.

The importance of empirical knowledge over learned discourse was also clear in a note from the editors accompanying a different article. A correspondent writing under the pseudonym *Medico de Capa Blanca* had submitted information on treating the bites of venomous snakes and scorpion stings, which the editors enthusiastically accepted and printed. However, they had only printed part of the contributor's letter in this article, since he had prefaced his report with long excerpts from Pliny and Galen, in Latin. Despite praising his erudition in principle, the editors rejected this part of the letter, and even explicitly asked him to omit Latin texts from his letters in the future. If the correspondent might go to the trouble of providing them in Spanish translation, the editors explained, they would not have to leave them out. 'We have ended up being very economical in [our use of] Latin', the editors explained, 'because we are all laypeople, and we suspect that so are many of our readers of this newspaper'.[35] The information would only be useful if it could be understood by the newspaper's dispersed readership, and Latin erudition made it unsuitable for this purpose. The editors' insistence that they would keep their descriptions practical also seemed justified in the face of some readers clamouring for solutions appropriate to their circumstances. For instance, a contributor from New Spain disparaged the idea of the newspaper wasting space on articles which contained medical solutions to snake bites so complicated that one seemed to need specialised medical knowledge or equipment to use them.

---

[34] 'Elogio de Cabanilles', *Gazeta*, Vol. 8, no. 277 (21 January 1805), 537. These hopes did not come to fruition, since Mociño's botanical work on the expedition was not published in his lifetime.

[35] *Gazeta*, Vol. 7, no. 290 (14 February 1803), 23–4.

'Should one walk around holding a vial, with a doctor by one's side?' the reader asked mockingly. Instead, he suggested an even simpler solution, a herbal remedy.[36] The *Gazeta*'s editors embraced this bellicose attitude towards theoretical knowledge as an approach they could agree with, since it was relevant to the local context and more accessible to a wider audience.

Economic Society members were well aware that goods and information often arrived in Central America after a long journey. However, pre-empting any potential problems with their application, they sometimes sought to adapt them 'to the circumstances of the country'. The prioritisation of empirical knowledge and scepticism of theoretical knowledge was related to this impulse and applied across different fields of study. In this, the Economic Society and *Gazeta* were no doubt influenced by Creole naturalists in other parts of the empire. They would have happily agreed with the New Granadan botanist Francisco Antonio de Zea's statement 'First the useful, then the scientific, if desired', or the practical ideologies of Francisco de Caldas or Jerónimo Torres in the same viceroyalty.[37] In Central America, this perspective was so fundamental to the Economic Society's mission that notes from its earliest meetings in December 1794 already mentioned it. The founders proposed to create an academy of mathematics to further their aims of spreading Enlightenment through education, but made it clear that the curriculum for this intended school was to be 'adapted to the circumstances of this country'. It was to teach 'useful' ideas, and make valuable discoveries that would lead to the implementation of projects that had hitherto been considered impossible. In everything, the school would 'prefer the practical to the theoretical or the speculative'.[38] President Domás y Valle also mentioned this quality of locally specific knowledge in his letter of support for the Economic Society as a whole in 1795: its entire plan of action was 'most useful' because it was 'in all its parts accommodated to the circumstances of the country'.[39] Two years later, the first issue of the *Gazeta* announced that the authors were interested in all manner of political topics, but would not be drawn into discussions on the 'grand science of government, or the laws of war, or peace, of nature, of men'.[40] The Society set out the parameters of enlightened knowledge carefully. Their application of principles of distance and reliability to different types of knowledge showed that their distrust of theory was not grandstanding, but pragmatism related to their practical aims.

---

[36] 'Contravenenos', *Gazeta*, Vol. 7, no. 293 (18 February 1803), 38.
[37] Quoted in Silva, *Los Ilustrados*, 468; for Torres, see e.g. p. 420; further practical ideologies 451–92.
[38] 'Plan de la instruccion que se ofrece dar en la Academia de Matematicas, dispuesto por el Capitan de Ingeniero Ordinario R.s Ex.tos Dn Josef de la Sierra', AGI, Estado, 48, N.7, 2v.
[39] 'Presidente de Guatemala sobre establecer una Sociedad Económica', AGI, Estado, 48, N.7, 5.
[40] 'Política', *Gazeta*, Vol. 1, no. 1 (13 February 1797), 1.

Even the highest medical authorities shared this emphasis on pragmatic approaches. For instance, in reaction to an 1804 outbreak of the measles in Guatemala City, the interim *protomédico*, José Antonio Córdova, wrote an 'Instruction' on how to recognise and treat the illness, which the government sent out to all Central American provinces. The full title of the work was 'A simple method for recognising and curing the measles adapted to the knowledge and lack of assistance in Indian lands', emphasising the practicalities of treatment and available resources over the latest European medical knowledge.[41] In fact, it mostly included nothing more complicated than herbal infusions and chicken soup, reflecting perhaps the meaning of the word *curar* as 'caring for' a patient rather than 'curing' an illness in the modern sense.[42] Infusions of the rare and medicinally valuable tree-barks *quina* and *copalchí* were only mentioned as a last resort. The circular ended on a dry note, pointing out that Córdova deliberately omitted a discussion of the possibility of inoculation against this illness, apparently because it would contradict the instruction's central aim of simplicity. Córdova signalled his knowledge of the international world of science and his formal education with this note, but stayed true to his promise to focus on simple remedies that were locally relevant. This approach to treating diseases had also appeared in the work of José Felipe Flores, Córdova's predecessor as *protomédico*. Martha Few has described how in the fight against a smallpox epidemic, the first quarantine measures were explicitly framed as being adapted to the medical committee's 'understanding of the country'. In subsequent outbreaks, Flores additionally adapted his instructions for vaccinating rural communities to accommodate indigenous resistance to the procedure.[43] Of course, the idealised vision of grateful communities of indigenous people happily receiving the Crown's medical provisions presented in government communications was often quite different in reality, as reports of the government's medical officers chased out of communities show.[44] These examples do, however, demonstrate that even the Crown's medical officers were sensitive to the idea that it was not worth promoting knowledge, especially medical knowledge, that could not be replicated across Central America.

Travelling Knowledge

Enthusiasm for successful overseas networks was tempered by the concern that distance would distort information. This was visible even in the universally

---

[41] 'Sarampion', *Gazeta*, Vol. 8, no. 338 (19 March 1804), 9. Manuscript circular sent to the *corregidor* of Totonicapán: AGCA, A1.1, Leg. 6091, Exp. 55306, 2 January 1804. It is possible that this circular was related to the earlier 1769 handbook of *Curing Measles and Smallpox* discussed by Martha Few (*For All of Humanity*, 49), since that document alluded to country-specific instructions that would be distributed across Guatemala in manuscript form.

[42] Few, *For All of Humanity*, 50–1.   [43] *Ibid.*, 136–49.   [44] *Ibid.*, 90–3.

praised project on exotic plants that appeared in the *Gazeta* in 1801 (discussed in Chapter 2). When Alejandro Ramírez proudly introduced the Sumatran seeds that he had imported, he made it clear that they were only relevant as far as they could be adapted to Guatemalan soil. He was pleased to be able to offer a 'scrupulous translation' of the description of the seeds by the British botanist Campbell, who had originally compiled the collection. He did not seem to doubt Campbell's authority, speaking reverently of his qualifications as a doctor and scholar. Even so, Ramírez made a conscious choice to omit the description of one of the seed packets, that of the Asian 'varnish-tree', although Campbell provided it in the English original. He explained that this was because 'this seed has arrived spoilt, lost, useless. *Ergo* its description is useless'.[45] This blunt statement demonstrates the limits of what type of knowledge could be considered useful. Even the description of a trusted botanist could become 'useless' when it was tied to botanical samples that would never grow. For Ramírez, the presence of the collection in Guatemala was a triumph of useful knowledge only to the extent that it could be applied. Knowledge about the Asian varnish tree, therefore, had no practical usefulness for the Guatemalan context if there were no seeds to grow it to harvest its lacquer-like sap.

Long journey times were not just a danger to physical materials, like the seeds of plants, but they also made it much more difficult to ascertain the veracity of knowledge as it appeared in print. Communication routes were key in determining whether a location counted as 'distant' or manageably close, as Sylvia Sellers-García has shown, and so relatively unproblematic (in peacetime) postal and mercantile connections to specific cities such as Havana or Cádiz became sources of generally reliable knowledge.[46] By contrast, a reader of the *Gazeta* gave an exasperated description of a letter on agricultural matters by a North American plantation owner which had been printed in a Spanish newspaper before travelling back to the Americas and Guatemala: 'although the distance between where [the letter] left from and here is very short, it is difficult, and with as few or fewer modes of communication than exist between Tibet and Germany'.[47] The editors also encountered this problem when they published a description of 'a remedy for the bite of venomous snakes' which had originally appeared in the *Papel Periódico de Santafé de Bogotá* newspaper in New Granada. The *Gazeta*'s editor saw himself in a difficult position, because the description of the plant from Bogotá 'was not as complete or clear as we might like it to be'.[48] There was no simple way of clarifying this

---

[45] 'Carta', *Gazeta*, Vol. 5, no. 195 (16 March 1801), 421–2.
[46] Sellers-García, *Distance and Documents*, 7.
[47] *Gazeta*, Vol. 8, no. 369 (22 October 1804), 476–7.
[48] 'Remedio para la mordedura de culebras venenosas', *Gazeta*, Vol. 6, no. 270 (13 September 1802), 223–5.

information. Indirect transport connections meant that the article from Bogotá only arrived in Guatemala because it had been reprinted in the Spanish newspaper *Mercurio de España*, which in turn was brought to Guatemala by post. The editor lamented that 'our means of communication with Santa Fé are quite difficult'. It was therefore unlikely that more detailed written information about the plant could be acquired in any useful timeframe. The solution, for the *Gazeta*'s editors, was instead to ask for the collaboration of the readers, 'especially the parish priests'.[49] Rather than embarking on the seemingly impossible task of asking for details from Bogotá, the *Gazeta* decided to use the circuitously acquired notice as a mere starting point for its own researches. In this case, the *Gazeta* made a virtue out of the inconvenient distance that separated Bogotá from Guatemala by declaring that its network of correspondents and priests would be more than able to make up the shortfall in information provided by the original article. Where Guatemala's long-distance connections failed, the community of priests, merchants, and officials who had first come together in the Society's correspondence networks and the *Gazeta* remained reliable sources.

It is important to remember that for some of the Economic Society's members, the very process of the diffusion of treatises written in Europe could form a key part of 'doing Enlightenment', almost regardless of their content. However, this argument was not always convincing to those reformers who strove to bridge intellectual information transmitted through print with its practical application. The *Gazeta*'s editors specifically touted agricultural treatises and practical guides as 'useful', but were not quite able to follow up on their practical implementation. The archbishop of Guatemala, for instance, provided 130 copies of the Spanish journal *Semanario de agricultura* to parish priests so they might emulate 'the advances [made] in other countries', seemingly replicating on a small scale the 'spirit of emulation' that historian Sophus Reinert has observed across European empires.[50] The mere availability of knowledge was not the same as its implementation. After all, as Peter Jones has stressed, we should distinguish between the availability and the use of knowledge.[51] A priest in Nicaragua, for instance, who had discovered a supposed improvement for the process of indigo extraction was seen by his contemporaries in the Economic Society as 'enlightened'. Yet in practice, 'not

---

[49] *Ibid.*, 224.

[50] The archbishop was reported to have paid for 130 examples to be imported to Guatemala, but it is unclear whether these ever arrived, or if a subscription was maintained: Real Sociedad Económica, *Quarta junta pública*, 6–7. Sophus Reinert, 'The Empire of Emulation: A Quantitative Analysis of Economic Translations in the European World, 1500–1849', in Reinert and Røge (eds), *The Political Economy of Empire*, 105–28.

[51] 'Economia rural', *Gazeta*, Vol. 2, no. 80 (24 September 1798), 261; Jones, *Agricultural Enlightenment*, 6.

one' of his parishioners had shown themselves willing to take up the new method.

Another reformer eager to persuade farmers of new methods was President Gonzáles Mollinedo. In a circular of 1805 to the *alcaldes mayores* of various Central American provinces, he recommended the two recently imported 'distinguished' books, translated from French, to his provincial governors: Brisson's *Diccionario universal de física* (published in Spain in 1796) on the topic of mineralogy; and Rozier's *Diccionario universal de agricultura* for a widely applicable guide to agriculture. The latter was especially sought-after, having been published in eight volumes from 1797 on behalf of the Madrid Economic Society, but both books had been listed three years previously as books whose arrival in Guatemala the *Gazeta*'s editors were impatiently awaiting.[52] President Gonzáles noted that he had ordered various articles from Rozier to be copied 'in order to send them to different parts of the Kingdom'. He hoped that the regional governors would help to popularise these books to help to improve existing crops and introduce new ones. He expected that the articles would be 'read by all classes of people, those who need to be instructed the most, and who can take the most advantage of these two works ..., replacing other [books] which are pernicious or useless, but that tend to be abundant among the people of the countryside'.[53] The Economic Society was suspended at the time Gonzáles was writing, and he wrote in this case simply as captain-general of Guatemala. The ideas which later led him to re-establish the Economic Society are visible in his comments on Rozier and Brisson, although it is also worth recalling here that the Economic Society was the main proponent of these debates about local applicability, but it did not have a monopoly on such ideas. Gonzáles, for instance, lest he be accused of meddling, hastily added that this agricultural work was also in His Majesty's spirit. Whether Society enthusiast or high-ranking administrative official, it was clear to Gonzáles that popularising the most locally useful passages from these books would mean the propagation of 'enlightened and useful knowledge' (*las luces y conocimientos útiles*). However, there were many obstacles to the translation of such book knowledge into practical improvement. Gonzáles supported his argument by underlining the specific importance of these books for Guatemalan agriculture, and identifying precisely the rural populations for whom he thought they would be most useful. Through such specific selection, the president hoped to overcome some of the obstacles that the travel of knowledge across spaces might face. However, there is no clear

---

[52] 'Variedades', *Gazeta*, Vol. 6, no. 278 (27 September 1802), 241–2. See also Luque Alcaide, *Sociedad Económica*, 27.

[53] 'Circular', 18 August 1805. AGCA, A1.1, Leg. 6091, Exp. 55307, fol. 118.

evidence, as so often happened with the enlightened elite's projects, that the hoped-for popularisation of the practical guides ever materialised.

The insistence on locally relevant knowledge, especially where it came from foreign sources, continued to be a sticking point for the Society through its different incarnations. In 1812, the leadership of the Society (here represented by José Cecilio del Valle) discussed this problem with respect to political economy. Although the political context was now different, against the background of the liberal Cádiz Cortes in Spain, the weighing of information against local usefulness continued earlier concerns. Valle explained that education in matters of civil and political economy was vital to the ability of Central Americans to contribute to the country's progress. However, he was dissatisfied with the well-known authors on the matter. The Spanish statesman and philosopher Jovellanos, for instance, had limited himself to the agriculture of Spain, while Condillac had only written about commerce in its relation to government. Hume, although 'deeply learned', had not written a 'complete course' of the type that could instruct beginners in the matter. The *Encyclopédie méthodique*, finally, he found unsuitable because its 'dictionary-style' presentation was not sufficiently comprehensive (one of many Central Americans to object to the *Encyclopédie*, a topic which is discussed in Chapter 4). In a few sweeping sentences, Valle had dismissed all European authorities, even a Spanish one. He concluded that a local scholar such as himself should draw up a course of instruction in ideas of political economy suitable for Central Americans, especially those with little background knowledge in these matters. Although the resulting lecture course was influenced by the writings of Adam Smith, Jeremy Bentham, and the French physiocrats, his intention was to present a locally appropriate reinterpretation of these ideas.[54] Not all political-economic writing was rejected. For instance, the Economic Society's 1797 essay competition on the 'usefulness' that would come from Indians wearing Spanish-style clothes and shoes was in part inspired by the writings of the Irish-Spanish economic thinker Bernardo Ward. The *Gazeta* published an excerpt from Ward's *Proyecto económico* to give the 'less learned part of the public' some crucial background information on the theory that Indians would make good consumers of Spanish goods, and also end up producing such clothing, in addition to printing friar Matías de Córdova's prizewinning essay.[55] Such schemes to incorporate the Indian population more closely into the Spanish

---

[54] Real Sociedad, *Novena Junta Pública*, 46. On Valle's Discurso on the matter, see Bonilla Bonilla, *Ideas económicas* en la Centroamérica ilustrada 1793–1838 (San Salvador: FLASCO Programa El Salvador, 1999), 129–39; Ralph Lee Woodward, *Class Privilege and Economic Development: The Consulado de Comercio of Guatemala*, 1793–1871 (Chapel Hill, NC: University of North Carolina Press, 1966), 113.

[55] 'Economía política', *Gazeta*, Vol. 1, no. 7 (27 March 1797), 56. Bernardo Ward, *Proyecto económico* (2nd ed., Madrid: Ibarra, 1779), 266–9. Córdova, *Utilidades*, unpag. An argument that Indians might start buying Spanish cloth appears in 'Discurso', AGI, Estado, 48, N.7, 5.

economy were usually intended to help the Spanish (including explicitly the imperial) economy primarily, with the secondary implication that it might also benefit Indians. Here, the specificity of Ward's ideas, designed to be applied to the Americas, made them welcome. Rather than working with the blunt categories of 'non-Hispanic foreigner', 'American', or 'European', improvers in Central America made it clear that local relevance should be the ultimate standard by which to judge knowledge, and started to construct as 'local' those Central American landscapes to which they imagined their reform projects would apply.

The Great Plantain Debate

The ambiguities of importing botanical knowledge from different parts of the world came to the fore in a somewhat curious debate that demonstrated exactly why knowledge drawn from distant networks could be contentious, and the difficulties of finding the right parameters for the local application of published texts. In 1798, the *Gazeta* published a report by José Celestino Mutis of New Granada on the topic of the dangers to public health posed by banana or plantain trees [*platanares*].[56] In the report, Mutis argued that the trees, and the swampy puddles of water and other waste which accumulated underneath them, produced 'pestilent air', which in turn was acknowledged as the cause of many diseases. In this, he participated in a discourse about trees, environment, and public health widespread throughout the Americas (also discussed further in Chapter 5).[57] Mutis's report had been submitted to the *Gazeta* by a citizen (*El Desengaño*) who had apparently long been concerned about the health effects of plantain trees in Guatemala City. He shuddered at the thought of the poor nuns in the Capuchin and Concepción convents right in the centre, who lived

---

A similar argument had been put forward by a sixteenth-century governor of Peru, who argued that if Indians dressed like Spaniards, 'much more Spanish merchandise will be sold, which will all be to the benefit of the treasury': Juan Matienzo, *Gobierno del Perú, 1567* (ed. by Guillermo Lohmann Villena, Paris and Lima: Institut Français d'Études Andines, 19), 69–70. Many thanks to Rebecca Earle for this reference.

[56] 'De los platanares', *Gazeta*, Vol. 2, no. 51 (5 March 1798), 19. Accounts of this report can also be found in Maldonado Polo, *Huellas de la razón*, 257–9, and José Hernández Pérez, 'Medicina y salud pública: su difusión a través de la Gaceta de Guatemala (1797–1804)', *eä* 2 (2010), 19–21. *Platanar* may refer to the trees known respectively in English as 'plantain' or 'banana' trees, but due to the emphasis on their use as an important food source in the *Gazeta* articles, 'plantain' seems more appropriate here. A previous version of this case study has also been published in Brockmann, 'Retórica patriótica y redes de información científica en Centroamérica, 1790–1814', *Cuadernos de Historia Moderna*, Anejo XI (2012): 165–84, at 177–8.

[57] Anthony E. Carlson, 'Vast Factories of Febrile Poison: Wetlands, Drainage, and the Fate of American Climates, 1750–1850', in Sarah Miglietti and John Morgan (eds.), *Governing the Environment in the Early Modern World: Theory and Practice* (London and New York: Routledge, 2017), 156–60.

amid such trees. Finally, he had had a chance to voice his concerns, in response to a letter previously published in the *Gazeta*.[58] The letter he was responding to seemed perfectly innocuous, and well in line with the *Gazeta*'s mission: a reader from León in Nicaragua (*El Provinciano*) had suggested that in order to make the city's lands useful, they should be planted with useful plants like plantains, *zapote*, or avocados. In an emergency, they would make good subsistence for the city, and the citizens of León would not have to part with their hard-earned money and pay the nearby city of Chinandega for food.[59] The Nicaraguan *Provinciano* spoke of believing in enlightened ideas and the 'fire of real patriotism' fuelling his imagination.

The Guatemalan *Desengaño*, fearing that León was about to commit a grave mistake, used Mutis's report in support of his position, which was that the Nicaraguan city would be putting its inhabitants in danger if they carried out their plantain plans. He additionally invoked a Spanish volume on public health – a translation (and presumably anthology) of works by the European authors Joseph-Aignan Sigaud de la Fond, Andrés Piquer, and William Cullen on the putrefaction of plants and 'bad air' as causes of disease. He argued that it was necessary to pay attention to the writings of scholars like Mutis in decisions of governance, since it was 'not that easy or obvious' to prescribe the correct norms of public order without 'vast knowledge of certain arts and sciences [*facultades y ciencias*]', making clear the close connection between true and useful knowledge, and its application to governance.[60] The suggestion to apply a scientific treatise to the art of governance and public health as it manifested in the environment of a regional capital may appear to have been an uncontroversial statement in line with broader aims to spread practical and applied Enlightenment around Central America. However, the report's publication in the *Gazeta* also submitted it to the judgement of a broad audience who had their own opinions on the matter, and who questioned the applicability of this knowledge to their local situation, and their own views of political economy. Rather than taking Mutis's word and authority for granted, many readers tried to refute his evidence on the basis of their own observations of these plants in their own country. A veritable outcry in readers' letters followed the publication of Mutis's article – pages and pages were written in response, mostly in defence of the plantain tree.

The eminent member of the Economic Society, Antonio Liendo y Goicoechea, produced an 'apology for plantain trees', writing under the pseudonym *Licornes*. Although he was a respected naturalist himself and his reply made clear that he had considered various theories of air and causation of disease, his reply appealed

---

[58] 'Carta', *Gazeta*, Vol. 2, no. 50 (26 February 1798), 13–14.
[59] 'León y 8 de Octubre', *Gazeta*, Vol. 1, no. 41 (13 November 1797), 227.
[60] 'Carta', *Gazeta*, Vol. 2, no. 50 (26 February 1798), 13–14.

to readers' emotions as much as a scientific argument. Mutis's report would make anyone scared of plantain trees, Goicoechea wrote, but he himself would never be scared of them. They were 'beautiful' and 'beneficial', and he had strong feelings about them: 'I know these beneficial plants well, and the love I profess for them makes me take up the pen in their favour.' He described the plantain trees' remarkable resilience to pests, their roots that allowed them to gather nourishment from deep in the ground, and their ability to bear fruit year-round, which by extension was an economic benefit that would always 'enrich' the people who planted them. In addition, Goicoechea stressed the plantain's every-day importance in rural Guatemalans' nutrition in particular, and mentioned that the juice of this useful tree might even have medicinal qualities. 'Poor plantain trees!' he exclaimed. These misunderstood plants were truly a gift from God. Concluding his letter by declaring that 'plantain trees, and all other trees, are always good', he displayed an appreciation for the nutritional, economic, and aesthetic benefit of the plants.[61] In this, he seemed to firmly follow the original Nicaraguan's argument, who had not gone so far as to argue for plantains' beauty, but was certainly convinced of their economic and nutritional benefit.

Just as Mutis and the Guatemalan *Desengaño* argued within the framework of enlightened concern about public health that they thought equally applicable to any location, Goicoechea and the Nicaraguan argued on the basis of another Enlightenment ideology, that of 'public usefulness'. The debate in the *Gazeta* highlights the potential clash between these related concepts. Mutis and Goicoechea agreed on the importance of considering vegetation as a factor in public health, and both thought that concerned 'patriots' had a role to play in studying these topics before making interventions in the landscape. Mutis lauded the zealous service of the governor of a small town who had initially pointed out the alleged problem with banana trees, and held him up as an example of an administrator who worked to spread Enlightenment ideas within the limited scope of this office. Goicoechea argued that this was misplaced patriotism, because the public happiness of a well-nourished population was more important than these – in his view isolated – problems with bad air. Proper management rather than outright dismissal of these plants should be the solution in Guatemala, he argued. For Goicoechea, plantain or banana trees were an integral part of the Guatemalan landscape, and his response, written 'from my plantain field in Petapa', implied that he considered them something on which no outsider could decide. This episode, in which naturalists from two different parts of the Spanish empire turned their attention to the relative utility of plantain trees, highlights the varied approaches to nature that fell under the umbrella of 'enlightened thought', and were not easy to reconcile. Although the

---

[61] 'Apologia por los platanares', *Gazeta*, Vol. 2, no. 54 (26 March 1798), 43–5. On pseudonyms, see Lanning, *Eighteenth-Century Enlightenment*, 87, n. 29.

two naturalists operated within the same framework of the importance of patriotism and public happiness, and maintaining public health, their interpretation of the report differed.

As the story of these plantains shows, the same enlightened concepts could also be used to support opposing interpretations, as each argued that they knew best what would apply to Guatemalan landscapes. The *Gazeta*'s reading public carefully weighed up knowledge produced by scholars such as Mutis and applied it to an ever-shifting framework that pitted local utility against utility as interpreted by an international scholarly world. A focus on local circumstance did not mean an outright rejection of knowledge from elsewhere: several contributions to the great plantain debate of 1798 invoked European botanical treatises to support their position. Yet while readers accepted Mutis's authority to speak on plants in general, not everyone agreed that his treatise should be applied to Central American ones. A reader who went by the name of *Mutilchos* countered *Desengaño*'s arguments with his own knowledge of botany. He pointed out that a thorough knowledge of the trees in question as well as the location and context of plantains was important, turning the debate's attention to the physical spaces in which one was likely to find plantains. In his view, Mutis was not necessarily attacking *all* plantains, just those that had been planted too closely together and therefore produced putrid air from the puddles that developed around their base. If anything, all plants generally expelled noxious air, as Jan Ingenhouz's experiments had shown. Plantains as such did no particular harm.[62]

In fact, most defenders of plantain trees did not reject 'modern botany' outright, but preferred to apply it only in ways that could be considered in line with their observations of Central American plants and their uses. Yet another contributor named *Eudiofilo* (probably the Mexican botanist José Mariano Mociño) also defended plantains. His favourable opinion was partly influenced by his own observations – plantains were 'precious to Americans' and should indeed be regarded as literal 'bread fruit trees' (*arboles de pan*) for their ability to provide nourishment to the populace. The learned botanist's perspective, however, differed from the Guatemalan patriots'. His was a view that explicitly welcomed the contributions of European scientists to the debate, because he was convinced that the latest discoveries supported a positive verdict on plantains. Mociño stressed his knowledge of the composition of the earth's atmosphere, in particular of experiments on 'fixed air' which had shown that leaves of plants purified the air around them. He cited the experiments of Joseph Priestley, Ingenhousz, and Jean Senebier, which he stated he

---

[62] 'Carta', *Gazeta*, Vol. 2, no. 83 (28 May 1798), 123–4. Goicoechea also discussed Ingenhouz's experiments in his edition of Mociño's treatise *Tratado del Xiquilite y Añil de Guatemala* (Guatemala, 1799), 38–9.

had read about in a treatise by the French physician Sigaud de la Fond (specifically, M. Rouland's 1785 edition of the same). However, even he made sure to clarify the relevance of these experiments to Guatemala, drawing a direct comparison between experiments on plants' capacity for purifying air, and plantain trees. He argued that each individual leaf of the plantain tree should be seen as a 'type of laboratory in which this healthy chemical operation takes place', with special emphasis on the large surface area which increased the potential for this reaction. Armed with such chemical knowledge and the names of famous northern European scientists, he challenged Mutis's implied scientific authority with another botanical explanation centred on Central American experience.

Mociño had even found a passage in the French treatise he had been reading which dealt specifically with plantains, a trump card which seemed to entirely contradict Mutis's interpretation. In the Persian city of Ispahan, Sigaud de la Fond assured his reader, *plátanos* were specifically planted around unhealthy places such as hospitals or prisons, because they were considered to purify the air and to guard against the bubonic plague. For Mociño, the solution was clear. Just as in Ispahan, the plague was not known in America, so 'why can we not attribute this benefit to our plantains, as they do in Ispahan?'[63] Universal knowledge derived from Persian plants was clearly perfectly acceptable to this scientifically trained author, if clear parallels could be drawn between the two places that suggested the comparison was valid. There was, however, a slight problem: the Persian *platanes* that Sigaud de la Fond spoke about referred to plane trees rather plantains. The distinction was lost in translation, since both were known as *plátano* in Spanish.[64] Most of the debate in the *Gazeta* was certainly about the latter, using specific terms such as *platanar* (a group of plantain trees) and referring to their nourishing fruit which could be eaten cooked as well as grilled. We can assume that Mociño read the treatise in the original, and in the great comparison of plants across three continents, the very thing that botanists always cautioned against happened: they were debating a plant without agreeing on its identification, and compared plants across geographical distances without taking into account the local context. Unwittingly, the plantain debate of 1798 therefore underlined the very criteria of useful knowledge that the *Gazeta* had started to develop. A rigorous examination of sources and their practical applicability to the local context was needed to arrive at knowledge that would truly be suitable to drive the goals of *utilidad y felicidad pública* forward.

[63] 'Concluye la carta concerniente platanares', *Gazeta*, Vol. 2, no. 57 (16 April 1798), 76–7. Saladino García, *Ciencia y prensa*, 227, identifies Mociño as the author.
[64] Joseph Aignan Sigaud de la Fond, *Essai Sur Différentes Especes D'Air-Fixe Ou De Gas* (Paris: P. Fr. Gueffier, 1785), 411–12.

Imperial Comparisons

Two worlds of useful knowledge therefore existed in the pages of the *Gazeta*: that of natural history and its economic applications studied on a local scale, often with patriotic undertones, and the international world of scientific knowledge that traversed long distances. These two worlds collided in the search for domestic equivalents of foreign plants, which explicitly required a comparison of Central American nature with other places. One feature of patriotism as it was explicitly tied to the natural world was the idea of fertile soils. The fertility of American soil, and its ability to accommodate European as well as American crops, had been seen as a sign of divine providence from Columbus onwards.[65] The seventeenth-century Franciscan friar and scholar Martín Lobo published a treatise on the possibilities of exploiting this providential nature specifically for Guatemala, entitled 'Measures to ensure that all the fruits, herbs and plants of Europe and of all the world be harvested in the Kingdom of Guatemala'.[66] The rhetoric of abundant land as a source of more specifically patriotic Guatemalan pride had also already been found in other earlier writings, such as the 1722 *Historia Natural del Reino de Guatemala* by the Dominican friar Francisco Ximénez, and in the writings of exiled Jesuits.[67] By the late eighteenth century, the description of Guatemala as 'a vast terrain ... a soil where nature by itself produces what it cannot produce elsewhere even with the most diligent cultivation' in the petition for the Economic Society's establishment might have almost sounded generic.[68] The interpretation of landscape as bountiful meshed with the idea of the distinctive nature of a Central American locality in the pages of the *Gazeta*. Despite a decline in indigo prices in the 1780s, the worth and quality of Guatemalan indigo, for instance, was taken to be beyond dispute. In a botanical description of the indigo or *xiquilite* plant as it was found in the Kingdom of Guatemala in general and El Salvador specifically, the report's author stressed that from these plants' twigs 'the best indigo in the world is made'.[69] Local pride could reside in either the province or the kingdom, but it certainly referred to Central America in this case.

In the late eighteenth century these sentiments were also drawn into discussions about the worth of Guatemalan nature and imperial competition. If soil-based patriotism provided vague assurances about the special status of Central

---

[65] Rebecca Earle, *The Body of the Conquistador: Food, Race and the Colonial Experience in Spanish America, 1492–1700* (Cambridge University Press, 2012), 93–110.

[66] Antonio Batres Jáuregui, *La América Central ante la historia* (Guatemala: Sanchez y De Guise, 1920), 520.

[67] See Saint-Lu, *Condición colonial*, 129–35.

[68] 'Discurso sobre las utilidades que puede producir una Sociedad Económica en Guatemala', 1795, AGI, Estado, 48, N.7, 1.

[69] 'Relación hecha por el teniente coronel del Ejército y capitán de Dragones, Felipe de Sesma, sobre la simiente de siquilite de Guatemala y S. Salvador', 1788, AGI, Indiferente General, 1545, 169.

American nature and its productions, reformers now also applied this principle by searching for and acclimatising specific products within the kingdom. This signified a departure from local descriptions: to compare or export plants, they were likely to have to subscribe to some extent to a globally recognised system of nomenclature. The creation and use of plant knowledge within the geographically restricted area of Central America obeyed its own logic, which would need to be modified to merge local knowledge with participation in imperial worlds. A search for foreign plants' equivalents was a feature of many eighteenth-century societies, from Carl Linnaeus's Sweden to Tokugawa Japan, and often happened in tandem with acclimatisation projects. Botanists working in the Spanish empire's botanical gardens both in Europe and the Americas were well aware of the economic potential of botanical commodities from the colonies, and focused especially on spices, herbs, and tea that might rival Asian products monopolised in Europe by the Dutch.[70] Members of the Guatemalan Economic Society joined in this search for equivalents, well aware of the desirable status of spices and tea. José Domingo Hidalgo, a resident of the Guatemalan town of Quetzaltenango, wrote to the Society in 1797, claiming that on the slopes of the Santa María volcano the tea herb grew. Hidalgo claimed to have verified its identity 'according to its description by Dn Felix Palacios on page 688 of the *Palaestra Farmacéutica*', referring to the Spanish pharmacist's famous 1706 treatise. The vice-secretary of the Society noted his doubts about the legitimacy of this comparison between the Guatemalan herb and true tea, but nevertheless forwarded Hidalgo's report to the director, Villaurrutia, since in his estimation, Hidalgo was a learned man who deserved to be taken seriously. If Hidalgo were to send a sample of the tea, the Economic Society would analyse it.[71] The undertaking was unsurprisingly unsuccessful, but it demonstrates that Society members were alert to the possibility of Central American plants acquiring a new status as valued treasures for export. Another correspondent sent samples of allspice from Honduras, where this tree grew, to the secretary of the Society. The *Gazeta de Guatemala* eagerly reported on this promising development. The article quoted the Spanish economist Nicolás de Arriquibar's complaints about the Dutch pepper

---

[70] Robert Liss, 'Frontier Tales: Tokugawa Japan in Translation', in Simon Schaffer, Lissa Roberts, Kapil Raj, and James Delbourgo (eds.), *The Brokered World: Go-Betweens and Global Intelligence, 1770–1820* (Sagamore Beach, MA: Watson Publishing, 2009), 1–47, at 15; Lisbet Koerner, *Linnaeus: Nature and Nation* (Cambridge, MA: Harvard University Press, 1999), 113–39; Gonzáles Bueno and Nozal, *Plantas americanas,* 19–20; Manuel Lucena Giraldo, 'Los experimentos agrícolas en la Guyana española', in Antonio Lafuente, Alberto Elena and María Luisa Ortega, *Mundialización de la ciencia y cultura nacional* (Madrid: Ediciones Doce Calles, 1993), 251–58; Bleichmar, *Visible Empire*, 25, 137–8, and 229–33.

[71] Tomás de Moreda to Jacobo Villaurrutia, Guatemala, 17 May 1797, AGCA, A 1.1, Leg. 2007, Exp. 4347, fol. 35.

monopoly, and his suggestion that allspice might be a suitable alternative to pepper. The editors latched onto Arriquibar's idea and its promise that cultivation of allspice might 'replace the eastern one, which we unnecessarily pay for with gold coins', if the Honduran samples were indeed as good as their initial promise.[72] While there was apparently never an attempt to carry out the projects of growing allspice on a larger scale either, the idea of a native plant replacing a foreign one was a common one. *Gazeta* readers and Economic Society members regularly sent in such information within the broader project of compiling information about the Central American natural world and extracting prosperity from Guatemalan soil.

If the search for the equivalents of tea and spices was directed at rivalling the Dutch monopoly on these Asian products and would benefit the Spanish empire as a whole, other natural productions stoked internal imperial competition. In an example of a search for local products to rival specific imports, a priest from Granada in Nicaragua sent information about cinchona bark, or *quina*, to the newspaper. This was a direct response to a treatise on *quina* by Mutis, which had appeared in the *Gazeta de Guatemala* the previous year – a description which used Linnaean nomenclature but provided a wealth of more discursive detail to guide the botanist to distinguish species.[73] The correct identification of the bark and disputes over the number of *quina* species in existence were of great concern to botanists and administrators in the Spanish empire, particularly in the wake of the establishment of a cinchona monopoly in Loja, Peru.[74] The Nicaraguan priest now wanted to prove that Central America was also home to true and effective *quina*, even though it was more commonly associated with Peru. The priest compared the information in the botanist's treatise to locally occurring plants with which he was familiar, and, like Mutis, used the language of all senses including taste, not just visual description, to verify the plant: 'Regarding *quina* … we can affirm that we have the four species described by Dr Mútis, even though we call them different names. One of them exists on the coast of Managua, its fibres have bitterness, astringency, and some sharpness.'[75] And despite the focus being on identifying 'true' *quina*, the

---

[72] 'Sobre un arbol de toda especia', *Gazeta*, Vol. 4, no. 177 (6 October 1800), 342.

[73] 'Salud publica', *Gazeta*, Vol. 2, no. 81 (1 October 1798), 269. 'Errores inevitables en el uso de la quina', *Papel Periódico de Santafé de Bogotá*, no. 89 (10 May 1793). '*De las diferentes especies de Quina, y sus virtudes medicinales.* Por el Dr. D. José Celestino Mutis, celèbre medico y botanico de Sta Fè de Bogotá' – reprinted in *Gazeta*, Vol. 6, no. 270 and 275 (7 and 13 September 1802), itself a reprint of *Mercurio de España* of January 1802.

[74] Matthew Crawford, *The Andean Wonder Drug*: Cinchona Bark and Imperial Science in the Spanish Atlantic, 1630–1800 (University of Pittsburgh Press, 2016), 151–75; Bleichmar, *Visible Empire*, 146–7.

[75] 'Plantas medicinales de Nicaragua', *Gazeta*, Vol. 7, no. 297 (28 March 1803), 71–2. On the use of taste in identifying quina, see Crawford, *Andean Wonder Drug*, 82. Later botanists criticised this discursive approach: Clements Markham, *The Cinchona Species of New Granada* (London: Eyre and Spottiswoode, 1867), reproduced Mutis's *Arcano de la Quina* but insisted that only

priest could not resist listing local alternatives to *quina* which would be 'just as good' as the authentic species, for instance, a tree known locally as 'almond tree ... which because of its similarity in its bark, even though it is thicker, seems to me a type of perfect *quina*'.[76] The Nicaraguan pretenders to the Peruvian *quina*'s throne, like its other challengers across the empire, did not supplant the Peruvian bark, but locally, these alternatives continued to be considered an equivalent. As late as 1833, a French explorer noted in his journal a plant by the name of 'Cinchona Caribbeoa, which has the same property as quinquina'.[77] Local beliefs about the ability of Central America to produce such an array of economically and botanically useful plants in parallel to imperial and Linnaean worlds of knowledge persisted.

A particular opportunity to consider the place of Central American knowledge within an imperial framework appeared in the very first year of the Economic Society. The Society collaborated with the Spanish naturalist José Longinos Martínez to establish a cabinet of natural history in Guatemala City in 1795. Some members of the Royal Botanical Expedition to New Spain also travelled to the Audiencia de Guatemala, where they spent three years (1795–98), much of it travelling around the Audiencia. Longinos, as a member of the expedition, was an influential figure in the ambit of the Society during the months that he spent in the capital. Guatemalan reformers welcomed the botanists and engaged with their aims, but they also saw it as a way to fulfil their own scientific and commercial priorities. Longinos compiled detailed instructions on how to collect specimens, echoing instructions regarding the collection of herbs sent from Madrid to Spanish officials in the colonies, and perhaps with input from Mexico City's José Alzate.[78] In this case, it was, however, the Economic Society and *Gazeta* who jointly took responsibility for requesting contributions from their members. Signed by both Longinos and the Society's vice-secretary Tomás de Moreda, up to 150 copies of a letter of instruction were sent to members of the Society, in addition to the *Gazeta* drawing attention to it.[79]

Yet again, Society members were to be the link between the cabinet and the rural population, this time on behalf of the Royal Expedition. Members should

---

Part Four, the Latin Linnaean description, was of any use. He also criticised Mutis as having an 'extreme love of generalising on insufficient and false data' (p. 10).

[76] 'Plantas medicinales de Nicaragua', *Gazeta*, Vol. 7, no. 297 (28 March 1803), 71–2.

[77] 'Note-book, in French, of Baron Jean Frédéric Maximilien de Waldeck', British Library, Add. MS 41685, fol. 14.

[78] Maldonado Polo, *Huellas de la razón*, 271–2, notes that José Felipe Flores had contacted Alzate as early as 1790 to ask about the practicalities of establishing a natural history cabinet. Expedition members travelled around the region separately and extensively. Their itineraries are reconstructed in Taracena Arriola, *Expedición científica*, 26.

[79] AGCA, Leg. 2008, Exp. 13844, fol. 15.

read the instructions out 'a few times to those subjects who walk across many lands, because from them knowledge is always acquired, and some news of resources which chance usually presents to them'. To make sure it was suitable for this purpose, the instructions had therefore been written so as to be 'intelligible to all kinds of people'.[80] The Society in this case appeared to be an instrument of an explicitly imperial collection of specimens, one which made use of the Society's networks but in which knowledge collected 'by chance' needed to be formalised in a structured, scientific, and Spanish way, in order to fit into the imperial cabinet. To encourage submissions, the Spanish naturalist Longinos sponsored a prize for the best collection of 'natural productions' submitted to the museum. The response was disappointing, with no submissions of sufficient quality to merit a prize received three months later, and the museum itself was short-lived, closing after less than five years.[81] The Economic Society's correspondence suggests that it was not for the association's lack of trying. One Economic Society member, Julian Hernández, replied to the secretary of the Society from the Honduran administrative capital of Comayagua in early 1797, noting that he had distributed the printed instructions for the collections of natural products among 'appropriate individuals'.[82] Another member, José Domingo Hidalgo from Quetzaltenango, was apparently responding to the line in the instructions that referred to collecting the 'productions of volcanoes' (under the heading of the 'Mineral Kingdom') when he accompanied his letter to director Villaurrutia, with samples of 'productions of the Sta Maria de Jesus volcano' in the Guatemalan highlands.[83] In addition to this submission from Quetzaltenango and one from the nearby town of Texutla, someone, perhaps Hernández, sent contributions from Comayagua by the end of the year, as did another member or reader from Omoa in Honduras.[84]

Despite its ultimately lacklustre results, this initiative shows that the Economic Society and its members were quite engaged in this project. They encouraged submissions of 'any sort of thing in a good condition' for conservation in the cabinet, adding that it might be sent to Madrid if it was exceptional for its 'novelty or exquisiteness'.[85] Since the best specimens were to be sent to Spain, imperial claims to the most unique American plants appear to have taken precedence over local interest in them. However,

---

[80] 'Compendio Instructivo', Guatemala City, 1 January 1797.

[81] 'Nueva Guatemala 19 de Febrero', *Gazeta*, Vol. 2, no. 49 (19 February 1798). Maldonado Polo, *Huellas de la razón*, 302.

[82] Julian Hernandez to Tomás de Moreda, Comayagua, 25 February 1797, AGCA, A 1.1, Leg. 2007, Exp. 4347, fol. 11.

[83] Tomás de Moreda to Jacobo Villaurrutia, Guatemala, 17 May 1797. *Ibid.*, fol. 35; Real Sociedad, *Quinta junta pública*, 16.

[84] Real Sociedad, *Tercera junta pública*, 12.

[85] 'Gabinete de historia natural', *Gazeta*, Vol. 2, no. 49 (19 February 1798), 7. See also Maldonado Polo, *Huellas de la razón*, 272–303.

Guatemalan reformers even carved out a niche for a more multi-layered patriotism when they encouraged Central Americans in 1798 to see the cabinet as more than an imperial obligation:

The Society hopes that good patriots will contribute to this useful enterprise ... in which they will perform a distinguished service to the public and to their *patria*, because of the general usefulness which will result from these submissions, insofar as they can contribute to the good of humanity, and to the advancement of natural sciences, whose study is currently encouraged with the greatest effort by our glorious Monarch.[86]

While the *patria* in this case was clearly Spain and its empire, the invocation of the 'good of humanity', 'general usefulness', and 'advancement of the sciences' indicates a broader ambition for the natural specimens collected within Guatemala. If they were distinctive enough to contribute to the richness of imperial collections, this would in turn elevate the collections to living proof that Guatemala was a jewel in the king's crown. Throughout the Guatemalan cabinet's short existence, this dual ambition was visible. The cathedral canon García Redondo had no problem calling the Spanish naturalist Martínez Longinos a 'wise, enlightened, and generous patriot' at the Economic Society's general assembly in 1796. However, he immediately also refined the meaning of the cabinet within the empire. In addition to giving 'all the appropriate glory to the beloved *patria*' (which might here be the whole empire), the purpose of the cabinet was to send a message to the king about the much more specific locality of the Audiencia: 'to our august sovereign a convincing proof that Guatemala is neither the last nor the most inferior of his many reigns'.[87] Despite the project's imperial roots, the Economic Society's patriotism aimed to crystallise more localised feelings of pride in the productions of the country. The cabinet would serve to promote Guatemala's place in the wider empire. The Central American intelligentsia did not co-opt the expedition into an entirely local enterprise, but they clearly envisioned ways in which the empire's collecting practices would work for them, instead of the other way around.

Reformers did not expect the royal expedition to do all the work of promoting Guatemalan nature for them. The Economic Society's members also attempted to popularise local plants beyond the Audiencia in another bid to gain recognition and commercial success for Central America. While the newspaper's immediate concern was to exchange useful knowledge within Central America, reformers also had lofty ambitions for medicinal plants beyond just healing locally occurring illnesses and bites. The *protomédico* José Felipe Flores had undertaken several projects in which he sought to prove the place of Central American medicines in a global world of knowledge.

---

[86] *Ibid.*   [87] Real Sociedad, *Junta pública*, 34.

For instance, he had conducted experiments into the medicinal properties of certain lizards at the Chiapas hospital earlier in his career.[88] Moreover, in a report on Guatemala's medicinal plants sent to Madrid in 1788, he had couched the process by which indigenous people extracted a febrifuge from the bark of the *bálsamo* tree in explicitly scientific terms, describing them as acting 'like able apothecaries' performing a 'chemical operation ... so appropriate ... that they could be located in a laboratory'.[89] Flores in these formal reports had shown that enlightened knowledge and universally applicable modern science could already be found in the Guatemalan countryside, if one only knew where to look. However, this was not the type of formal and abstracted knowledge to be found in the *Gazeta*, which would be more difficult to translate.

José Rossi y Rubí, the contributor to the *Gazeta de Guatemala* who had been involved in Peruvian enlightened circles before his move to Guatemala, suggested new uses for the *algalia* plant. As long as it was considered a local cure for the bites of local spiders and local snakes, it had less obvious value as an export, or as the sort of botanical knowledge that might be internationally significant. However, Rossi believed that it might also be used to treat the much more universally recognisable diseases of epilepsy and rabies. He noted that he had tried to export the seeds to 'Peru, Valencia, Messina, and other European regions, whose climate is the least different to this one', seeking to establish recognition for the herb beyond the confines of the Kingdom of Guatemala and the reach of the newspaper. In fact, he was convinced that the *algalia* would be 'as appreciated in that part of the world as *quina*'. He reported that, unfortunately, none of his remittances to Europe had arrived (ongoing wars disrupted Atlantic traffic). He nevertheless appealed to the 'patriotism' of the Economic Society to persuade them to make new attempts to popularise the seeds in Europe.[90] Rossi's approach demonstrates that additional ambitions of 'universal' utility could lie behind Guatemala's enlightened reformers' interest in medicinal plants, but also the incongruity of these different forms of knowledge. Rossi's plant seeds in this case were lost en route, but the *Gazeta*'s own descriptions lacked an established taxonomy that would make the translation of knowledge more difficult in any case. Even Rossi's own description, which was titled 'botanical' but listed only the root's uses, was more reminiscent of a remission to Spain within an imperial framework of collecting useful productions of nature from the countryside and sending them to be evaluated in

---

[88] Miruna Achim, *Lagartijas medicinales: Remedios americanos y debates científicos en la Ilustración* (Mexico City: Consejo Nacional para la Cultura y las Artes, 2008).
[89] 'Razón de lo que contienen los dos Cajones remitidos de Goatemala', AGI, Indiferente General, 1545.
[90] 'La Algalia', *Gazeta*, Vol. 5, no. 234 (22 October 1801), 610. See also Hernández Pérez, 'Medicina y salud pública', 25.

Madrid, than establishing an autonomous bank of knowledge.[91] Despite their best rhetorical efforts, Central Americans were more frequently importers than exporters of such botanical material and 'enlightened' knowledge. The reluctance to employ Linnaean descriptions that was entirely suited to a local paper was more problematic here. The knowledge of the plant in the *Gazeta* itself was not readily exportable. Instead, it was the seeds, to be grown, acclimatised, and made local in a different geographical location, that would signify the successful translation of Central American local knowledge to the wider world. The Central American project of enlightened knowledge drawn from the Central American countryside could serve to promote the Kingdom of Guatemala as a significant centre of knowledge within the empire, but it required a stretching of the usual categories in which Guatemala's local knowledge economy traded.

Conclusion

Central American reformers were eager to use the wide communication networks they had already established to source knowledge that would form the basis of applied projects. However, the translation of knowledge within and across continents was not without its pitfalls. Even beyond mistranslations, the interpretation of scholarship from Europe, North America, or other parts of the empire in a local context was never simple. Reformers and readers of the *Gazeta de Guatemala* explained their doubts in persuasive detail, raising issues that would have been familiar to any scholar residing outside the world's best-connected scientific metropoles. They voiced concerns that the lack of communication routes with other parts of the Spanish empire, such as Bogotá, made it more difficult to obtain reliable detailed knowledge. As a result, they set out to create their own Central American epistemologies because they could not even rely from knowledge within the Spanish empire, even while patriots in places such as New Granada were building their own theories of American knowledge. Concepts of utility, for these reformers, should ultimately be anchored in, and tested against, observations of Guatemalan nature.

Reliance on empirical knowledge was most pronounced in the case of medicinal plants, where local applicability was key. The debates around plantain trees that played out in the pages of the *Gazeta* further put this local lens of interpretation into focus. They show that the emphasis on local observations in the making and application of natural knowledge was not just a result of limited knowledge networks available to the editors and the Economic Society, but a matter of weighing up different priorities within sometimes contradictory principles of Enlightenment, applying different geographical and human reference points to it. While this did not negate a broader possibility of this knowl-

---

[91] 'Algalia', *Gazeta,* Vol. 5, no. 216 (3 August 1801), 531.

edge being useful beyond Central American shores, it did complicate the matter. Reformers such as Rossi y Rubí schemed to export medicinal plants from Guatemala to Europe, for the benefit of humanity. However, the rejection of Linnaean and other universal systems of taxonomy also complicated the transmission of such knowledge across the world.

These considerations led to the *Gazeta* pondering not just the situatedness of knowledge, but also the relationship of their project to indigenous knowledge, especially when it came to medicinal plants. They made it clear that knowledge needed to function within whatever geographical space reformers had currently defined as their territory of intervention: from the specific locations of the Capuchin convent of Guatemala City that needed to be saved from plantains, to villages on the shores of Lake Atitlán, where medicinal plants might be chopped with a machete. If, as President Gonzáles Mollinedo argued, knowledge of the 'locality and the circumstances which are present in it' was the key to progress, and utility was measured against its practicality in the Guatemalan context, then notions of what Enlightenment was in Central America also relied strongly on an understanding of what this 'Guatemalan nature' was that enlightened progress would transform.

4 Useful Geography in Practice

In 1803, the *Gazeta de Guatemala* tried to answer a question about a Mexican mountain known as *Volcán de Orizaba* that quickly developed into a broader critique of geographical knowledge. Did the eruptions of the Orizaba, a snow-covered peak in the Veracruz region, produce a strange rubbery material that could be used to waterproof all kinds of fabrics? The question should have been easy enough to answer, but editors and readers disagreed about a fundamental detail: whether the mountain in question was a volcano at all. A reader, who called himself *El amigo de la verdad*, protested that the mountain may have been known 'vulgarly' as *Volcán de Orizaba*, but it had not been known to erupt in living memory. The rubbery waterproofing material was, in fact, merely a tree resin. The editors apologised for reproducing erroneous information. They knew that they had left themselves open to a challenge from American readers when they printed this article, sourced, as they explained, from two Spanish newspapers. They quickly agreed that by virtue of writing from the village of Orizaba itself, the reader was in a far better position to verify this information.[1] To some extent, this was a familiar discussion about the ramifications of information that had travelled a long way. However, the absurdities of relying on knowledge imported from Spain were acute in the matter of identifying an American mountain, and questions of applying and authenticating knowledge seemed especially stark to the editors when it came to discussions of geographical knowledge. After all, if the geographical context of knowledge production and application had been relevant when reformers questioned the validity of botanical or medicinal knowledge, in debates about geography there was an even clearer spatial reference point that privileged the local observer.

---

[1] 'Orizava 27 Sept 1803', *Gazeta*, Vol. 7, no. 327 (31 October 1803), 411–12. This debate may be partly explained by a note on a *relaciones geográficas* map (copied in the eighteenth century) which suggests that, historically, there had been a tendency within Guatemala to simply blur the categories of mountain and volcano: 'in these provinces, they call the tallest mountains volcano, but they are not all of fire or sulphur'. See Figure 4.1, 'Costa de Zapotitlan, y Suchitepeques', BL Add. MS 17,650e.

Exasperation with the availability of geographical information was a recurring theme among high- and low-ranking government officials, *Gazeta* editors, and contributors alike. Of course, as we have seen, there was a lively geographical tradition within the Spanish American bureaucracy that had found particular expression in debates around volcanoes and archaeological ruins in the previous decades. However, these were not publicly available documents, and their ideas about the relationship between landscapes and societies was implicit rather than explicit. To the Economic Society, this formed a geographical vacuum. As a response, several geographical surveys, designed to reassess the existing knowledge about the coasts and interior of the kingdom, and test received wisdom about the navigability of particular roads and rivers, occupied reformers in the 1790s and early 1800s. For the *Gazeta*'s editors, taking decisive action about this unenlightened state of disinformation meant the publication of a series of articles that would eventually form a new and authoritative 'geographical description of Guatemala'. In the course of this project, they set out a framework of geographical knowledge and its sources which energetically doubled down on the paper's guiding principles of prioritising local practice over foreign theories.

The creation of new geographical information was tied up with interventions in the landscape itself. Reformers personally carried out reconnaissance projects with the dual intention of gathering and publishing information (itself seen as an enlightened act), and to reframe Central America's transport network in a way that would be suited to local priorities. The most pressing application for geographical knowledge were infrastructure projects, on land as well as water. It was a truth universally acknowledged among residents of the Audiencia de Guatemala that the road networks of this province were 'perhaps the most uncomfortable and the most rough'.[2] This stood in the way of improvement because it hindered trade. Reformers countered with projects that were designed to improve transport: not always along the existing trade routes that connected Central America with Spain, but new connections within the Audiencia's interior, coastal routes, and rivers that had hitherto not been successfully navigated. The much-debated relationship of Guatemala to global forms and networks of knowledge was therefore paralleled within Central America by a re-examination of the relationships between city and countryside, between coastal regions and the interior.

---

[2] 'Puente de Zamalá', *Gazeta*, Vol. 4, no. 174 (15 September 1800), 330. Other complaints, e.g. Cortés y Larraz, *Descripción geográfico-moral*, Vol. i, 292; 'Camino de Suchitepéquez', AGCA, A1.21 Leg. 207 Exp. 4171, fol. 47v. Road transport in Central America: Ramón Serrera Contreras, *Tráfico terrestre y red vial en las Indias Españolas* (Madrid and Barcelona: Ministerio del Interior, 1992), 53–63; Sellers-García, *Distance and Documents*, 79–137.

The *Descripción Geográfica de Guatemala*

In a 1797 article entitled 'Notes for creating a precise description of the Kingdom of Guatemala', the editors of the *Gazeta de Guatemala* unveiled an ambitious plan for compiling a new kind of geography. It was inspired by the well-known *Descripción topográfica* in the *Gacetas de Literatura de México* in November and December 1791, and had a counterpart in other newspapers, echoing in particular the arguments put forward in favour of a 'patriotic' description of Peru in the *Mercurio Peruano* in 1791 and 1792.[3] In the hands of the *Gazeta*'s editors, the plan for such a patriotic description of Guatemala combined their concerns with the applicability of foreign knowledge to Central American lands, and a thoughtful discussion about the sort of knowledge which would be truly 'useful' on a local scale. Having first alluded to the necessity of a geographical description of Guatemala in the prospectus of the newspaper in early 1797, the editors elaborated their stance over four issues in May and July of the same year. They saw it their task to 'gather different materials, acquire authentic news, documents, and declarations, in one word, to bring together all the learned men who are dispersed in these provinces that we are trying to describe'. In their search for information, they would value useful recent information that might be gleaned from city archives, but they also asked readers to contribute knowledge to the project. They had a rather pessimistic view of the practices that currently fell under the umbrella of 'geography'. As they saw it, travellers might pass off as complete topographic descriptions of a town information that they had gleaned from the waiter at a brief rest-stop, perhaps while they consumed 'a light refreshment'. Others might declare that a distance amounted to a certain number of miles, but a closer inspection would reveal that they had arrived at their text 'without examining this, or doing a practical reconnaissance'. This was not the sort of knowledge the *Gazeta* was after. The proposed *Description* would focus on reliable knowledge, and on those aspects which would contribute to the 'happiness' of a province: the 'number of useful men ...., the possibilities offered by the land to advance agriculture, arts, and commerce'.[4]

The proposed *Description* was carefully defined in terms which the editor found useful. Local descriptions, the editor argued, would serve the patriotic project better than grand encyclopaedic ambitions. The 'slow' and 'thorough' method of chorography was presented as distinctly tailored to the yet undescribed expanses of Central America. The second-century scholar Ptolemy had defined 'chorography' as detailed, small-scale descriptions, the geographical study of an 'individual locality'. The *Real Academia*'s 1791 dictionary defined

---

[3] Peset, 'Ciencia e independencia', 205; Meléndez, 'The Cultural Production of Space', 177–83.
[4] 'Memorias para hacer una descripcion puntual del Reyno de Guatemala', *Gazeta,* Vol. 1, no. 13 (8 May 1797), 14 (15 May 1797), 24 (17 July 1797), and 25 (24 July 1797).

it as a 'description of a particular Kingdom, Country, or Province'. In contrast, geography was a global genre, the 'science which deals with the universal description of the whole earth'.[5] The *Gazeta*'s project would cover only a manageable scale. The editors repeatedly stressed the simplicity of the *Description*, which would be in line with the general intentions of the newspaper:

> We have not promised to give a general and complete description of the Kingdom of Guatemala. Such a pompous offering would be contrary to our simplicity, and to what the public can expect from us on the basis of our prospectus. ... We only offer some good intentions, alongside the customary dose of patriotic zeal, and love of humanity.[6]

For well-known places such as New Spain or Peru, the editors explained, a few words might suffice to provide an update on their situation, but in an 'almost unknown' kingdom such as Guatemala, 'the most inconsequential things must be touched upon' in order to present an exact description which might replace earlier, imprecise reports. This required the detailed approach of chorography, and the peripheral situation of Guatemala itself conditioned their choice of approach.[7]

The *Gazeta*'s editors, faced with different possible geographical formats, chose the density of local descriptions and of 'physical, historical and political geography' over the all-encompassing measurements of 'astronomical geography', a mathematical abstraction of space which they saw as less accurate and certainly less relevant on a local level. The *Description of Guatemala* must not be 'scientific', the editor (most likely Ignacio Beteta) argued, preferring the 'plain and simple' local knowledge of chorographic compilations over abstract and scholarly works. In his eagerness to make the description locally relevant, he stated:

> I would also relieve the author of the necessity of noting longitudes and latitudes ... May the task of considering the earth in relation to the sky remain for the navigators and the learned men ... This description, plain and simple whenever possible, should not contain any scientific vocabulary, not one thing that might be worthy of being discussed by learned men.[8]

This dismissal of 'learned vocabulary' and 'considering the earth in relation to the sky' shows a rather particular vision of what kind of geographical

---

[5] Lennart Berggren and Alexander Jones, *Ptolemy's Geography* (Princeton, NJ and Oxford: Princeton University Press, 2000), 57. Real Academia Española, *Diccionario de la lengua castellana* (Madrid: Viuda de Joaquin Ibarra, 1791), 262, 453.

[6] 'Memorias', *Gazeta*, Vol. 1, no. 24 (17 July 1797), 185.

[7] 'Memorias', *Gazeta*, Vol. 1, no. 14 (15 May 1797), 107–8 and no. 24 (17 July 1797), 187.

[8] *Ibid.*, no. 14 (15 May 1797), 106. Beteta sent what appears to be a draft of the *Gazeta*'s articles arguing for a 'geographical description' to a municipal official as part of his request to see archival documents, see also Chapter 2. Ignacio Beteta to Regidor Syndico Joseph Manuel Laparte, 17 February 1797, AGCA, A1 Leg. 5361, Exp. 45314.

description was of use to Guatemala. To some extent, these phrases probably betrayed Beteta's doubt about the capacity of his readers to understand any complex science. There was also a practical component, given that even trained engineers sometimes chose to reflect travel times and estimates in their reports and cartography. For instance, one engineer's map showed the 'Plain of Jalapa', but only gave its distance from the capital city 'according to the common estimate of this country'. Another promised 'geometrical observations' in his explorations, but nevertheless seemed to rely mostly on estimates than on measurements of latitude and longitude for this report.[9] A third weighed up two conflicting distance measurements, explaining that only the 'common leagues' reflected the terrain that the road traversed.[10] 'Mathematical' maps, from this perspective, were neither particularly useful nor trustworthy. Once again, forms of knowledge that tended towards the universal or theoretical were dismissed.

It also followed a more deeply engrained scepticism of maps, an argument which the editors sustained elsewhere, too. Intervening in a dispute about the size and latitude of Russia, the editors felt compelled to remind their readers that even 'the most acclaimed maps' were 'not irrevocable testaments as far as the exactitude of their geographical calculations is concerned' – they were not trustworthy.[11] In general terms, Anne Godlewska points out that most eighteenth-century geographers might have agreed with this argument. If geographers wanted to make maps, they would simply have to accept some information of 'lesser quality' within their sources.[12] The origin of such scepticism within Guatemala was even clearer. Reformers could not generally expect maps to provide accurate information about the interior of the isthmus, since a long-standing Spanish policy to jealously guard and regulate cartographic information of the Spanish American interior (mirrored by similar secretive strategies of other European states) meant that most early modern published maps of the region were created by British or Dutch cartographers, often on the basis of incomplete spies' reports. The government in Madrid started relaxing its policy of cartographic secrecy by the end of the eighteenth century, but no project equivalent to the Jesuits' or Juan Cruz Cano's maps of

---

[9] José María Alejandre, 'Mapa que comprende el Llano de Jalapa situado en la Provincia de Chiquimula de la Sierra', 1773, AGI, MP-Guatemala, 200; 'Relacion e informe del reconocimiento de la Costa de Mosquitos, para tratar de su fortificación y defensa. Por el Ing. Porta y Costas', 1790, AGCA, A1.17.3, Leg. 4501 Exp. 38303, fol. 1v.

[10] 'Demarcación y medida de los volcanes de la ciudad antigua de Guatemala realizada por Luis Diez Navarro y José Gregorio de Rivera', 1791 [cited text from 1774 within a later compilation of documents], AMN Ms 0339/033, fol. 75. See also Sellers-García, *Documents and Distance*, 91–3; Sophie Brockmann, 'Surveying Nature: The Creation and Communication of Natural-Historical Knowledge in Enlightenment Central America' (PhD dissertation, University of Cambridge, 2013), 144–64.

[11] 'Criticas hechas', *Gazeta*, Vol. 1, no. 16 (22 May 1797), 121.

[12] Godlewska, *Geography Unbound*, 39.

South America existed for Central America.[13] Some 'professional' maps were produced by members of the Royal Corps of Engineers, who were trained in specialised Spanish academies. However, they mainly focused on strategically important areas of the Nicaraguan and Honduran coastline, on fortifications, and some city plans for the newly built capital Guatemala City.[14] As a result, much of Central America's interior remained unmapped throughout the colonial period. Several historians have suggested that maps, especially maps of the interior, were primarily vehicles to make Central America visible to far-away politicians, rather than a way of conveying geographical information within the region.[15]

In addition to expressing long-held views about the futility of maps, the *Gazeta*'s statement about longitude and latitude signalled the adaptation of a philosophy about the uses of science that was firmly centred on the Audiencia de Guatemala and its circumstances. This attitude had parallels in other parts of Spanish America. Francisco José Caldas, the Colombian naturalist and geographer, saw a role for astronomy in geography if it was linked to improvements in 'geography, our roads and our commerce', and stressed the importance of making science 'useful'. By contrast, he considered astronomy for the sake of 'observation', that is science in isolation from an earthly purpose, 'fruitless'.[16] The Central American authors of the *Gazeta* also insisted that the enlightened criterion of knowledge 'useful' to Guatemala simply left no time for heavenly observations, though they may have been willing to follow Caldas in conditionally accepting some 'mathematical' geography. The Economic Society itself established a school of mathematics, and its founding documents included a reference to young Central Americans being trained in the useful science of geography. However, this art of mapping must

---

[13] Richard Kagan, *Urban Images of the Hispanic World, 1493–1793* (New Haven, CT and London: Yale University Press, 2000), 77–81; Alison Sandman, 'Controlling Knowledge. Navigation, Cartography, and Secrecy in the Early Modern Spanish Atlantic', in Nicholas Dew and James Delbourgo (eds.), *Science and Empire in the Atlantic World* (New York: Routledge, 2008), 31–51. Matthew Restall, 'Imperial Rivalries', in Jordan Dym and Karl Offen, *Mapping Latin America* (paperback ed., University of Chicago Press, 2011), 79–83. David Buisseret, 'Spanish Colonial Cartography, 1450–1700', in Brian Harley and David Woodward (eds.), *The History of Cartography*, Vol. III, Part 1 (Chicago, IL and London: University of Chicago Press, 2007), 1143–71, at 1148; Tang, *Geographic Imagination*, 141–2.

[14] Ignacio Gonzáles Tascón, *Ingeniería española en ultramar: siglos XVI–XIX*. Vol. I (Madrid: Colegio de Ingenieros de Caminos, Canales y Puertos, 1992), 79–90; Karl Offen, 'Edge of Empire', in Dym and Offen (eds.), *Mapping Latin America*, 88–92.

[15] Sylvia Sellers-García, 'The Mail in Time: Postal Routes and Conceptions of Distance in Colonial Guatemala', *Colonial Latin American Review* 21 (2012): 77–99, at 86. Kagan, *Urban Images*, 60, makes a similar case for eighteenth-century Spain. See also Dym, *From Sovereign Villages*, 50–3.

[16] Quoted in Antonio Lafuente, 'Enlightenment in an Imperial Context: Local Science in the Late-Eighteenth-Century Hispanic World', *Osiris* 2nd Series, 15 (2000): 155–73, at 170.

have been carefully circumscribed to promote utility. After all, it was the school's founder, Josef de Sierra, who had so vehemently declared that utility 'adapted to the circumstances of the country' rather than theory must be the foundation of this education.[17]

The *Gazeta*'s project never covered more than a few provinces of the kingdom, since it depended on the initiative of its members outside the capital to contribute these descriptions. Nevertheless, the few that were submitted by readers were exceptionally thorough and ran over several issues. The contributors were parish priests or regional secular officials – men with access to the latest local records or familiarity with entire provinces that facilitated compiling such texts. The first one to be published, in July 1797, was the 'Chorographic Description of the Province of Quetzaltenango' by José Domingo Hidalgo. A prominent resident of the highlands city of Quetzaltenango, Hidalgo was a member of the Economic Society and subscriber to the *Gazeta*, and participated in several Society projects, for instance submitting specimens to the cabinet of natural history in Guatemala. He was also a self-taught civil engineer, a scribe for the regional government and a *juez subdelegado de tierras*, that is, a judicial official tasked with overseeing land measurements and property claims.[18] The description of the Province of Quetzaltenango, Hidalgo explained, would be divided into 'chapters of which there will be as many as there are parishes, which will be subdivided into articles or paragraphs, in order to talk in detail about each village'.[19] This format of listing information parish by parish was reminiscent of *relaciones geográficas* or *visitas*, the geographical reports compiled by Spanish secular officials, priests or bishops, which collated responses from questionnaires or followed the direction of the compiler's travel. Official documents such as censuses and tax records mirrored this format, too, and as a land arbiter and scribe Hidalgo would have been familiar with these. The members of the Economic Society, embedded as they were in the region's governing circles, translated bureaucratic geographical conventions directly to the *Gazeta*. The new 'civic' geographical description that the editors envisioned showed that its authors' geographical imagination was shaped by bureaucratic experience of administering the countryside as much as any distinctly 'Creole' vision of the land.

Despite these formal resemblances, the *Description* was designed to surpass currently available information. Hidalgo's next contribution to the *Description*, that of the province of Totonicapán, prompted the *Gazeta*'s editors to reflect on

---

[17] 'Plan de la Ynstruccion que se ofrece dar en la Academia de Matematicas', Josef de Sierra, 6 December 1794, AGI, Estado 48, No. 7.

[18] Real Sociedad, *Tercera junta pública*, 12; Jorge Gonzáles Alzate, 'Hidalgo, José Domingo', in *Diccionario AFEHC*.

[19] 'Descripción corografica de la Provincia de Quesaltenango', *Gazeta*, Vol. 1, no. 26 (31 July 1797), 201.

the improvements that their newspaper provided to the realm of geography. They defined these in a direct comparison between the new description of Totonicapán and the treatment of the same *alcaldía mayor* by established authorities. First, they stated that José de Alcedo's geographical dictionary wrongly attributed it to the jurisdiction of neighbouring Chiapas and did not provide any details about it. Thomas Gage's account of the same lands was superficial at best. Beyond these printed authorities, they had also gained access to the geographical manuscripts of the prolific military engineer Luis Diez Navarro, who had surveyed the region in 1743 and produced a series of route descriptions. While they were happy to acknowledge that the engineer had been thorough when it came to the 'northern and southern coasts', his work on the interior of Central America left much to be desired. For instance, he had only seen the province of Quetzaltenango 'in passing'. The editors further had before them two sets of census records of the province from the 1750s, and deemed that while one of them seemed relatively accurate, Hidalgo's newer figures were more reliable.[20] Having evaluated a range of sources, including those originally created for the government's limited network of correspondence only, they were satisfied that the *Gazeta*'s own *Description* updated and refined existing information about the province's human and physical geography. In addition, it also made official documents publicly available that were otherwise not easily accessible, such as Diez Navarro's reports or population records.

The purpose of the newspaper's *Description* was not to entertain, but to compile facts in the most familiar format available. Textual descriptions rooted in bureaucratic processes provided the opportunity to create multi-layered representations of physical and human geography as well as natural history. The detailed, enumerative format that drew on bureaucratic models was not to the taste of all readers – the editors noted that they had received a number of complaints about such a 'dry and monotonous' subject taking up so many pages. The editors rebutted these arguments by pointing to their mission to publish any topic which might be 'useful with respect to the country [*pays*], and to the general state of its enlightenment [*sus luzes*]'. In a rather patronising tone which reflected their ethos of the *Gazeta* as a medium for dispersing enlightened knowledge among a generally un-enlightened population, they went on to explain that all readers would eventually see the value of these dry descriptions:

Even those short-sighted readers who do not find any value in this paper except that of providing a moment of entertainment on Mondays, will thank us later for our work, when after some time, once these individual articles are united and ordered, they will find that, imperceptibly, a complete description of this Kingdom, province by province, has been made.[21]

---

[20] 'Descripción de la provincia de Totonicapan', *Gazeta*, Vol. 1, no. 31 (4 September 1797), 242.
[21] 'Descripcion de la provincia de Totonicapan', *Gazeta*, Vol. 2, no. 65 (11 June 1798), 137.

Each individual description in itself was valuable, but put together, the whole would be greater than the sum of its parts. This perspective on the meaning of the *Description* showed that the *Gazeta* valued the potential for individual reformers' work in provinces, but was not giving up on the hope of creating an account of the Audiencia as a whole. The great economic potential of Guatemalan nature would be unlocked through studying geography, as an editor dreamt: 'What an immense multitude of riches Nature generously presents to our eyes everywhere! How many commodities for a great and lucrative trade!'[22] The *Description*, serving as a basic record of hitherto unknown economic and agricultural possibilities, would encourage improvements of the land and therefore contribute directly to 'men's happiness'. Unlike traditional forms of mapping, or the superficial approaches of engineers, who considered earth in relation to the sky but ignored the inland provinces, a new geography would bring new wealth to Guatemala.[23]

Sources of Geographical Information

In addition to the familiar but rather vague aims of economic progress and public happiness, the editors used the geographical project to reaffirm the place of knowledge created in Guatemala in relation to wider networks of scientific information. Just as they had done in the context of botanical questions, they related the knowledge they gathered to the sort of geographies that were presented as 'universal' in Europe. The comparison was not favourable for European knowledge. A good part of the introduction to the *Description* was taken up by an extensive criticism of knowledge about Guatemala as it was available to the rest of the world. Most of it, the editors argued, was superficial and out of date. While they attacked different authorities, they singled out the *Encyclopédie méthodique*, declaring its three volumes on geography to be completely 'futile'.[24] The *Encyclopédie* rankled Central American reformers because its articles about American geography were 'for the most part a work of Mr. Masson de Morvilliers'. Masson, as the author of a controversial 1785 article which asked 'What has Spain ever done for Europe?', was a reviled figure.[25] The *Gazeta*'s first argument against the *Encyclopédie* was therefore a patriotic one, chiming in with countless publications in which inhabitants of Spanish America railed against theories of American inferiority. Cornelius de

---

[22] 'Comercio', *Gazeta*, Vol. 1, no. 9 (10 April 1797), 67.
[23] 'Memorias', *Gazeta*, Vol. 1, no. 14 (15 May 1797), 105.
[24] *Ibid.*, no. 25 (24 July 1797), 193. Luque Alcaide, *Sociedad Económica*, 148, erroneously transcribes the editors' opinion of the *Encylopédie*'s geography as 'useful' [*útil*] rather than 'futile' [*fútil*], although she agrees that the *Gazeta* attacked these geographical articles.
[25] Matthieu Raillard, 'The Masson de Morsvilliers Affair Reconsidered: Nation, Hybridism and Spain's Eighteenth-Century Cultural Identity', *Dieciocho* 32 (2009): 31–48, at 33.

Pauw and the Comte de Buffon in particular had argued that the Americas, especially the flora, fauna, landscapes, and peoples of Central and South America, were degenerate. Spanish Americans, including exiled Jesuits, were quick to react with outrage to any publication which did not seem to take their continent seriously, and countered by demonstrating the richness and utility of American nature.[26] More than any other part of the *Gazeta,* the proposed *Description of Guatemala* was explicitly conceived as a counterpoint to these 'petulant philosophers' who had 'blackened America without knowing her'. Geographical knowledge, that ultimate expression of local expertise, would be a defence of 'this exceptionally vast and rich part of the globe' against the 'ridiculous and extravagant system' of anti-American philosophers.[27] Local experience would provide a sound empirical footing to rectify erroneous claims that were based on faulty sources. The *Description* would 'suffice to refute the futile arguments of Pauw, and all anti-Americans, to whom one should not respond with anything but facts [*hechos ciertos*], taken from the very same nature of the lands which they concern'.[28] Most readers seemed to agree. A contributor had previously insisted that to refute the theory of debilitating climates it was enough to simply 'to live in America, in these lands which Montesquieu did not know, and of which no European can form an idea from his study [*gabinete*]'.[29] Local, empirical, and up-to-date facts about American nature would show de Pauw and Buffon's claims to be untrue, while serving the enlightened project of progress through geographical knowledge.

But the *Gazeta*'s critique of the *Encyclopédie* was not merely a polemical piece about foreigners' attempts to belittle Spain and its empire. The editors instead launched into a thorough criticism of the format and source material of European encyclopaedias. 'Beware of any book that pompously titles itself as universal, however grand and famous its authors may be', advised the *Gazeta*'s editors, in stark contrast to the popularity of encyclopaedic works in Spain and Europe at the time.[30] Such works left out local details and represented a superficial universalism – the complete opposite of the 'slow and thorough' approach they expected of their own chorographic description. Not everyone agreed with the grandstanding – one reader's letter attacked the editors for their arrogance in challenging 'that great book' and the 'society of scholars whose respectable names are on the *Enciclopedia*'. The reader conceded that the French version had its problems, but he found the additions of the Spanish

---

[26] Antonello Gerbi, *The Dispute of the New World: The History of a Polemic, 1750–1900* (University of Pittsburgh Press, 1973); Cañizares-Esguerra, *Nature, Empire and Nation,* 99–103; Cowie, *Conquering Nature,* 114–48.
[27] 'Memorias', *Gazeta,* Vol. 1, no. 13 (8 May 1797), 97–8.
[28] *Ibid.,* no. 14 (15 May 1797), 106.   [29] *Ibid.,* no. 8 (3 April 1797), 58.
[30] *Ibid.,* no. 24 (17 July 1797), 185; Clorinda Donato, 'The Enciclopedia Metódica. A Spanish Translation of the Encyclopédie Méthodique', in Clorinda Donato and Robert Maniquis (eds.), *The Encyclopédie and the Age of Revolution* (Boston, MA: G.K. Hall, 1992), 73.

translators Juan de Arribas y Soria and Julian de Velasco 'invaluable'.[31] It is true that on some practical issues such as agriculture the *Gazeta*, too, was happy to refer to the *Encyclopédie*, but the editors drew the line at using it for geographical information. Their issue was with the genre represented by the *Encyclopédie* as much as the debate about the value of the New World.

This became apparent when the editors dismissed another work with encyclopaedic aspirations recently published in Spain: José de Alcedo's *Diccionario Geográfico de América* (1785–89). Editor Ignacio Beteta had already taken aim at Alcedo's geography a few months before, in a letter to the city government in which he explained the shortcomings of existing geographies of Guatemala, and requested access to the city records in order to find more reliable sources. He now repeated these arguments publicly in an editorial for the *Gazeta*. Alcedo may have been a Spaniard, which meant that Beteta credited him with 'honest intentions'. However, the editor could not look past Alcedo's deeply flawed sources. He concluded that 'few truths are found in these articles, and they are embedded in a thousand anachronisms', because Alcedo took his information about Guatemala from the British seventeenth-century traveller Thomas Gage, and other 'inaccurate reports'. It was a 'shame' that Alcedo should have been forced to 'beg information from foreigners ... in matters of Spain's dominions', he lamented. Despite these criticisms, Beteta showed himself sympathetic to the difficulties that Alcedo must have had in writing about Central America. Like other 'modern authors', he had simply 'copied from the old ones', and the scarcity of existent sources made it 'impossible' for such a geographical dictionary not to contain errors. Getting geographical information from 'books, which lie, or which talk of things already past' was geographers' mistake. If Thomas Gage really was the best available authority on Central America, this situation needed to be remedied. The imperative to rectify geographical knowledge that was 'unjustly' reported incorrectly could be a question of moral duty as much as intervening in debates about de Pauw was.[32] Predictably, Beteta offered the *Gazeta*'s own *Description* as a correction to Alcedo, and was indeed granted access to the city records to help him fulfil this task.[33]

The problem with European geographers – who were inevitably the authors of such 'pompous' encyclopaedic volumes – was then that they were using the wrong sources, and may not have been able to judge the veracity of any source

---

[31] 'Carta', *Gazeta*, Vol. 1, no. 39 (30 October 1797), 309–11.

[32] Compare Dean W. Bond, 'Plagiarists, Enthusiasts and Periodical Geography: A.F. Büsching and the Making of Geographical Print Culture in the German Enlightenment, c.1750–1800', *Transaction of the Institute of British Geographers* 42, no. 1 (2017): 58–71, at 64.

[33] Ignacio Beteta to Regidor Syndico Joseph Manuel Laparte, 17 February 1797, AGCA, A1, Leg. 5361, Exp. 45314; 'Memorias', *Gazeta*, Vol. 1, no. 14 (15 May 1797), 106; no. 24 (17 July 1797), 186; no. 25 (24 July 1797), 193.

without the experience of America that it demanded. This critique included Spanish authors as well as non-Hispanic ones. If, due to difficulties in communication and verifying information, the *Gazeta* and Economic Society had an ambivalent relationship with foreign knowledge, they were resolutely certain that there was but one acceptable reference point for geographical knowledge, and that was local knowledge and local perspectives. To these reformers, geographers seemed to be insufficiently concerned with the quality of their sources, leading to errors and confusions. They might not even notice that their sources were hopelessly outdated. Central American geographical sources themselves could well be accused of such problematic histories of copying and obfuscation. For instance, one of the few maps of an inland province in circulation in the eighteenth century was in fact a copy of a 1579 map from a *relación geográfica*. The original sixteenth-century mapmaker had drily included a warning that its scale of distance 'will not be correct'. And yet, in the absence of newer large-scale mapping projects, a copy of this 200-year-old map still appeared to be in use in the second half of the eighteenth century. The newer copy brought the cartographical conventions of representing lakes and mountains in line with eighteenth-century customs and refined representations of roads, but otherwise replicated the topographical information and all the original text, including the mapmaker's own complaint that it lacked accuracy (Figure 4.1). It did not include any reference to the new map being a copy, and was marked up in pencil by a different hand, suggesting that it was considered to be relatively current, and in active use.[34] We must therefore conceptualise geographical information in the eighteenth century not just as the coexistence of bureaucrats' and reformers' approaches, of narrative and mathematical measurements, but also as the material copying of older geographical documents.

As far as written geographical knowledge was concerned, the editors accused authors of 'particular and universal geographies, whether in the form of dictionaries, treatises or travel narratives', of being a mindless crowd who merely copied from each other and frequently 'confused the historical state of things with the present one'.[35] The inability of foreign authors to know which facts were ancient and which were current compounded the issues that arose

[34] 'Costa de Zapotitlan y Suchitepeques', BL Add. MS 17,650e. The copied map is undated, but is clearly a later copy and bound with several maps from the 1770s to the 1790s, from the collection of the Venezuelan Francisco Michelena and Rojas before 1848. It is possible that it is related to the map once catalogued as 'Mapa de la zona de la alcaldía de Suchitepeques, 1799', AGCA, A3.30 Exp. 37764 Leg. 2578, which appears to be lost. The original map is described in Manuel Morato Moreno, 'El mapa de la Relación Geográfica de Zapotitlán (1579): una isla de racionalidad en un océano de empirismo', *Journal of Latin American Geography* 10, no. 2 (2011): 217–29 and is located at the University of Texas at Austin, UTX, JGI xx-9, digitised at https://legacy.lib.utexas.edu/benson/rg/zapotitlan.jpg.

[35] 'Noticias historicas de la antigua Truxillo', *Gazeta*, Vol. 6, no. 284 (8 November 1802), 285.

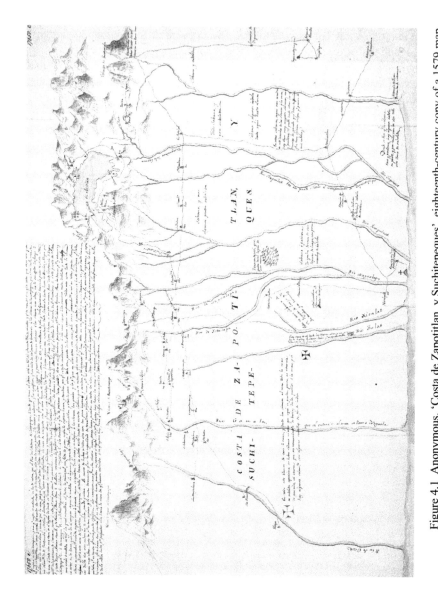

Figure 4.1 Anonymous, 'Costa de Zapotitlan, y Suchitepeques', eighteenth-century copy of a 1579 map, British Library, Add. MS 17,650e. By permission of the British Library.

from the lack of sources available in Europe. The *Gazeta*'s readers continued to find problematic examples of such practices. In 1803, a contributor criticised the French encyclopaedists' misconception that the abundant cacao harvest known as 'Caracas' came from Guatemala. This was doubly wrong. Firstly, there was the geographical error of confusing the two places. Secondly, it was true that Guatemala had once been the dominant exporter of cacao, but its position had been usurped by Caracas, so these writers were conflating two historical epochs. The foreign writers 'confuse[d] the times as well as the names', the contributor sadly concluded.[36] Just as Alcedo had problematically assumed that the century-old superficial descriptions of an English traveller were a trustworthy source of information about Guatemala, the editors concluded that all these geographical formats and the way in which they were reproduced encouraged a 'mindless' and static vision of geography.

It was not that they rejected history, quite on the contrary. Granted, history was not useful if it occupied itself only with wars and politics: the history of the 'noisy efforts of some men intent on destroying each other'.[37] However, archival material that recorded geographical or economic information could serve as a useful comparison for a particular region's potential. The *Gazeta*'s editors actively encouraged the use of history. For instance, they noted that the record books of the old capital Santiago from 1524 onwards would be a good potential source for their description.[38] Beteta asking for access to the city archives was just one example of a Society member consulting manuscripts. The leader of Trujillo's *junta de correspondencia* Juan Ortiz de Letona referred to a 'very old document' in his possession, which provided data about the previous century's sugar harvest up to 1642.[39] Ortiz de Letona was encouraged by this manuscript, since it claimed that Trujillo's sugar harvest was on a par with Cuba's famously plentiful yield in those years. Historical reports of 'thousands' of bags of cacao being exported to Peru or Oaxaca from Guatemala's Pacific provinces could also identify specific places such as Trujillo or Suchitepéquez, as suitable for growing this particular crop.[40] Similarly, an article in the *Gazeta* referred to historical commerce and marine traffic reaching Granada in Nicaragua and Matina in Costa Rica between the years of 1685 and 1733, using such previously established patterns for trade as a template for what was possible in the present.[41] The contributors to the

---

[36] 'Apuntamientos', *Gazeta*, Vol. 7, no. 313 (25 July 1803), 295.
[37] 'Consulados', *Gazeta*, Vol. 8, no. 342 (16 April 1804), 44. These were the words of a Consulado rather than Economic Society official, but the editors added a note to say that they agreed with the article and that the author was someone 'close' to them.
[38] 'Memorias', *Gazeta*, Vol. 1, no. 25 (24 July 1797), 195.
[39] 'De Truxillo', *Gazeta*, Vol. 2, no. 76 (27 August 1798), 226.
[40] 'Cacaos', *Gazeta*, Vol. 7, no. 295 (14 March 1803), 54.
[41] 'Del antiguo Comercio de Nicaragua y Costa Rica', *Gazeta*, Vol. 5, no. 233 (19 October 1801), 605.

Gazeta, as well as government officials on whose archival practices these ideas drew, found a template for the future potential of the region in historical data, much as explorers of Palenque had envisioned the ancient city as proof of possible future trade.

An anonymous contributor who called himself *el copista*, possibly again Beteta, also looked at the intersection between human history and the history of agriculture and landscapes when he reproduced a document from 1607 that he had come across in the Guatemala City archives. He gathered from the document that herds of cows were widespread in Guatemala at that time. He acknowledged that Guatemala having livestock 'equivalent to Buenos Aires' might seem strange to its current inhabitants, but explained it in a grand historical arc. It 'had to be like this because of the natural order of things, and the nature of these lands [*la naturaleza de estos terrenos*]', he explained. All peoples were herders before they were farmers, and after the conquest the rearing of livestock was the first industry to grow in the Spanish colonies (*en nuestras colonias*). 'It grew in all of them without another limit than the fertility of their inexhaustible pastures', he concluded.[42] This was not just another affirmation of a belief in the 'fertility and expanse' of Guatemala. While it represented a simplistic view of economic and environmental history that further ignored all indigenous agriculture before and after the conquest, the *Gazeta* showed a real curiosity for the ways in which Guatemala's agricultural and economic basis had changed on a large scale over the years of Spanish habitation. This close relationship between history and geographical and environmental information continued to be important in Central America, even into the independent period. The statesman and intellectual José Cecilio del Valle explained in 1830 that one of the functions of the history of any state was to 'indicate its climate, waters, winds, productions, etc' – the history of people's coexistence with, and management of, natural phenomena and resources, while the independent government of Guatemala also declared it obligatory for history students at the Academy of Sciences to also take courses in geography.[43]

However, the *Gazeta* editors also went further. The *Gazeta*'s geographical description formalised these traditional archival practices into a theory of geographical knowledge that explicitly relied on local archives and contributed to their overall intellectual project of contextualising Central American knowledge within global knowledge. The editors ventured their own theoretical approach to the relationship between history and geography, and their relevance to the present. At the same time that scholars in Madrid stressed the importance of primary sources in their historiographical debates, the Society

---

[42] 'Abundancia de ganados en lo antiguo', *Gazeta*, Vol. 7, no. 297 (28 March 1803), 70–1.

[43] 'Historia', *Mensual de la Sociedad Económica*, no. 3, 63–4. José del Valle was the author of the previous article on geography: his authorship of this article on history is possible, but not certain. 'Al C.no Secretario del S Gobno del Estado', Guatemala, 27 April 1836, AGCA, B. Leg. 1075, Exp. 22843.

and its newspaper (possibly unaware of the debates in Europe) also put trust in the historical value of archival material.[44] Quoting what was apparently a draft manuscript of the Guatemalan scholar Domingo Juarros, who was at the time engaged in writing a history of Guatemala, they insisted on the value of local historical sources. Since they found encyclopaedic geographies too superficial, it is unsurprising that they considered general histories and Chronicles of the Indies to be full of 'mistakes'. Instead, they insisted that much better sources could be found in 'city and provincial chronicles, authentic and trustworthy manuscripts'.[45] This solid historical base was important for the geographical project because the two disciplines were intertwined. Having debated geographical epistemologies and representations at length, it now became important to understand history as well. The editor theorised that history and geography should be seen as 'the same thing, and one should be considered part of the other'. He immediately qualified this statement by explaining that the newspaper's *Description* would 'not be historical or geographical of the past, but of the present'. That is, geographical descriptions could be a feature of history, and historical descriptions a feature of current geographies, because the present land itself was a result of historical processes.[46]

Central Americans had no trouble in accepting landscapes and geography as historically contingent. At the time, theoretical questions about the historicity of the landscape were also increasingly debated in Europe, as Chenxi Tang has shown.[47] Unlike the German geographers of Tang's narrative, however, Guatemalan reformers did not reach the conclusion that an objective geography (either physical or political) was unachievable. Recognising the changeability of landscape did not throw them into an epistemological crisis because the fixity of landscapes had never been assured in the first place. On the most basic level, the context of volcanic eruptions, earthquakes, and mudslides represented a reminder that landscapes might be physically altered at any time. At sea, fishermen reported that swelling rivers in the rainy season changed the location of a small harbour's entrance.[48] On land, the two moves of the capital in the sixteenth and again in the eighteenth century and the redrawing of administrative boundaries in the Bourbon intendancy reforms further cemented the idea that the political geography of this kingdom was also not fixed. Raymond Craib has used

---

[44] Cañizares-Esguerra, *How to Write the History*, 174–96.

[45] 'Descripcion geografica del reyno de Guatemala', *Gazeta*, Vol. 6, no. 279 (2 October 1802), 245–6. Juarros's history was published from 1808 in two volumes as *Compendio de la historia de la ciudad de Guatemala* (Guatemala: Beteta, 1808) and in English in 1823. While the comments on the value of sources are Juarros's rather than the editors', they highlighted that Juarros's geographical description was a sister project of their own *Description*.

[46] 'Memorias', *Gazeta*, Vol. 1, no. 14 (15 May 1797), 105.

[47] Tang, *Geographic Imagination*, 35–8.

[48] Consulado de Comercio, 'Sobre la exploracion de las barras de Iztapa y de Michatoya', 1798, AGCA, A1, Leg. 1691 Exp. 3421, fol. 6v.

the term 'fugitive landscapes' to describe places and territories that were disputed in their extent and location, and has shown that beyond topography, there was a lived landscape of local uses and markers that surveyors had to become familiar with it in order to map it.[49] In colonial Mexico City, as Vera Candiani has shown, knowledge of seasonal variations in water levels, draughts, and floods was also a crucial component of the management of Lake Texcoco over centuries.[50] Given these widely accepted ideas about changeable landscapes, it is unsurprising that the *Gazeta*'s editors accused European geographers of being incapable of piecing together American landscapes from the limited source material available to them. A wealth of local observations and experiences already existed within Central America, and could be employed instead of erroneous, outdated, insulting, and needlessly grand geographies.

Putting the *Gazeta* to Work

While the *Description* was the *Gazeta*'s flagship geographical project, readers also used the newspaper to drive forward their own projects. They were quick to utilise its pages as a platform of communicating with each other in a less mediated way about the practical geography of Guatemala. One member of the Society, the governor of Chiapas Agustín de las Quentas Zayas, spent much of the 1790s attempting to found a 'new town' by the name of San Fernando de Guadalupe along the banks of the river Tulijá. Its benefits would include facilitating communications and transport through that province towards New Spain, which the editors of the *Gazeta* welcomed.[51] While he sent a rather lengthy text and grand watercolour image (we might call it a map-view) to the king in Spain to introduce this project, depict its progress, and ask for a pension, given that he had sustained a serious injury when Indians attacked him during the early days of this colonisation project (Figure 4.2), none of this really mattered within Guatemala.[52] What his fellow reformers would need to know is how to find this town, and how they might trade with it. To this end, an article about the town in the *Gazeta* included crucial details, such as how it could be reached from Campeche via boat (a boat captain 'who has made the journey

---

[49] Craib, *Cartographic Mexico*, 69–90, 141–51.
[50] Vera S. Candiani, *Dreaming of Dry Land*: Environmental Transformation in Colonial Mexico City (Stanford University Press, 2014).
[51] 'Ciudad Real de Chiapa', *Gazeta*, Vol. 1, no. 14 (15 May 1797), 111. Dym, *From Sovereign Villages*, 39. Although this was a project with deeply traditional roots – similar to the establishment of a *reducción* to compel Indians who had fled from Spanish towns to pay tribute and live under the Spanish state and church's power – the intendant consciously used the wording *nueva población*, more closely associated with the Bourbon reforms, e.g. the Nuevas Poblaciones at La Carolina, Andalucía: Maria Luisa Martínez Salinas, *La colonización de la costa centroamericana de la Mosquitia en el siglo xviii. Familias canarias en el proyecto poblador* (Valladolid: Ediciones Universidad de Valladolid, 2015), 18.
[52] Map-view: Kagan, *Urban Images*, 2–6.

twice' gave details), where to hire Indians to carry loads (the village of Tumbalá), and where mules were for hire (Comitán).

In addition, it gave an account of the roads that led from the new town to the regional trading hubs of Ciudad Real, capital of Chiapas, and Santa Ana Huista in highland Guatemala. (Figure 4.3). Details of the villages *en route*, and the distances between them, are arranged in the same spatial way as they were in the manuscript pages of postal and other administrators. As Sylvia Sellers-García has explained, these route descriptions were much more important than maps and more accurate in their representation of distance as it would be experienced by a traveller. They were the format in which all manner of colonial officials conceptualised space and distance, representing the idea of knowing the territory through travel.[53] In the pages of the *Gazeta*, it acquired the form of a travel guide. In the left column is the list of villages connected by the San Fernando-Santa Ana road, with the distances in leagues between each place. After 19 leagues, at Chilon, the road forks. This is not made clear in the text, but implied in the list by the dash which connects the name of the town 'Chilon' to the sub-heading of *'Camino para Ciudad Rl'*. This is a crucial element informing the reader of the relationship of these two roads, transforming it from a list into a spatialised representation on the printed page. While in the left-hand column, the Santa Ana road carries on, the right-hand column (starting in the middle of the page) now lists the villages from Chilon to Ciudad Real. The distance between San Fernando and Chilon, 19 leagues, is added up and carried over into the right-hand column, where the count continues from village to village until arriving at Ciudad Real. Both columns finish with the total distance, in leagues, from San Fernando to Santa Ana and Ciudad Real respectively. While such route descriptions appeared in manuscripts of the period, its appearance in the press was a novelty. Even more than the *Gazeta*'s *Description*, this was a truly useful guide which reflected the templates of everyday descriptions of travel in Central America.

Another member of the Economic Society (and head of the Guatemala City merchant association) concerned with the practicalities of finding quicker connections for trade, Juan Bautista Irisarri, shared his own reconnaissance projects with the remainder of the *Gazeta*'s readership. He extended the debate about geographical information from land to sea. In 1798, he was engaged in a series of small exploration projects along Guatemala's southern coast, attempting to find out whether minor old harbours such as Iztapa, now silted up, could be made useable again.[54] Perhaps most notably, he attempted

---

[53] Sellers-García, *Documents and Distance*, 47–53, 96–101.

[54] 'Consulado', 1798, AGCA, A1 Leg. 169 Exp. 3421, fols. 6v–17v. Woodward, *Class Privilege*, 64, also concludes that Irisarri's work at Iztapa was an 'invidual effort'. The aim of these explorations was the *habilitación* of new harbours, which in the Guatemalan context referred specifically to the wholesale trade conducted on credit up to a year in advance and controlled by the Guatemala City merchants: Rojas Lima, *Diccionario histórico-biográfico*, 479.

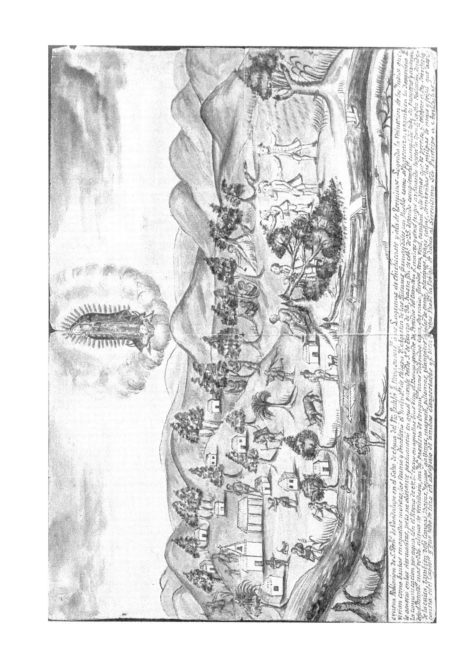

Figure 4.2 'Nueva Población de San Fernando de Guadalupe en el Salto de Agua del Río Tulijá' (1795). Ministerio de Cultura y Deporte. Archivo General de Indias. MP-Guatemala, 257.

Figure 4.3 Spatial representation of the roads from San Fernando de Guadelupe. 'De Chiapa', *Gazeta de Guatemala*, Vol. 1, no. 14 (15 May 1797), p. 111.

to send a small ship to navigate up Guatemala's Pacific coast from Sonsonate to San Blas in New Spain. His intention was to back goods imported to New Spain from China, with the ultimate aim of establishing a regular trading route. However, the ship was promptly damaged in a storm, in what was reported in the *Gazeta* in several issues as the 'failed expedition to San Blas'.[55] The journey had been a dangerous undertaking because Irisarri apparently lacked basic information on weather and shipping conditions for maritime traffic in the Pacific. These waters were home to important shipping routes between San Blas and El Callao in Peru, but Guatemalan harbours such as Sonsonate may have lacked persons with the right expertise to conduct

---

[55] 'Comercio del Sur', *Gazeta*, Vol. 2, no. 75 (20 August 1798), 223. See also Brockmann, 'Surveying Nature', 191–5.

trade from them. After all, longer (legal) trade journeys could for centuries only depart from authorised harbours, but harbours of Central America's southern coast had been given special permission for untaxed trade with Mexico and Peru in 1797. Trade with Mexico at Acapulco meant tapping into the Pacific trade with China, prompting a new interest in exploring other harbours of the southern coast.[56] Undeterred by the failure of his first expedition, Irisarri looked for ways to source information to make his next journey successful.

The editors of the *Gazeta* supported this enterprise. They believed the expedition was worth another attempt, since a successful navigation of these waters would mean 'utility not just for one individual for one moment, but for posterity and an entire kingdom'.[57] One reader even saw significance in the mere fact that the first expedition's outcome was reported in the newspaper at all, despite its failure, distributing realistic information about the capacities of regional harbours. He found Irisarri's honesty in chronicling even the less successful parts of his enterprise refreshing, and quite in contrast to his usual expectations of merchants' behaviour. He was impressed that Irisarri had 'put his own interests after the interest of the public (*causa publica*)'.[58] The appreciation was mutual. Irisarri understood the potential of the newspapers' readers well. After the failed journey, Irisarri now turned to the extraordinary avenue of communication between various segments of society provided by the *Gazeta de Guatemala* to solve his pressing geographical problem. Just as others had asked or contributed information about plants, Irisarri asked what the crew of the boat should have done differently, what time of the year and which routes they should have chosen to reach their destination safely. Irisarri saw the ability of local knowledge to solve problems of transport as instrumental to the development of international trade. He was sure that 'in San Blas, Acapulco, or some other place on this vast coast, or in the centre of the capital cities, there will be no lack of mariners who have experience in the navigation of the Pacific Sea' who might help answer this question.[59]

Had Irisarri been able to share in the information of the Spanish naval ministry or the viceroyalty of New Spain, he might have encountered a report about this very same coast that a Spanish military captain had compiled just a few years before, alongside a detailed map for navigation. In 1794, a Spanish military brig, on orders of the viceroy of New Spain, explored the coast between San Blas, Acapulco, and Sonsonate, the same route that Irisarri was

---

[56] 'Consulado 20 de Marzo de 1798', AGCA, A1, Leg. 169, Exp. 3421. Troy S. Floyd, 'The Guatemalan Merchants, the Government, and the Provincianos, 1750–1800', *The Hispanic American Historical Review* 41, no. 1 (1961): 90–110, at 95. Woodward, *Class Privilege*, 62–3; see 55–69 for the Consulado's harbour explorations.

[57] 'Comercio del Sur', *Gazeta*, Vol. 2, no. 75 (20 August 1798), 219.

[58] *Ibid.*, no. 84 (22 October 1798), 289.   [59] *Ibid.*, no. 75 (20 August 1798), 219–24.

interested in. While the coast was therefore known to some parts of the Spanish imperial bureaucracy, this information had not reached the officials of the Audiencia de Guatemala, and certainly not its resident merchants. The military expedition had examined the coast in a global context: it took place around the time of Spanish expeditions to Juan de Fuca Strait and Nootka Sound in modern-day Canada, and more broadly represented part of the Spanish reaction to British and French assertions of power in the Pacific, including the voyages of Captain Cook and Louis de Bougainville. The captain of the ship who carried out the coastal exploration from Acapulco to Sonsonate was very conscious of this. He explicitly put his voyage into dialogue with these foreign explorers, mentioning the journeys of La Pérouse, 'Boulambille', 'Cooc', and Vancouver in the Pacific.[60] By contrast, Irisarri saw the coastal journey as a way to increase the local profits of merchants – an opportunity for Guatemala City importers of goods, and exporters of indigo from Sonsonate. The merchant association he headed, Guatemala City's Consulado de Comercio, had previously tried to recruit imperial knowledge to this cause when it asked José Mariano Mociño, who had spent time in San Blas on his botanical travels, whether trade with San Blas was a promising endeavour. Mociño was able to give some advice, but did not have extensive mercantile knowledge, or experience of that maritime route.[61] 'The Empire', in theory, possessed plenty of geographical information about this coast, but, in practice, it was not accessible to Irisarri and other merchants of Guatemala who stood to profit from it.

The *Gazeta*'s geographically dispersed readership once again showed itself remarkably effective at addressing pressing concerns regarding local land-scapes (or in this case, waterscapes) rapidly. Barely a month after Irisarri's request was published, a reader from El Salvador responded under the pseu-donym '*El próximo*'. He provided answers to the questions with the caveat that he was 'not a mariner [*nautico*] by profession', but instead answered with a mixture of his own knowledge, and 'what he has been able to find out from some books'. He concluded that January to March would be the best time to realise the journey about which the owner of the *Marte* sought advice. *El próximo* stressed that he hoped others would build upon this knowledge, since the matter was of interest to 'humanity, the State, and the trade of this kingdom'.[62] The book to which he referred to was Admiral George Anson's account of his failed 1740–44 voyage around the world, which, although not

---

[60] Salvador Meléndez y Bruna, 'Diario del viage al puerto del Realejo para reconocer y levantar planos del troso de costa comprehend.o entre el puerto de Acapulco y el surgidero de Sonsonate: amas la exploración del golfo de Conchagua con el bergantin de Su Magestad el Activo', 1794, The Newberry Library, Chicago, Ayer MS 1135, fols. 2–5. For the context of Pacific journeys, see also Engstrand, 'Of Fish and Men'.

[61] Taracena Arriola, *Expedición científica*, 54.

[62] 'Comercio del Sur', *Gazeta*, Vol. 2, no. 81 (1 October 1798), 270–2. Minutes of a Consulado meeting suggest that Irisarri's boat may have actually already made it to San Blas from

translated into Spanish until 1833, circulated in Spain in various English and French editions.[63] The reader gave a relatively thorough account of the book, its contents, and the information on the winds and navigation around New Spain's southern coast provided by the 'admiral, and the reports of ship's pilots and imprisoned navigators [*prácticos*] whom he had on board'.[64] It is likely that the Salvadoran was in possession of this book, which provides a snapshot of some of the foreign-language information which was available to some literate individuals even in a peripheral part of the Spanish empire. Anson's book was judged 'useful' by the *Gazeta*, even though in New Spain, Alzate had taken issue with its depiction of Spanish Americans as cowards.[65] The Salvadoran reader nevertheless stressed the need to build upon this knowledge. He warned that he was not a mariner, implying that such practical experience would have made his knowledge more reliable. He evidently saw himself as participating in an enlightened exchange, citing his belief that this matter was of interest to 'humanity' as the reason for taking up a pen to contribute to the discussion. Disinformation on geographical matters was an individual inconvenience to merchants, but also a collective concern of enlightened reformers.

Cartographic Thinking: Roads and Waterways

As the *Gazeta* explained, drawing up a chorographic description of Guatemala had the practical aim of cataloguing the possibilities and productions inherent in the land. This potential, however, was held back by the lack of transport that prevented goods from being traded and exported from Central America's provinces. The *Gazeta*'s editor did not mince his words when he explained the link between roads and trade in Guatemala:

In addition to the [roads'] natural ruggedness, the abundance of dangerous and deep rivers that everywhere divide them present the impression of a miserable kingdom, which can only be transited with danger to one's life, poor by necessity, and which can never prosper as long as this most powerful obstacle is not overcome, which has been, and will always be the most grievous weight on its commerce.[66]

More than an inconvenience, the lack of appropriate transportation impeded progress, since it stopped merchants from accessing the country's natural productions. Reduced transit time through improved transport would directly lead to a more prosperous Central America. The editors of the *Gazeta*, for

Sonsonate in September, but it is also possible that this information was initially mis-reported in the capital. 'Junta no. 194', 20 September 1798, AGCA, A1. leg 169 Exp. 3421.

[63] Marta Torres Santo Domingo, 'Un bestseller del siglo XVIII: El viaje de George Anson alrededor del mundo', *Biblio 3W* 9, no. 531 (2004).

[64] 'Comercio del Sur', *Gazeta*, Vol. 2, no. 81 (1 October 1798), 271.

[65] Cañizares-Esguerra, *How to Write the History*, 282.

[66] 'Puente de Zamalá', *Gazeta*, Vol. 4, no. 174 (15 September 1800), 330.

instance, blamed the transport network for the crisis that Central America's most valuable export, indigo, was undergoing. Land transport was slow, they explained, only to be followed by delays in harbours where ships were 'endlessly subject to shipworms', and mariners to diseases. All of these created a 'considerable surcharge' for the end product, which finally made it impossible to compete with prices on the European market.[67] New transport routes, by contrast, promised access to the 'the astonishing fertility of our privileged soil'.[68] While the idea of building roads to improve trading connections was part of Bourbon rhetoric in Spain and the Americas, and had parallels in the British and French empires, Guatemala did not see much imperial funding for such road-building enthusiasm. In one instance, in 1793 the priest of Matatescuintla in Guatemala organised a measurement of the distance between the town and the new capital city, and submitted a plea for funding to the government, but it does not seem to have been converted to reality.[69] Exceptions were mainly of a military nature, such as the new road to connect the capital city with the military fort of Omoa, of obvious strategic and imperial importance.[70] Inland connections were often left to regional officials, individual enthusiasts, and local Consulado chapters.

A road project designed and constructed within inland Central America and for the benefit of regional trade was therefore a localist intervention to make it possible to extract economic benefit from a particular place forgotten by large-scale imperial undertakings. Constructed in 1801, this new road was designed to integrate the western Guatemalan province of Suchitepéquez more closely into the trading networks of the capital, and improve the province's connections with the Pacific coast as well as New Spain. The province's two larger towns, Suchitepéquez and Mazatenango, had been centres of indigo trade in the past, but had seen a decline in previous decades. Designed and mostly paid for by the *alcalde mayor* of Suchitepéquez, José Rossi y Rubí, it was a project dripping with paternalistic Enlightenment rhetoric that projected the new road as a small but significant step in the project of exploiting the dormant usefulness of the Central American countryside. As one of the founding members of the

---

[67] 'Navegación del Motagua', *Gazeta*, Vol. 4, no. 159 (2 June 1800), 268.
[68] Real Sociedad Económica de Amantes de la Patria, *Octava Junta Pública de la Sociedad Económica de Amantes de la Patria de Guatemala* (Guatemala City, 1811), 24.
[69] 'El cura de Mataquesq.ta y otro vecinos sobre havrir camino en derechura p.a esta capital', 1792, AGCA, A1.21.5, Leg. 2357, Exp. 17815. Michael Crozier Shaw notes that Spain itself was also seen as lacking sufficiently good roads, but even there, many projects never went beyond the planning stage: 'El siglo de hazer caminos', 413–34.
[70] Sellers-García, *Distance and Documents*, 60–1 and n. 13; Bruce Castleman, *Building the King's Highway. Labor, Society, and Family on Mexico's Caminos Reales, 1757–1804* (Tucson, AZ: The University of Arizona Press, 2005), 7–8. Omoa: Mathias de Galvez, San Fernando de Omoa, 7 July 1778, AGI, Guatemala, 451. British and French: Jonsson, *Enlightenment's Frontier*, 259; Lissa Roberts, '*"Le centre de toutes choses"*: Constructing and Managing Centralization on the Isle de France', History of Science 52, no. 3 (2014): 319–342, at 322–4.

Economic Society, a co-founder of the enlightened newspaper *Mercurio peruano*, and a participant in the *Gazeta*'s debates about medical plants, Rossi needed no introduction among Guatemala City's enlightened elite. He presented bad roads as a problem of Guatemalan backwardness and proposed to solve it by personally seeing to the construction of a new road. Most of the American continent, in his experience, had bad roads, but Guatemala stood out even in this company for its 'very bad and derelict' ones. Rossi added darkly that he knew this from personal experience, having 'crossed the whole kingdom in all directions, from Tumbalá to Trujillo'.[71]

The main part of the new road connected Guatemala City and Suchitepéquez via Patulúl, following the slopes of the volcanoes Fuego and Acatenango closely. This new direct connection was important from Rossi's point of view because the main route from Guatemala City to Chiapas and New Spain at the time, the *camino real*, headed north past Lake Atitlán instead and bypassed Suchitepéquez. Getting to Rossi's *alcaldía mayor* was therefore a circuitous undertaking on the 'highlands' road. Like an arrow, Rossi y Rubí's new road, marked *camino nuevo*, instead strikes an almost straight line on the map accompanying the project, next to the two longer alternatives marked *camino de los altos* and *camino de la costa* (Figure 4.4a). An additional branch road connected the village of Tolimán on the southern edge of Lake Atitlán with Patulúl, shortening an otherwise lengthy mountainous route (Detail, Figure 4.4b). Suchitepéquez, shunned by two main roads, was now to be the centre of a new regional geography. The new road was said to be between 18 and 20 leagues long.[72] Like most roads across the kingdom, it remained quite simple. It was not a cart-road, but perhaps rather a wide path designed for the mule trains that carried indigo from the coast. Rossi's achievement consisted in determining the route, clearing it of trees and shrubs, fortifying some ascents and descents with stones or logs, as well as building wooden bridges, and rest-stops at half-day intervals.[73] In March 1802, as his time as *alcalde mayor* of Suchitepéquez was coming to an end, Rossi made sure to collect testimonials from eight members of the clergy, militia, and local administration, who 'unanimously' affirmed that the new road was finished, and that it was 'comfortable, and spacious, with several wooden bridges and rest stops'.

---

[71] 'Camino de Suchitepéques', AGCA, A1.21, Leg. 207, Exp. 4171, fol. 34.

[72] Taking a league to be 4.2km, this gives us a road of 75–85km, which on a modern map is roughly the distance between San Antonio Suchitepéquez and San Pedro Martir (probably the starting point of the new road), if we imagine a straight road to be possible (there is no modern road on Rossi's exact route). The branch road Patulul-San Lucas Tolimán would be approximately 30km on modern roads.

[73] The words used are *abrir*/open and *desmontar*/clear, which left ambiguity compared to *componer*, most frequently used for cart-roads. State of the road: see *Gazeta*, Vol. 6, no. 272 (16 August 1802), 191 and AGCA, A1.21, Leg. 207, Exp. 4171, fols. 5, 55–55v.

In some ways, the format of this road project was less radically new than the simple but experience-based descriptions published by Irisarri and Quentas Zayas. As a project designed to both fulfil the aims of Enlightenment and also rescue the administrator's stalling career, it fits into the description of *proyectismo* as a chance to propose improvements, but also climb the ladder of administrative appointments.[74] However, it also presented several novelties of intervening in, and presenting the Guatemalan landscape. While this was a hybrid project that drew on the Audiencia's authority and ability to order people to work as labourers (further discussed in Chapter 5), the newspaper as well as administrative documents confirmed that Rossi financed the road himself. The *Gazeta* reported approvingly that he constructed it with as few as twenty labourers on some days, and as many as a hundred on others, all 'well looked after and paid by him', and further paid for overseers and materials. He even offered to pay an overseer's wages to help complete the last part of the road after he had been transferred to a different province.[75] Rossi's personal physical and financial involvement was impressive to his contemporaries. They praised his personal dedication to the 'public good', his active personal involvement (*agilidad*), and the 'laborious and dangerous' measurements he personally undertook.[76] Contemporaries saw real value in such personal application, contrasting it with the red tape that government projects brought with it. While they referred to Rossi as the governor of Suchitepéquez, they also saw the road as going above and beyond, separate from his regular responsibilities. Despite Rossi presenting this road as a function of his role as dutiful servant of the Crown ('de orden superior'), the project was certainly not particularly driven by Guatemala City – one clerk could not find any reference to the existence of the project and had to ask his superiors for information.[77] Instead, contemporaries seemed to perceive his work at least as much as a function of his Society membership and identity as enlightened reformer, since one official sent a report about the road within the regular bureaucratic channels, but referred his superior to '*la gazeta*' for further evidence of the particulars of the road construction.[78] Indeed, President Gonzáles Mollinedo a few years later was of

---

[74] Muñoz Pérez, 'Los proyectos', 181.
[75] 'Camino nuevo de Toliman, abierto y costeado por Don José Rossi', *Gazeta,* Vol. 6, no. 272 (16 August 1802), 189, 191. Rossi y Rubí, Comayagua, 17 June 1802, AGCA, A1.21, Leg. 207, Exp. 4171, fol. 35. Castleman, *Building the King's Highway,* for labour practices in road-building in mid-eighteenth-century Mexico.
[76] AGCA A1.21 Leg. 207, Exp. 4171.
[77] 'Ignacio Guerra, Palacio, Guatemala', 30 July and 10 September 1800, AGCA, A1.21, Leg.207, Exp. 4171, fols. 7–10. An additional order affirmed that the Economic Society would support the establishment of the branch road, but the file appears to be missing in the AGCA: 'Oficio del capitan general Jose Domas y Valle, al alcalde Mayor de Suchitepequez, sobre que la Sociedad Economica patrocinara la hechura de un camino en San Lucas Toliman. 1800', AGCA, A1.6.5, Leg. 4071, Exp. 32198.
[78] Pantaleon Ysidro del Aguila, Mazatenango, 9 September 1802, AGCA, A1.21, Leg. 207, Exp. 4171, fol. 45. Pantaleon de Aguila was listed in AGI, Estado, 48, N.7 as a founding member of

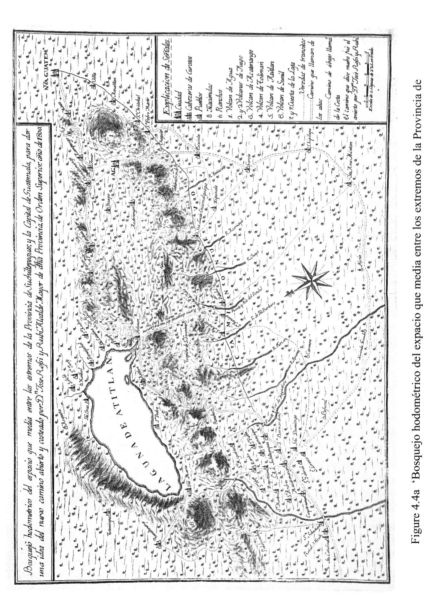

Figure 4.4a 'Bosquejo hodométrico del expacio que media entre los extremos de la Provincia de
Suchiltepeques y la Capital de Guatemala, para dar una idea del nuevo camino abierto y costeado por
D. José Jossi y Rubí, Alcalde mayor de dicha Provincia, de orden superior, año de 1800'. Engraved by

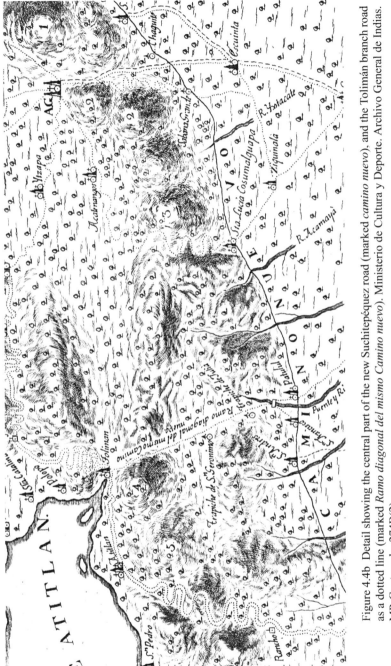

Figure 4.4b Detail showing the central part of the new Suchitepéquez road (marked *camino nuevo*), and the Tolimán branch road as a dotted line (marked *Ramo diagonal del mismo Camino nuevo*). Ministerio de Cultura y Deporte. Archivo General de Indias. MP-Guatemala, 274B(2).

the opinion that road-building was one of the great strengths of the Society. By contrast, effective reform could not come from bureaucratic projects which relied on unwilling individuals who were tied down by 'established norms and reports [*formulas y expedientes*], which complicate and confuse everything'.[79] He may have been right about the potential for administrative morass and disputes – compare, for instance, the 20-year-long quarrel that a project to build a bridge in El Salvador descended into.[80] As an individual, Rossi y Rubí embodied that willingness for initiative which the Society collectively represented.

At first glance, the decision to present the project as a map seemed to sweep aside the Economic Society's rejection of mathematical geography. It is worth noting that the title of the map, *bosquejo hodométrico*, makes it clear that this was still a time-distance based survey, a local mode of perceiving space rather than the 'mathematical geography' of longitudes and latitudes. However, its level of detail was far greater than any other locally produced maps of the region, and its sophisticated look paralleled the maps of government engineers usually reserved for coastal and defence projects. Given the 'forgotten' location Suchitepéquez occupied within the road network, such a professional-looking map was a real novelty in terms of the dedication of resources to an area far away from strategically important coasts. Forms of representation and modes of communication normally reserved for defence projects were being used for a project designed to develop the interior from a local perspective. The very existence of a printed map suggests that Rossi saw this road as something of a prestige project, and wanted to advertise its completion. However, it did not follow normal channels of distribution. Normally, a map would simply have been sent to Madrid as one manuscript copy, as Agustín de Quentas Zayas's *vista* had been. This map, however, was more mobile. At least six copies reached Madrid alongside reports of the road, which suggests that there was a glut of available copies. It seems that Society channels promoted this road and map alongside more traditional bureaucratic correspondence. After all, the draughtsman and engineer Pedro Garci Aguirre produced the fine engraving together with Rossi. Although Garci Aguirre had arrived in Guatemala in the wake of the 1773 earthquake, it was probably his role within the Society (he was a founding member and head of its School of Drawing) that connected these two reformers.[81] It even appears that one issue of the *Gazeta* included the

the Economic Society, but was writing here clearly in his capacity as holder of the secular office of interim *alcalde mayor*.

[79] Gonzáles Mollinedo, 'Sobre restablecimiento de la Sociedad Económica', AGCA, A1, Leg. 4035, Exp. 3118.

[80] 'Puente sobre el río Nexapa, llamado Sucio, en la provincia de San Salvador', 1795–1816, AGI, Guatemala, 892.

[81] Belaubre, 'Garci-Aguirre, Pedro de', *Diccionario AFEHC*; Garci-Aguirre as 'engineer': AGI, MP-Guatemala, 306(B). The term Garci Aguirre used was *ingeniero voluntario*, inviting a comparison to the more formal military offices of *ingeniero ordinario* or *extraordinario*.

map, and a local governor also mentioned that it was 'distributed throughout the province in various copies'.[82] Contrary to almost every other previous map of inland Guatemala, this was not just a chance to represent Guatemala to Madrid – it was a chance to represent a new possibility for Guatemalan transport networks within the region itself.

While the *Gazeta* envisaged their chorographical project to be a detached and factual record of the places and resources that could form the basis of improvement, any application of this knowledge would inherently be political. Useful knowledge was never envisaged to be abstract, but it could not be neutral either: its applications affected the Audiencia's political economy. One practical result of the application of useful geography was new infrastructure. Under the cloak of 'progress', reformers presented it as a neutral and universally positive state of affairs, but new infrastructure challenged the established physical backbone of the Audiencia's trading networks and its associated monopolies, and was therefore inherently political. Even among those who could broadly be described as reform-minded, the Consulado merchant association and the Economic Society represented two different political attitudes. They shared an interest in promoting road-building but were often at odds over the beneficiaries of these projects, with the Consulado likely to promote an improvement of current established trade routes from the capital to Santo Tomás and Omoa, and on across the Atlantic, and the Society more likely to be representing a diversification of routes (see map, Figure 0.2). President Gonzáles Mollinedo scathingly commented that the Consulado's constituency was a narrow group of merchants, while the Economic Society represented the much wider 'agricultural class' of producers.[83] These divisions were not set in stone, since the Economic Society and Consulado's

---

[82] P Y del Aguila, Mazatenango, 9 September 1802. AGCA A1.21 Leg. 207, Exp. 4171, fol. 45. The map arrived in Madrid alongside the article 'Camino nuevo de Toliman, abierto y costeado por Don José Rossi', *Gazeta*, Vol. 6, No. 272 (16 August 1802), 189–92. The *Gazeta*'s editors refer to the map in passing, and the copy of the *Gazeta* in Guatemala's National Library contains a note referring to the map as missing. The cost of printing such a map, twice the size of a regular page of the *Gazeta*, makes it uncertain whether it was really included with every copy, but Shafer, *Economic Societies*, 220, seemed to work with a copy that did include the map. The AGI preserves six copies (MP-Guatemala, 274). There are also copies in the AGCA (A1, Leg. 207, Exp. 4171) and the British Library (Add. MS 17650f), providing further evidence for wide circulation.

[83] 'Representacion que el Presidente de Guatemala dirije a S.M.', February 1802, AGI, Guatemala, 481. Consulados in general: Paquette, *Enlightenment, Governance, and Reform*, 134–5. Guatemala's *Consulado*: Serrera Contreras, *Tráfico Terrestre*, 59–63. The Consulado was funded partly by member contributions, but since it was also a powerful regional lender and probably managed some of the Audiencia's spending on public works, it always commanded a more secure budget than the Economic Society, who relied on voluntary contributions. See Floyd, 'Guatemalan Merchants', 99; Regina Grafe and Alejandra Irigoin, 'A Stakeholder Empire: The Political Economy of Spanish Imperial Rule in America', *The Economic History Review* 65, no. 2 (2012), 609–51.

memberships overlapped to some extent. Irisarri directly challenged the Guatemala City merchants' monopoly of Atlantic transport by exploring Pacific harbours, even though he was a representative of the Consulado. His family and associates' ownership of indigo-growing lands in Sonsonate explain his interest in the Pacific littoral.[84] Less combatively, even Rossi y Rubí was making a statement by literally putting Suchitepéquez 'on the map', a statement about the access people in different parts of the Audiencia had to its economic life beyond their own province, and their spatial relationships with the capital. With the unusual decision to display the results of his project on a map, Rossi y Rubí contextualised it, opening up the possibility about thinking of this reformer's territory of intervention within a larger cartographic space.

New localisms in geographical thinking reopened old arguments between people living in different parts of the Audiencia, and different views of political economy. A group of merchants from Nicaragua registered their opposition to the 'despotic' Consulado merchant association, who derived their power and wealth from entrenched trading routes. Aware of the highly political nature of the matter, the Nicaraguans sent this petition directly to the king in Madrid, 'in order not to put our reports at risk with the interested parties in the capital of this kingdom, who could have feelings about this, since the [reports] are contrary to their particular interests'. In the Nicaraguans' opinion, the only reason that the Guatemala City Consulado was able to maintain this monopoly was the arduous lands that connected the northern and southern provinces. It directly profited from the hundreds of leagues 'of exceedingly bad roads, and dangerous rivers' that separated Nicaraguan merchants from the nearest export harbour. If bad roads and inefficient transport were enlightened issues, the implication was that the Guatemala City merchants actually stood in the way of progress. The Nicaraguans also implied that it was in fact Omoa and Santo Tomás that were peripheral. In order to argue for a Nicaraguan local perspective, their language employed a distinctly cartographic and spatial perspective on the Audiencia. They pointed out that 'all the big cities' except for Comayagua were near the *Mar del Sur* or Pacific. Why, then, was it only through the ('deadly') harbours of Omoa and Santo Tomás that overseas traffic could arrive in the kingdom? Should ships not be permitted to travel directly via the San Juan River from the Caribbean to Lake Nicaragua? They equated the fate of interior trade (*comercio interior*) as it took place in Nicaragua and other provinces with the fate of the whole Audiencia. It followed that if the Nicaraguan issue were solved, this would benefit all of Central America: 'this miserable and abandoned Kingdom could flourish', as the merchants put it.[85] Geographical localism coupled with

[84] 'Junta número 173. 20 de Marzo de 1798.' AGCA, A1, Leg. 169, Exp. 3421, fol. 5v.
[85] 'Informe dirigido al marqués de los Hormazos por los Comerciantes y Hacendados de la Villa de Nicaragua en Junio de 1798 expeniendo los motivos que tienen tan entorpecido el Comercio del Reyno de Goatemala', AMN 0258 Ms 0484/005, fols. 99v–101.

the language of public happiness therefore helped to construct a new vision of the Audiencia as a geographical unit whose prosperity depended on its different parts. These unifying thoughts were not popular with everyone.

The Economic Society, given that its director Villaurrutia mentioned the San Juan river as one of the potential outlets for prosperous Central American trade in the future, was generally sympathetic to an Audiencia-wide concept of progress.[86] Neither the Consulado nor the Crown, however, were likely to welcome the Nicaraguan plan. For the Crown, such trade via new Caribbean harbours would have represented too much of a risk. The River San Juan and its surrounding coastal areas had been the site of British invasions and settlements over decades, and the eastern coast of Nicaragua was finally abandoned by Spain to the independent Miskitu in 1800.[87] The Crown and Consulado also saw a risk of contraband in such harbours. Illegal trade with foreigners and contraband trade had been taking place throughout the century. Trade with the United States had been briefly legal in 1797 under the short-lived liberalisation of trade with 'neutral countries', and continued afterwards: in 1798 a horrified Consulado official reported to the king that a frigate from Boston had anchored in Trujillo unchallenged, apparently by pretending that it was in distress, and peddled its wares to Honduran merchants. Audiencia officials even sometimes turned a blind eye to such activity. Merchants outside Guatemala City in particular benefited from more direct international trade by circumventing the monopoly of the merchants from the capital, leading to Consulado officials eager to protect the monopoly to declare the alternative harbours of Trujillo and the San Juan river to be 'good for nothing'.[88] However, even the Consulado softened their stance in difficult times: in wartime, some of its members explored the possibility of contraband. By the late 1810s, faced with an increasingly bleak economic situation, some of the Consulado members even supported the idea of encouraging taxed trade directly with the British. Others explored the possibility of re-opening the harbour of San Juan del Sur, which the Consulado had dismissed as a 'den of thieves' twenty years previously.[89] Nevertheless, the geopolitics of Central America and the merchants' and Crown's fiscal interest were diametrically opposed to much of localist thinking.

---

[86] Real Sociedad, *Quinta junta pública*, 3.

[87] Doug Tompson, 'The Establecimientos Costeros of Bourbon Central America, 1787–1800', in Dym and Belaubre, *Politics, Economy and Society in Bourbon Central America*, 157–84.

[88] Consulado, 3 July 1798, AGI, Guatemala, 888.

[89] Gustavo Palma Murga, 'Between Fidelity and Pragmatism. Guatemala's Commercial Elite Responds to Bourbon Reforms on Trade and Contraband', in Dym and Belaubre, *Politics, Economy and Society in Bourbon Central America*, 101–27, at 115–23. 'Demarcacion del puerto de Sn Juan de la Mar del Sur, 20 enero 1817 por Manuel Rodriguez', AGCA, A1.5 Leg. 267 Exp. 5860

The need to request permission for many journeys due to fears about contra-band not only slowed down trade in many parts of the colony's interior, but also impeded the acquisition of geographical knowledge. So a ship's captain, José Estévez Sierra, found his licence to trade along the Polochic river and with the fort at Omoa revoked in 1776 to avoid contraband trade on this route. It was, however, reinstated in 1777, and he was named '*piloto mayor de la Costa del Norte*' in 1779, a sign that he had earned the Guatemala City bureaucracy's trust.[90] One individual was reprimanded in 1790 for 'having navigated the River Polochic, which is forbidden', but in 1794, another entrepreneur, a resident of Santa Catarina on the banks of the Polochic, was given a licence to trade agricultural products with the Gulf of Omoa along that river, with the caveat that precautions must be taken to avoid contraband.[91] Creating new trade routes was easier said than done. The Consulado, Madrid government, and Economic Society spoke the same language of Enlightened public happi-ness, but had their own visions for how this would look in practice.

For every successful project of practical geography there were several failed ones. Between 1792 and 1841, the Consulado and the Economic Society repeatedly tried to solve the puzzle of whether the Motagua river, which flows from central Guatemala to the Caribbean coast, was capable of carrying cargo boats. To reach the Caribbean, merchants currently had to transport goods from the capital to the inland harbour of Izabal by mule-train, where goods were transferred to small boats, taken along the Golfo Dulce to the Caribbean, and then to Omoa via another transfer point at Santo Tomás. This trading route was the most important one for Guatemala City merchants, but was only fully established in the eighteenth century, when Omoa, constructed in the second half of the eighteenth century as a military outpost, was authorised to receive European trading vessels. For much of the colonial period, goods had travelled to Europe via the port of Veracruz in New Spain instead.[92] This was therefore a new addition to Central America's mercantile geography, and one which reformers were eager to base on practical knowledge of the land. In 1792, the engineer Antonio Porta y Costas had surveyed the river for the Consulado, and was quite optimistic about the enterprise, even suggest-ing that a canal might be built to connect Gualán directly with the harbour at Santo Tomás (the engineer had experience with building 'the great canals of

[90] 'Pretension de Dn Josef Estevez, sobre permiso para navegar el Rio de Polochic con Piraguas, introduciendo viberes y demas efectos permitidos a Omoa, y retornandos iguales, satisfaciendo los Derechos correspondientes', 1775–77, AGCA, A1.5., Leg. 382, Exp. 3706; 'Nombramiento a Piloto Mayor de la Costa del Atlantico, Real Palacio, Noviembre 28 de 1777', AGCA, A1.39, Leg. 1754, fol. 179
[91] 'Contra Carlos Leal, por aber navegado el Rio de Polochy, que esta prohibivido. 1790', AGCA, A1, Leg. 382, Exp. 3721; 'Dn Ambrosio Montalvo solicitando permiso para conducir zarza, y otros frutas del Paiz al Golfo, por el Rio Polochic. 1794', AGCA, A1, Leg. 382, Exp. 3729.
[92] Hall and Pérez Brignoli, *Historical Atlas*, 162–3.

Castile and Murcia').[93] Predictably, there was no funding for such an ambitious project, but even the more basic question of the river's accessibility to different types of trading boats remained disputed. A 'Motagua Navigation Company' was formed in 1796, with the initial list of subscribers including influential merchants and members of both the Consulado and Economic Society. One of them was Tomás Calderón, the parish priest of the village of Acasaguastlan, who was a great enthusiast for Enlightenment ideas and the Economic Society's and *Gazeta de Guatemala's* undertakings. The commissioners of the proposed trading company lauded President Domás y Valle's enthusiasm for the project, noting that the undertaking 'demonstrate[d] an enlightened understanding' (*un entendimiento ilustrado*). They pointed out that in his zeal to see the project fulfilled, the president was acting 'like a father' to his people (*mas Padre que Gefe de sus subditos*).[94]

These references to an enlightened (and explicitly paternalistic) framework of public utility, happiness, and progress provided a bridge between the officials and merchants from different political camps and show the important status that the exploration of old and new trading routes held in their geographical imagination. The *Gazeta de Guatemala* also supported the project, repeatedly publishing encouraging notes, for instance the report of a boat captain's successful journey and another by Bernardo Dighero, member of both the Economic Society and Consulado.[95] Several reports contradicted each other, and some of Dighero's fellow Consulado members doubted the honesty of his river guides.[96] And yet, a Consulado official who summarised many different reconnaissance reports about the Motagua for the king urged his fellow reformers not to lose sight of the larger picture. He stressed the utility of these successive reports as a geographical endeavour that could ultimately provide information about the country. His summary of several reports was designed to present the whole province of Guatemala 'at one glance'. The fact that he was writing to the king to represent Guatemala 'at a distance' rather than within the Audiencia explain some of this large-scale perspective, but it is nevertheless interesting that he claimed that the navigation issues of the Motagua river could be a stand-in for understanding the whole province.

[93] José de Isasi and Juan Antonio Araujo, *Ereccion de la Compañia de Navegacion del Rio Motagua* (Nueva Guatemala, 1796). Floyd, 'The Guatemalan Merchants', 93–5.

[94] Isasi and Araujo, *Ereccion de la Compañia de Navegacion*, 4, 28.

[95] 'Nota', *Gazeta*, Vol. 4, no. 161 (16 June 1800), 272 [wrongly printed as 178]. A later verdict on the Motagua navigation maintained that these journeys were simply fiction, because it was what the Guatemala City Consulado had wanted to hear: Remacha had in fact travelled from Omoa to Izabal or Sto Tomás, presumably along the coast, but he did not travel from the capital up to Gualán: 'Informe de José Munilla y Rafael Trullé', 14 February 1817, fol. 7. AGI, Guatemala, 892.

[96] 'Reconocimiento Río Motagua por Dighero, 1812', 7 August 1816, AGI, Guatemala, 892, fol. 15v.

Using a medical analogy, he saw this view of Guatemala as 'anatomised', exposing the 'intrinsic ailments of its constitution', to which the correct remedies could then be applied.[97] The physical geography of the isthmus, which currently suffered from problems of connection between the capital and the coast, was disadvantageous. But like a disease, it could be overcome. Geographical knowledge and, as a result of this knowledge, engineering projects, would remedy Guatemala's infrastructural ailments. Like the *Gazeta*'s own *Description*, the Consulado's vision of geography was one of patiently accumulating knowledge, until one day it could provide the foundation for practical landscape interventions to draw different parts of the Audiencia closer together.

Conclusion

If the fundamental problem that the Economic Society's members sought to solve was that the Kingdom of Guatemala was a 'vast' land whose riches had hitherto not been properly exploited, geography was perhaps the most promising discipline that could counter the vast unknown with concrete knowledge and solutions. Countering the misinformation about Guatemala that they saw everywhere from travellers' reports to the *Encyclopédie Méthodique*, they proposed an alternative *Description of the Kingdom of Guatemala*, a chorographic description that would be compiled by the *Gazeta*'s readers. In gathering geographical knowledge, they did not always seek to reinvent the wheel. Where they had access to the state's censuses and geographical reports through local archives, or indeed the state's resources when it came to the building of infrastructure projects, they made full use of them. The *Description* would represent Guatemala appropriately to the outside world, but most importantly define the region to its own inhabitants. The *Gazeta*'s networks played a key role in gathering geographical information and promoting applied projects, such as new roads or coastal navigations. While relying on the rhetoric of improvement through scientific or systematic knowledge, there were also more local and more political projects. Agustín de las Quentas Zayas and Juan Bautista Irisarri published guides to navigating the land- and waterscapes of Central America, while José Rossi y Rubí organised a project to build a road to improve trade connections and other journeys. These were applications of reconnaissance and geographical knowledge to practical aims for economic improvement from a local perspective, which also made them political.

---

[97] Real Consulado de Comercio, *Informe aprobado por la Junta de Gobierno del Consulado de Guatemala sobre el Objeto y Cumplimiento de las Rles ordenes de 13 Sept y 22 oct de 1812* (Guatemala: Beteta, 1814), 39. Medical analogies in European political thought: Paquette, *Enlightenment, Governance and Reform*, 58.

The Economic Society and Consulado's geographical and infrastructure projects represent perhaps the most ambitious colonial-era attempt by Central Americans to assess and reshape the region according to local priorities. Local and regional geographies also rose to prominence in the reformers' work. This allowed the emergence of the sort of territorial imagination that María José Afanador has described as 'blending political economy discourse and local territorial descriptions'.[98] Projects of chorographic description, but even more so Quentas Zayas's travel guide or Irisarri's crowd-sourced navigation advice, helped to envisage parts of the interior and coast not just as connected with the capital, but integrated through the practical actions and physical movements of reformers. While there was no territorially defined Guatemalan 'homeland', in these geographical reimaginations there were some examples of a definition of Guatemala as a 'territory of patriotic intervention' upon which reformers acted. Rossi y Rubí found the impetus for his road reform in his travels 'from Tumbalá to Trujillo'. When director Villaurrutia emphasised the advantages of navigating on the San Juan river, he expressed it from the perspective of bringing different parts of the Audiencia together, and in the same breath as its southern reaches (Nicaragua) also mentioned its northernmost part, Chiapas. Trading on the San Juan as well as the Chiapan Tulijá river would allow 'the resident of León as well as the one of Chiapa [to exploit] the advantages offered by nature, not any less than the resident of the capital of the Kingdom'.[99] As an organisation covering the whole area of Central America, the Society also acquired the potential to suggest political meanings for the whole of the Audiencia. The Society's director José de Ayzinena used the establishment of a branch of the Economic Society in San Salvador to restore order after he was sent there to put down an 1811 uprising, apparently trying to channel political energies into the extension of 'liberal and beneficient ideas' to the Audiencia.[100] The close connections between imagining the relationship of different parts of the Audiencia to the capital politically and geographically were also present in the Nicaraguans who asked the king to imagine the entirety of Guatemala as if on a map. These geographical imaginations show that reformers tied their landscape interventions not to generic 'Guatemalan' or 'American' landscapes, but to concrete cultural landscapes that involved not just territory, but trade and agriculture as well.

---

[98] Afanador Llach, 'Political Economy, Geographical Imagination and Territory', 5.

[99] Real Sociedad, *Quinta junta pública*, 3.

[100] 'Sobre ereccion de la Junta de Correspondencia de San Salvador, verificado en 5 de Febrero', 1812, AGCA, A1.6 Leg. 2008, Exp. 13856; Shafer, *Economic Societies*, 227.

5   Transforming Environments

'Wanted: The secret to striking up a conversation without starting it by complaining that Guatemala is lost', read a satirical comment in the *Gazeta de
Guatemala* in 1798.[1] Parts of the region certainly suffered from economic
circumstances which might lead observers to consider it 'doomed', from locust
plagues to falling prices on the European market affecting the main export crop,
indigo. Enlightened reformers themselves regularly asserted that the country
was in a 'miserable' state, but immediately countered their own rhetoric with
assurances of Central American exceptionalism and great dormant potential.
Views of rural landscapes in eighteenth-century documents encompass both
these extremes. Within a general pattern of seeing landscapes as unexplored,
they appeared in two ways: as a vast and fertile land harbouring many as yet
unknown riches of nature; or alternatively a savage wilderness, a danger to
health, morals, and efficiency, to be overcome. Such dual interpretations of
nature, especially uncultivated lands, continued until well into the nineteenth
century in Central and South America, and beyond.[2] Even, or especially, in
these narratives of threat and desolation, there was promise to be found. In
addition to being common to Enlightenment ideologies, the rhetoric of optimism, transformation, and potential inherent in these narratives may also have
had something to do with the 1773 earthquake that had destroyed the old capital
and had created upheaval in the new city, but also political, commercial, and
social change.[3] It would be surprising if the Economic Society, Consulado, and
Gazeta, based in this city that had been established only twenty years earlier in
a place that was considered a 'blank' on the Spanish map of Central America,
had not been influenced by this decades-long process of transforming 'empty'

---

[1] 'Noticias sueltas', *Gazeta*, Vol. 2, no. 69 (9 July 1798), 174.
[2] Pratt, *Imperial Eyes*, 126–7; David Arnold, *The Tropics and the Travelling Gaze. India,
Landscape and Science* (Seattle, WA: University of Washington Press, 2006), 82;
William Griffith, *Empires in the Wilderness: Foreign Colonization and Development in
Guatemala, 1834–1844* (Chapel Hill, NC: University of North Carolina Press, 1965).
[3] See, for instance, Richmond Brown, 'Profits, Prestige, and Persistence: Juan Fermín de Aycinena
and the Spirit of Enterprise in the Kingdom of Guatemala', *The Hispanic American Historical
Review* 75, no. 3 (1995), 405–40; Twinam, *Purchasing Whiteness*, 395–7, links Bernardo
Ramírez's 1783 petition to the opportunities provided by the 'city in ruins'.

space into the political and intellectual hub of the kingdom. In the eyes of most late-eighteenth century reformers, the expanse of Central America had come to be seen as one of rich potential.

In 1798, director of the Society, Jacobo de Villaurrutia, judged the Audiencia to be almost as 'miserable' as ever. However, he had observed some changes. Not only was transformation possible, but 'the first steps have been taken' already. Results were appearing too slowly for his taste, but he was able to give some examples of the progress that had been made: good ideas were being propagated, [merchant] monopolies and egotism were losing ground, and the 'illustrious and select catalogue of lovers of the patria keeps increasing'.[4] The Society member in Nicaragua who had set off a storm of letters by suggesting that his hometown of León might plant plantain trees also showed himself quite content with his observations of society. The opinion that 'nothing could ever get done in this unfortunate Kingdom' was unfounded, since with the effort of many, men could come together to 'spread enlightenment [*las luces*]'.[5] These were small-scale changes, but reformers believed that they were already able to see proof of their designs working in small and incremental ways. Further useful actions and knowledge would enable them to realise projects which until now had been regarded as 'fantastical' and 'unattainable', as the Society had set out in their statutes.[6]

However, to reimagine Central America was not just to draw new lines on a map and plant a few crops. With renewed energy, reformers also schemed to intervene more profoundly in landscapes. They connected their interventions to ideas such as the quality of airs, water, and temperatures, generating discussions about the vegetation, healthiness, and climate of particular places. They became concerned with factors such as the specific location and layout of towns, which they believed exposed merchants and mariners to disease. Using categories such as *temperamento* and *clima* as factors in the potential of particular places, and referring to the belief derived partly from Enlightenment thinkers such as Montesquieu in the transformative power of people acting on landscapes, they hoped that a healthier population and a more efficiently managed soil would be part of the legacy of enlightened reform.[7]

---

[4] Real Sociedad, *Quinta junta pública*, 2.
[5] 'León y 8 de Octubre', *Gazeta*, Vol. 1, no. 40 (6 November 1797), 319–20.
[6] 'Plan de la Ynstruccion', Josef de Sierra, 6 December 1794, AGI, Estado 48, No. 7.
[7] The words *clima* or *temperamento* built on ancient ideas of 'climes' or different zones of the earth, and were used to describe characteristics of a place such as latitude, altitude, humidity, weather, temperature, and vegetation, and could also incorporate human settlement (and indeed the nature of human bodies in a particular *clima*). In the practical sense of the suitability of *climas* or *temperamentos* for particular aspects of agriculture and settlements used in the *Gazeta*, it does very roughly map onto the modern idea of 'climate' and is here translated as such; for a similar approach see Sara Migletti and John Morgan, 'Introduction', in Migletti and Morgan, *Governing the Environment*, 2–4. Mark Thurner points out that in 1790s Peru it was used in a metaphysical

Labourers, who had been merely implied as part of initial discussions of expanding cacao or cotton crops, now appeared more firmly as part of this equation. Director Jacobo Villaurrutia , was characteristically optimistic when he praised not just 'our soils and climate', but also 'the general aptitude of the natives' [*naturales*] talents, character, and good disposition for any sort of thing' as assets of the *patria*. Importantly, he also qualified this statement. These attributes were true 'as long as they are directed by good policies [*politica*]'.[8] He mentioned these qualities of 'natives' in conjunction with the landscape, in a conflation of natural and human resources as well as their management. This was typical of the 'biopolitics' of the eighteenth century, but also an extension of earlier European and Spanish-American thought about the links between landscapes and governance.

Untameable Lands and Their People

Visions of landscape as wilderness had been tied to the people who lived in it throughout colonial Spanish rule. Spanish and Creole views of the Central American countryside overwhelmingly maintained ideas of certain spaces as an Indian 'Other'. A forested area, criss-crossed by Indian footpaths, did not even merit mapping to a sixteenth-century author: he left the space blank on the map.[9] Spanish views linked Indians living in the 'wild' to loose morals and even cannibalism, while Indians in towns were portrayed as 'docile' converts to Catholicism and civilisation.[10] As late as 1802, the governor of Suchitepéquez in Guatemala linked dispersed communities with a variety of moral failings, including 'deaths, injuries, concubinage, robberies, &c'.[11] Older patterns of colonial governance here combined with a well-established trope of many Enlightenment-era regimes, equating lands that were difficult to map, govern, and tax with a lack of civilisation and barbarism.[12] Mountainous terrain in particular traditionally amounted to unusable territory in the Spanish imagination. Possible translations of the word *monte* include 'wilderness'.[13] Communities who

---

sense best not translated with modern terminology: *History's Peru*, 83–5. For a summary of the concept in eighteenth-century Spain and the influence of naturalists such as Buffon, see Luis Urteaga, 'La teoría de los climas y los orígenes del ambientalismo', *Geocrítica* 99 (1993): 1–55. For related ideas of vegetation as tied to altitude as well as latitude: Cañizares-Esguerra, *Nature, Empire and Nation*, 113–16.

[8]   Real Sociedad, *Quinta junta pública*, 2.

[9]   'Costa de Zapotitlan, y Suchitepeques', BL Add. MS 17,650e, see Figure 4.1.

[10]   Sellers-García, *Distance and Documents*, 61–3; Kagan, *Urban Images*, 36–8.

[11]   'Sobre reducción á Poblado los dispersos', Tadeo Cerdas, Mazatenango, 1802, AGCA, A.1, Leg. 207, Exp. 4176, fol. 8.

[12]   Sellers-García, *Distance and Documents*, 63; Blackbourn, *Conquest of Nature*, 37–46.

[13]   Real Academia Española, *Diccionario de la lengua castellana*, 574; Esteban Terreros y Pando, *Diccionario castellano con las voces de ciencias y artes*, Vol. II (Madrid: Viuda de Ibarra, 1787), 614.

had fled Spanish authority and lived in Guatemala's eastern lowlands were known as *montañeses*.[14] Spaniards who encountered such terrain, then, often revealed a sense of estrangement and suspicion. Although a survey for a new road through the mountains of the Sierra Madre in 1803 took place under the auspices of Society member José Domingo Hidalgo and was described as 'beneficial to the public good', the local surveyor's report betrayed little but suspicion of these lands, as well as demonstrating the coercive or violent aspects of the state when it came to gathering geographical information. Knowing that there was some way across the mountains, but not where those paths led, he went on a reconnaissance mission. Accompanied by a lieutenant and interpreter, he quickly found local guides:

taking with us Juan Zay and Manuel Belezuy, who live in the mountains and know all the paths that there are therein. We had taken them by surprise and given them beasts on which they were to go with us, and I told them that if they did not show me said road I would send them to the capital as prisoners.[15]

Not all officers of the state may have been so forceful, or so bluntly honest about their methods, but this threat of force affirmed a recurrent idea in colonial thinking: that knowledge of mountains, and more broadly wilderness, was the precious secret of the Indian population. The presence of an interpreter further suggests linguistic and practical barriers that the surveyor assumed to stand between him and the mountains. Indians seemed to physically and intellectually inhabit a parallel world of knowledge.

Specific notions of landscapes to which Spaniards might not have access could also serve to designate entire areas without Spanish settlements as 'wilderness'. Of course, areas of allegedly 'empty', 'useless', or 'wild' space often referred to places that were in reality inhabited by indigenous people. The engineer in charge of the 1792 report on the navigability of the Motagua river, for instance, considered the indigenous population ignorant about the possibilities of navigating it, claiming it was 'little known by the indigenous inhabitants [*naturales*]'.[16] In reality, the region was not as uninhabited as initial

---

[14] Héctor Aurelio Concoha Chet, 'El concepto de montañés entre los Kaqchikeles de San Juan Sacatepéquez, 1524–1700', in Robinson Herrera and Stephen Webre (eds.), *La época colonial en Guatemala. Estudios de historia cultural y social* (Guatemala City: Editorial Universitaria, 2013), 19–41.

[15] Francisco Xavier de Aguirre, 'Sobre apertura de un nuevo Camino con que se evita pasar la cuesta de aquel Pueblo, la de la hambre, y la de la Laja', 1801, AGCA, A1.21, Leg. 193, Exp. 3934. An influential member of the Economic Society, José Hidalgo was also involved in this project in his capacity as *subdelegado de tierras*.

[16] Antonio Porta y Costas, *Relacion del reconocimiento que ... practico el ingeniero ordinario D Antonio Porta, en la costa comprendida desde Omoa, hasta la punta de Manabique; y desde la barra del Rio de Motagua, hasta donde se le une el de Chicosapote, a 14 leguas de la ciudad de Guatemala* (Guatemala City, 1792), 23. This report did acknowledge that foreign contrabandists had previously navigated the river.

Spanish reports would have it. Héctor Concoha Chet has shown that indigenous Kaqchikel communities, fleeing from the Spanish authorities in western Guatemala, settled here to escape the harsh tax demands and abuses of power in Spanish-controlled villages. These communities were occasionally included in census figures and *visita* records, emphasising repeated attempts to incorporate them into Spanish religious life, tax, and labour demands.[17] First-hand Spanish reports from the river also showed that people very much lived on both sides of the river and used it to fish and transport their goods, not just in the Spanish settlement of Gualán and a few smaller Indian villages, but up and down the river, too. The planners of the Consulado seemed to sweep these observations aside. They could declare the Motagua as unknown even while making reference to the stone dams and staked-out enclosures (presumably for the purpose of fishing) that the river's residents had created.[18] In this, Spanish authorities followed a long tradition of seeing land cultivated by indigenous people (across different contexts of imperialism) as uncultivated and 'wild', and cemented the 'pristine myth' that Europeans ascribed to pre-conquest American landscapes.[19] In late-colonial Central America, this idea of undisturbed landscapes did not just affect views of the pre-conquest natural world, but translated into an understanding of all American landscapes being fundamentally wild without Spanish intervention.

Attempts at conquering such 'wilderness' could be risky and in the narrative of reformers required a strong belief in the possibility of man's dominion over landscape. Contemporaries were impressed with the governor Rossi y Rubí's willingness to personally participate in the process of building his road (discussed in Chapter 4), to 'live in the mountains [*montañas*] for many months', 'immersing himself in the uncultivated wilderness [*incultos montes*] in order to do good for the public'.[20] The consequences of engaging with uncultivated lands, however, became clear just a few years after Rossi had completed his road. By 1802, he had moved from his post as governor of Suchitepéquez to Comayagua, but he had barely arrived at this new posting when the new governor of Suchitepéquez complained that the famed road was, in fact, simply

---

[17] Concoha Chet, 'El concepto de montañés'; on Indians fleeing Spanish control, see Martínez Peláez, *Patria del criollo*, 463–8; Lovell, *Conquest and Survival*, 88.

[18] Isasi and Araujo, *Ereccion de la Compañia de Navegacion*, XXVIII.

[19] Earle, *Body of the Conquistador*, 78–83; William Denevan, 'The Pristine Myth: The Landscape of the Americas in 1492', *Annals of the Association of American Geographers* 82, no. 3 (1992): 369–85; Andrew Sluyter, 'The Making of the Myth in Postcolonial Development: Material-Conceptual Landscape Transformation in Sixteenth-Century Veracruz', *Annals of the Association of American Geographers*, 89 (1999): 377–401; Martyn J. Bowden, 'The Invention of American Tradition', *Journal of Historical Geography* 18, no. 1 (1992): 3–26; for an example in the British empire, see Philip, *Civilising Natures*, 135.

[20] AGCA, A1.21, Leg. 207, Exp. 4171, fol. 56. 'Camino nuevo de Toliman, abierto y costeado por Don José Rossi', *Gazeta*, Vol. 6, no. 272 (16 August 1802), 191. See also Sellers-García, *Distance and Documents*, 112–13.

not there. Another official, asked to act as an arbiter, concluded that the accusations were unfounded: the road had been built with great care. However, it had already been overgrown by dense vegetation. Rossi y Rubí admitted this possibility, saying that the 'fertility [*feracidad*] of the terrain will have caused shrubs to grow'. Rossi y Rubí recognised Central American environments as a destructive force ready to overpower even the best-made roads here. He explained that although the tropical lowlands had the most exuberant vegetation, the highland rains 'produce similar ruins, although more slowly'.[21] However, he did not blame nature for the 'bad and derelict' roads.

Roads through the wilderness could only exist through human commitment to them, and so it fell to humans to maintain them. Rossi homed in on provincial administrators, *jueces de provincia*, as the cause of these 'bad and derelict roads'. They only cared about trade and money, neglecting 'other branches [of government]', including the day-to-day governance that would keep infrastructure functioning. They were also at fault for thinking that they could simply give orders on paper from their offices, when roads needed action, and the physical presence of administrators, who should lead by example – yet another clear statement from a reformer emphasising that practical human interventions rather than empty words would be the source of transformation.[22] Rossi implied that the fault for the destruction of the road lay neither in its construction, nor the landscape which it had tried to conquer, but administrators' failure to maintain it. In this view, the natural forces at work in the Guatemalan highlands were simply a backdrop for the judicious actions of Enlightenment reformers ready to intervene in these landscapes. He thereby presented a qualified endorsement of Enlightenment optimism regarding man's control of nature: human action could temper Guatemalan landscapes and make them more European, if not completely tame them. The arbiter in Rossi y Rubí's case gave the practical advice that it would no longer be necessary to clear a road that was strengthened by frequent use, and in fact one part of Rossi's road had been maintained successfully. The inhabitants of surrounding villages were 'convinced by its advantages' and therefore helped to keep it clear of vegetation, but they were not using the other fourteen leagues of new road. Consequently, most of it looked about to disappear. The arbiter seemed to speak from direct experience with the landscape when he sighed that there had been nobody to give the road a *machetazo*, perhaps best translated as a 'good whack with a machete'.[23] Whether administrators or local communities, humans were needed to act on the landscape.

---

[21] Rossi y Rubí, Comayagua, 17 June 1802, AGCA, A1.21, Leg. 207, Exp. 4171, fols. 34–36v.

[22] *Ibid.* Rossi thought that the President was well placed to persuade these *jueces* to fulfil their roles, since his political leadership had in the past won over even 'the wild hearts of the Catalans'.

[23] 'Camino de Suchitepéquez', AGCA, A1.21, Exp. 4171, Leg. 207, fols. 35–36v, 53–6.

In his narrative of his 'personal dedication' to the cause of building the road, Rossi drew as much attention to himself as a 'leader' of the Indian workforce as a disciplined explorer and surveyor. Local knowledge of the landscape was not just required to stake out and map the road: it was also required to manage Indian labour, which Rossi was not shy to request from the Audiencia. His strategy was to request orders specific to each village so that, for instance, the people of Patulúl would only be in charge of maintaining the road between the Santo Domingo river and San Juan, two nearby locations which would, in Rossi's opinion, reduce the stress on these communities. While he stressed his intent to protect the Indian workforce paternalistically, and the *Gazeta* lauded him for 'treating the Indians with the greatest love', his concern was just as much with the practicalities of getting reasonably willing labourers to his road-building project without hiccups.[24] As David McCreery notes, by the late eighteenth century, towns regularly started legal protests or resisted orders for labour drafts that they considered dangerous or unfair.[25] Rossi wanted to take precautions to avoid such a situation, because 'the Indians get upset when they have to work in strange lands [*tierra agena*], and they turn to complaints and [legal] recourse, and the works are ruined'.[26] As Rossi's fears of mutinous labourers show, projects that intervened in the physical geography of the mountains required some degree of co-operation as well as physical labour from the Indian population. Rossi advertised himself as the connoisseur of local circumstances who would manage this labour force and enable construction.

In his advice for caring for the road after his departure, he made clear what place he accorded to Indians in this new enlightened landscape. He portrayed the Indians of Suchitepéquez, his own province, as trustworthy custodians of the road, stressing that they 'already know what they have to do, in which places, at what times of the year, in what way and so forth'. However, it quickly became clear that Rossi attributed these particular Indians' knowledge of how to maintain the road to the leadership he had provided. He demanded much more detailed written orders for the Indians of other provinces, whom he assumed helpless without such guidance. Rossi exhibited a paternalism typical of Central American Enlightenment reformers when he portrayed Indians as ultimately passive, even as he stressed elsewhere that he believed them to be the hardest-working inhabitants of the realm. Indians might have been 'useful' for their labour, but in the human interventions that conditioned landscape, some men were more 'useful' than others. Rossi opined that local officials should lead by example at the road works rather than just writing orders on paper, because 'the people of America [*el Pueblo Americano*] do not know how to

---

[24] 'Camino nuevo de Tolimán, abierto y costeado por D José Rossi', *Gazeta*, Vol. 6, no. 272 (16 August 1802), 191.

[25] McCreery, *Rural Guatemala*, 95–100.

[26] 'Sobre abrir camino por Costa del Sur', AGCA, A.1, Leg. 207, Exp. 4171, fol. 36.

lead, but they will follow in the most docile manner anyone who leads them'.[27] The management of the countryside, for Rossi y Rubí, may have relied on local knowledge, but it would have to be mediated through Spanish leadership.

This idea of trade routes being contingent on encroaching vegetation and humans maintaining them in accord with changing nature meant that trade routes were no longer just geographically defined, but a problem of nature. This idea even applied to rivers, and went beyond obvious differences in stream between the rainy and the dry season. One description of the Motagua river emphasised Central American nature-exceptionalism to the king: this was not like a European river that could be made navigable simply because someone had decided to apply engineering to it. Instead, to navigate it one would have to deal with the additional challenge of 'rough uninhabited wilderness, poisonous reptiles, and wild beasts'.[28] One section of the Consulado had not lost the hope that the Motagua river might be made accessible for mercantile boats, but it would require sustained human action once again. They noted obstacles in the form of trees, stones, or silt that might make this trade impossible. Unconcerned, the authors declared that this would be solved with an 'annual cleaning of the river, without which the navigability would be lost'. In this case, they 'could not expect individual boat-men to do it, nor entrust them with a task so important for the state and the public [al estado y la causa publica]'. Managing the landscape was too important a task to leave it to individuals. They therefore designated Consulado funds for the cleaning.[29] The colonial concept of a road or waterway was therefore a living construct, conditioned not just by the people who built and used it, but others who maintained it. Maintenance was dependent on the interaction of human action, the natural and the built environments. In traditional bureaucratic narratives, too, there was an element of considering the management of landscapes as a part of road-building: people who manned ferries and cleared roads were an important part of route descriptions.[30] Creating and maintaining roads and waterways repre-sented an ongoing human engagement with the vegetation of the region, and collecting geographical information also meant recording and understanding these interactions. The reformers' newly acquired geographical knowledge would therefore directly contribute to making new roads. Reconnaissance of trade routes was but a small part of it: it was equally important to augment

---

[27] 'Del trabajo de los Indios', Gazeta, Vol. 5, no. 232 (15 October 1801), 599; 'Sobre abrir camino por Costa del Sur, para facilitar comunicacion y transito de este Reino con el Mexicano y comision dada á el Alcalde Maior de Suchitepeques Dn Josef Rosi y Rubi', 1805, AGCA, A.1, Leg. 207, Exp. 4171, fol. 36.

[28] 'Informe sobre la navegacion del Rio Motagua en 14 de Febrero de 1817', fol. 6, AGI, Guatemala, 892.

[29] Isasi and Araujo, Ereccion de la Compañia de Navegacion, 21.

[30] E.g. 'Viaje que hizo el Ingeniero en Xefe don Luis Diez Navarro', Guatemala, 31 October 1758, AGI, Guatemala, 874.

society's knowledge of the management of the landscape that the roads traversed.

Custodians of the Land

Nowhere were considerations of the nature of the landscape and the effects of its undisputedly fertile soils more important than in the context of agriculture. The abundant fertility of the soil could pose a problem when it came to building roads, but could easily be reimagined as beneficial. In 1802, president Gonzáles contrasted his pessimistic assessment of the kingdom as a whole, whose dispersed population he judged to be living in 'misery', 'ignorance', and full of 'vices', with the land itself. Impressed, he noted the 'fertility of this terrain, which produces without cultivation, and without the help of any arts, the fruits of all zones, because there are climates analogous to each of them'.[31] As was typical of late-eighteenth-century reformers, Gonzáles here couched the land's fertility in terms of different climatic zones. At least one reader, signing his name *El patriota forastero* but possibly the editor Alejandro Ramírez himself, pointed out the obvious problem with this manner of thinking: why had agricultural productions and trading potential been lost in the first place if the soils were so rich? 'I hear a lot of people pondering the richness that the natural productions of this Kingdom contain, but little or nothing is said about their conservation', he complained.[32] What were once meant to be 'inexhaustible royal treasures' only existed in people's memory now. Where had these glorious past harvests gone? Unlike many, this reader tried to examine the root causes of the decline of production and waded into a well-established dispute about the dangers of a monoculture which had in turn driven up prices for basic foodstuffs, and also criticised the power of Guatemala City indigo merchants. In his opinion, the former self-sufficiency of El Salvador indigo growers had been disrupted when they came to depend on cash advances and loans from Guatemala city merchant-bankers or 'capitalists' (*el capitalista*).[33] Reformers, in his opinion, held no metaphorical gold nuggets in their hands that might prove an inherent superiority of Central American soils compared to those of competing producers. He pointed out with sadness that in Caracas, where indigo production was a new phenomenon, parts of the harvest were already better than Guatemala's. *El patriota forastero* acknowledged the idea of 'our'

---

[31] 'Representacion que el Presidente de Guatemala dirije a S.M.', AGI, Guatemala, 481, 2 February 1802.

[32] 'Carta', *Gazeta*, Vol. 1, no. 36 (9 October 1797), 283–4. *El patriota forastero* means the 'foreign' or 'immigrant' patriot. He refers to himself as an *español* (though not *peninsular*) later in the article. Speculation that he is Ramírez in Maldonado Polo, *Huellas de la razón*, 327.

[33] 'Carta', *Gazeta*, Vol. 1, no. 36 (9 October 1797), 286–8; Floyd, 'Guatemalan Merchants', 99; Troy S. Floyd, 'The Indigo Merchant: Promoter of Central American Economic Development, 1750–1808', *Business History Review* (1965): 476–9.

Guatemalan soils standing in competition with those of other Spanish American places as a patriotic issue, but refused to buy into the promise of magical lands if the evidence suggested that Venezuelan soil could produce indigo just as good.

Many articles in the *Gazeta* blithely ignored such realities. They emphasised the supposedly near-miraculous qualities of the soil independent of any other factors, and argued for the specific potential of particular places on its basis. In particular, they argued that Guatemala was blessed with 'different zones' – varied climates which would lead to varied agriculture, and bountiful exports in turn. The notion of one territory encompassing a variety of climates could be a symbol for divine providence as well as economic potential. Peruvian Creoles had elaborated such theories to argue for the Andes as a microcosm of creation for centuries. The concept was increasingly popular in the eighteenth century and went hand in hand with new scientific or pseudo-scientific approaches to the land and the expansion of climatic theories. It recalled the edenic themes of earlier centuries of European thought on America, and was prominent in the pages of patriotic newspapers across the empire. It would later also appear in independence-era writings across Latin America, from the Venezuelan poet Andrés Bello to the Guatemalan statesman José del Valle, who stressed that Guatemalan temperatures ranged from intense cold to extreme heat.[34] So entrenched were these ideas of ideal landscapes by the 1830s that a report (submitted by the Irish-Guatemalan Juan Galindo) in the *Journal of the Royal Geographical Society of London* also referred to Central America's 'advantageous scale of altitudes' reflected in a variety of climates, and stressed the bounty of its 'various vegetable productions'.[35]

And, what was more, apparently no cultivation or artifice was needed to make its fruits appear, with the logical implication that humans were almost superfluous in these bountiful landscapes. The leaders of the Society took it upon themselves to examine 640 grains of cacao from cacao trees that grew wild on the northern (Caribbean) coast of Honduras near Trujillo and declared the seeds to be of excellent quality. The reformers particularly highlighted the qualities of the earth producing 'excellent' fruits even from wild trees.[36] Enlightened reformers ascribed agency and vitality to a disembodied landscape

---

[34] Cañizares-Esguerra, *Nature, Empire and Nation*, 116–26; Thurner, *History's Peru*, 83–5; Jerry Hoeg, 'Andrés Bello's "Ode to Tropical Agriculture": The Landscape of Independence', in Beatriz Rivera-Barnes and Jerry Hoeg (eds.), *Reading and Writing the Latin American Landscape* (New York: Palgrave Macmillan, 2009), 53–66, at 57–60; Letter from Valle to Humboldt, 29 October 1829, in José Cecilio del Valle, *Cartas de José Cecilio del Valle*, ed. Rafael Heliodoro Valle (Tegucigalpa: Universidad Nacional Autónoma de Honduras, 1963), 49.

[35] Juan Galindo, 'On Central America', *Journal of the Royal Geographical Society of London* 6 (1836): 119–35, at 124.

[36] 'Calificacion de la calidad del cacao nuevamente plantado en Trujillo', 4 March 1797, AGCA, A1.6, Leg. 2006, Exp. 13811.

that would produce agricultural bounty without always explicitly explaining the role of existing inhabitants in this. 'Wild' productions of the country might be marshalled into economically productive crops (the Economic Society, for instance, conceded that the Honduran cacao would improve if the trees were cultivated and enclosed rather than left wild) even as reformers remained uneasy about the status of the (often indigenous) farmers who cultivated them. As the historian Severo Martínez Peláez put it, 'the more miraculous the land appears, the more the merits of those who work the land vanish'.[37] Martínez Peláez was referring to the seventeenth-century writings of the Creole chronicler Fuentes y Guzmán here, but eighteenth-century reformers only sharpened this distinction when they put forth detailed descriptions of the land's abundance, couched in the language of scientific detachment.

Here was a precious soil that produced a cornupia of fruits, but at the same time seemed divorced from the Indians who cultivated it. José Hidalgo's geographical description of Momostenango, for instance, informed the reader that the village's climate was temperate (*el temperamento es templado*): 'and this is why the soil [*terreno*] produces *anonas*, figs, *matasanos* and avocadoes: these Indians do not trade, and only maintain *milpas* for subsistence'.[38] Another contributor to the *Gazeta* used this idea of naturally abundant soils as proof for Indians' inherent laziness. He argued that Indians were work-shy because in this bountiful continent, maize and plantains sprang from the ground everywhere, so 'wild' and 'civilised' Indians alike could live off the land without even having to invest work in it. 'Wild' Indians might hunt, but mountains and rivers actually 'provided' them with animals and fish 'without any effort' on their part. The 'civilised' Indian who paid tribute, meanwhile, still did no work, because all he had to do is plant 'a kernel of maize', which produced 'without any other cultivation … more than he needs to sustain himself'.[39] The intendant of Chiapas, Quentas Zayas, upon establishing a new town was confident in the agricultural potential of the area and advocated the establishment of 'large haciendas of cacao, peppers, coffee, indigo, sugar-cane, cotton, rice, and whatever is desired'. However, his assessment of 'the natives' in this context was ambivalent. He mentioned that the natives 'knew' about the surrounding plains' 'famous' fertility, but it was the new Spanish settlers whose dedication and care in 'working the fertile terrains' he highlighted.[40] It was probably the intendant who wrote to the *Gazeta* on another occasion to note once again that Chiapas possessed different climates (*temperamentos*), leading to great diversity in its 'exquisite productions'.

---

[37] Martínez Peláez, *Patria del criollo*, 106.
[38] 'Descripcion de la provincia de Totonicapan', *Gazeta*, Vol. 1, no. 33 (18 September 1797), 258.
[39] 'Apuntamientos estadisticos del Br. Talcamáhida, sobre la agricultura, industria y comercio de este reyno', *Gazeta*, Vol. 7, no. 314 (1 August 1803), 301–3.
[40] 'Ciudad Real de Chiapa 7 de Abril', *Gazeta*, Vol. 2, no. 59 (30 April 1798), 96.

However, he also spoke of the 'inertia which . . . increases proportionally to the greater fertility of the land' as an impediment to successful trade.[41] He evidently subscribed to the idea that the tropics had a negative effect on bodies. Although most reformers would have recoiled at the mention of this theory of degeneration, this is one of the clearest expressions of the divorce of the region's natural and human potential.[42]

The idea of Indians as ineffective custodians of the land who did not realise its potential was pervasive. This dismissal of the Indian population, of course, also conveyed a message that reformers themselves had more enlightened plans for these lands and their people. As Richard Drayton and Kavita Philip have pointed out, British 'improvers' also invoked such claims that they would manage landscapes better than their current occupants as a moral justification for colonisation.[43] The intendant of Honduras, Ramón de Anguiano, pointed out the commercial advantages to be gained from establishing missions in the mountainous territory of the 'Xicaque' Indians. He cited the case of the British as an example of people who had enriched themselves by taking 'the fruits of this vast terrain' via the smuggling route of the river León to the British settlements on the Honduran coast. They had swapped 'trinkets' and items such as knives and scissors for a variety of natural products, including balsams, wax, tars, resins, vanilla, pepper, cacao, woods, and gold, 'which the Indians do not appreciate and of which there is great abundance'.[44] The description implies that all these riches were there for the taking of whomever endeavoured to exploit them, and could realise their proper potential. In another article on the commercial potential that the cultivation of flax and hemp offered to Central Americans, the *Gazeta*'s editors assured their readers that seeds were readily available in the capital city, Nueva Guatemala, 'thanks to the Indians of San Lucas and of Parrámos who have since time immemorial sown linseeds'. In this narrative, Indian villages might have possessed the seeds but did not realise their potential, since they 'did not know another purpose for it than selling it to apothecaries for a moderate price'. [45] The *Gazeta* article, by contrast, recommended the cultivation of flax for the purpose of spinning and weaving linen

---

[41] 'De Chiapa', *Gazeta*, Vol. 2, no. 91 (10 December 1798), 327; 'Ciudad Real 8 de Diciembre', *Gazeta*, Vol. 2, no. 94 (31 December 1798), 308.

[42] As discussed in Chapter 4, the *Gazeta*'s editors and many readers roundly dismissed these ideas. Creole scientists and philosophers such as Francisco Caldas in New Granada and Hipólito Unanue developed nuanced counter-arguments to these European theories (although Caldas did not reject it outright): Thurner, *History's Peru*, 92; Mauricio Nieto, Paola Castaño, and Diana Ojeda, "'El influjo del clima sobre los seres organizados" y la retórica ilustrada en el *Semanario del Nuevo Reyno de Granada*', *Historia Crítica* 30 (2005): 91–114.

[43] Drayton, *Nature's Government*, 50–55, 229–38; Philip, *Civilizing Natures*, 259–61.

[44] 'El Gobernador Yntendente de Honduras cuenta á VSS de un nuevo Proyecto p.a reducir á la Fé á los Yndios Xicakes, estableciendo con ellos el comercio p.a utilidad de la Rl Hacienda. Comayagua 1er de Julio de 1798', AGI, Guatemala, 587.

[45] 'Suplemento. Agricultura', *Gazeta*, Vol. 2, no. 55 (2 April 1798).

fabrics in addition to the more traditional use of producing linseed oil. The suitability of the Guatemalan soil for the cultivation of flax was not questioned, nor did the authors doubt that the indigenous inhabitants could grow the plant: they were, after all, already harvesting it. However, they saw large-scale cultivation for the production of fabrics as the desired outcome of this branch of agriculture, in order to eventually be less reliant on British textiles. Indians were, in their view, lamentably not working towards this most useful outcome.

The notion of unrealised potential sometimes even translated into viewing Spaniards and Creoles as complicit in the neglect of useful landscapes. An early edition of the *Gazeta* had encapsulated the editors' at least initially disdainful view of the population as unappreciative of the region's natural wealth with an invitation to all '*Guatemalenses*' to 'travel through these fertile lands with me, which you do not deserve to inhabit' while a 'multitude' of commercial avenues and natural riches remained unknown.[46] John Browning has argued that the exiled Guatemalan Jesuit Rafael Landívar's 1782 poem, *Rusticatio Mexicana*, also included an endorsement of the nascent policies of free trade and economic reform in his description of the natural history and landscapes of Mexico and Guatemala. Browning sees in Landívar's exhortation to his fellow Guatemalans to 'learn to esteem highly your fertile lands' and to 'abandon old ideas and adopt the new', a coded message about the political responsibility that came with such landscapes rather than mere literary reminiscence.[47] When the Spanish *Gazeta de Bayona* had reported on the waterproofing material found in the Orizaba province of Mexico, the author was not particularly concerned about whether it came from trees or volcanic stone. What worried him most was that the 'inhabitants did not pay the slightest attention' to such a useful product.[48] The author disdainfully commented that in Mexico, 'everyone [only] thinks about gold and silver', when there were other lucrative and useful trades to be done with the products of the land. All residents seem to stand accused here of mismanaging the natural potential. Since the *Gazeta de Guatemala* then discussed the uses of this resin over several issues, the authors and editors seemed to take the accusation seriously. They were, as always, attentive to any suggestions of unexploited natural riches that appropriate management of resources might turn into profitable economic activity. The *Gazeta* contributor who had so disparaged Indians' work ethic spoke of Guatemala as a land in which nature itself effortlessly sprung forth precious objects that might enrich the country, but at one point blamed all inhabitants for not appreciating this:

---

[46] 'Comercio', *Gazeta*, Vol. 1, no. 9 (10 April 1797), 67.
[47] Browning, 'Rafael Landivar's *Rusticatio Mexicana*', 22–6.
[48] 'Goma elastica', *Gazeta*, Vol. 7, no. 318 (29 August 1803), 339. This article is briefly discussed at the start of Chapter 4.

'no human hand has bothered to pick them up'.[49] Nature itself was Central America's greatest asset. The population was a more uncertain factor.

In Search of Useful Labourers

In contrast to the almost unanimous support for the idea that Central American landscapes were abundant, reformers had differing, sometimes pessimistic, views about the workers on whom the production of agricultural wealth would inevitably rest. When Creoles and Spaniards in Guatemala spoke of 'labourers', they often referred to Indians, who, as the majority of the population, provided the bulk of the agricultural workforce. Several accounts painted a picture of rural Central America as desolate of both labourers and agricultural productivity, to devastating economic effect. Interpretations of the cause of this decline varied. While some were preoccupied with population statistics and believed that there simply were not enough men available to make Central America prosperous, others questioned the personal qualities of available labourers. There might have been enough men in the population, but were they the kind of men who were also 'useful to the state'? As we saw in Chapter 2, reformers generally delegated the question of the labour force that would implement agricultural reform to priests and individual landowners, who they hoped would recruit Indians to work on the land. Large-scale ambitions for transforming landscapes, however, could not so easily ignore this question. The question of productive landscapes therefore became wrapped up in demographic, moral, and social as well as geographical and economic arguments.

Some reformers approached this question of the perceived rural decline in a moralising, but compassionate way. The *Gazeta*'s editors were inspired by Society member and enlightened scholar José Antonio Liendo y Goicoechea's treatise on vagrancy when they agreed that rural labourers were prone to abandoning their work, leading to desolation in rural Central America. However, they blamed absentee landholders for this situation. These landholders lived in the city but profited from the work of 'hundreds' of 'hungry' rural labourers without taking care of them, leading to a system in which neither agriculture nor industry could flourish. The authors argued that it was not surprising that agricultural labourers should tire of their work and abandon it if it did not pay an adequate wage, ending up in a life of vagrancy and vice.[50] As

---

[49] 'Apuntamientos estadisticos del Br. Talcamáhida, sobre la agricultura, industria y comercio de este reyno', *Gazeta*, Vol. 7, no. 313 (25 July 1803), 297.

[50] 'De la mendicidad', *Gazeta* Vol. 2, no. 51 (5 March 1798), 17–18; José Antonio de Liendo y Goicoechea, *Memoria sobre los medios de destruir la mendicidad y de socorrer los verdaderos pobres de esta capital* (Guatemala City: Ignacio Beteta, 1797); Dym, 'Conceiving Central America', 113–15.

discussed in Chapters 2 and 4, reformers generally represented a relatively egalitarian point of view regarding Indians' role in society and the Spanish economy, but their plans were often marked by paternalistic attitudes and an ambiguous relationship with systems of labour such as the *repartimiento* draft, which many saw as inevitable for economic productivity. There is plenty of truth in Francisco Sánchez-Blanco's argument that Economic Societies' local efforts to cultivate new plants and introduce new branches of industry were ultimately a mere distraction from the more profound problem of social inequalities.[51] Some Economic Society members proposed to reform Indians' status as a way of improving Central America's economy. Society member and canon of Guatemala City's cathedral, Antonio García Redondo, composed an influential treatise on the matter of cacao cultivation in 1798 which, as Christophe Belaubre explains, was one of the first serious proposals of agrarian reform beyond the current exploitative system.[52] By 1811, the Consulado de Comercio also proposed a reform of Indian participation in Central American economy. They denounced the common practice of calling Indians 'lazy' and envisaged giving them secure titles to their land to make them 'useful farmers' and encourage their full participation in the Guatemalan economy.[53] These remained mere proposals, but show that the issue of land rights was not entirely lost on reformers when they spoke of the factors that would transform Central American nature into ever more productive land.

The majority of responses, however, swept the question of working conditions and land ownership aside. Instead, reformers' often stereotypical views of Indian labourers were compounded by fears that a population decline might be to blame for the perceived economic decline of the late eighteenth century, that is, that a lack of available labourers led to potentially lucrative land being given over to wilderness. Spanish commentators persistently feared that population numbers might decline, even when census figures suggested population growth: as Gabriel Paquette has noted, a growing population in early modern political thought was considered a sign of good governance and a healthy economy.[54] The idea that Spanish America was under-populated was a persistent argument on both sides of the Atlantic. While contemporaries in

---

[51] Francisco Sánchez-Blanco, *El Absolutismo y las Luces en el reinado de Carlos III* (Madrid: Marcial Pons, 2002), 158.

[52] García Redondo, *Memoria sobre el fomento*; Belaubre, 'Lectura crítica'.

[53] Mario Rodríguez, *The Cádiz Experiment in Central America, 1808 to 1826* (Berkeley, CA: University of California Press, 1978), 25–7; Jorge Luján Muñoz, 'Guatemala en las Cortes de Cádiz', in Luján Muñoz and Herrarte (eds.), *Historia general de Guatemala* Vol. III (Guatemala: Asociación de Amigos del País, 1995), 409–18. For a recent analysis of the documents included in the instructions for Guatemala's representative at the Cortes, see Tania Sagastume Paiz, 'Las propuestas ilustradas sobre la propiedad corporativa, 1750–1811', *Estudios* 59, tercera época (2014), 41–70.

[54] Paquette, *Enlightenment, Governance, and Reform*, 63.

other European countries invoked 'Black Legend' histories to demonstrate how a combination of Spanish cruelty and emigration had decimated America's population, Spanish American authors also feared that a population decline was taking place, but were more likely to blame illness and inefficient agriculture as its root causes.[55] More efficient agriculture and a healthy population of the kind that the Economic Society was working towards might therefore provide a solution. The Guatemalan Economic Society's founding documents even pre-empted concerns about the lack of labourers by suggesting that a rational and enlightened approach to the countryside would render the 'problem' of labourers less pressing. They declared learned knowledge and technology a way of coping even with a reduced labour force. In a newly reformed agricultural economy, 'five men would be enough where in other circumstances five hundred would not be', because technological innovation and a scientific understanding of the world would increase agricultural yields.[56] They were, for instance, very interested in new machines for processing rice that had been developed in North America.[57] In this vein, they combined theories of population growth with the agricultural ideology of British improvers, such as Arthur Young or Lord Kames.[58]

Some Central American reformers were sceptical of the assumption that recent economic decline was linked to a declining population at all. While they admitted that Central America appeared to be less populous now than it had been around the time of the conquest, the *Gazeta* reported that the population had actually increased in recent memory: there were now more Spaniards, Creoles, and Indian tributaries than in 1740. Not content to base their sugges-tions on an unfounded fear of population decline, the editors felt they were in a position to correct false statistics. Like the geographical authors the news-paper had previously maligned, the *Gazeta* thought that authors who had recorded Central American population statistics relied on outdated estimates, which they repeated 'like stupid sheep'.[59] Consequently, the editors requested a list of the number of Indian tributaries directly from the Guatemala City government, and, in 1802, the *Gazeta* published an estimate of the total population of the kingdom based on recent parish registers, provided at least

---

[55] Adam Warren, *Medicine and Politics in Colonial Peru: Population Growth and the Bourbon Reforms* (University of Pittsburgh Press, 2010); Gabriel Paquette, 'The Image of Imperial Spain in British Political Thought, 1750–1800', *Bulletin of Spanish Studies* 81, no. 2 (2004): 187–214, at 188–96; Thomas Robert Malthus, *An Essay on the Principle of Population, as it Affects the Future Improvement of Society* (London: Johnson, 1798), 101–5, 109; William Godwin, *Of Population. An Enquiry Concerning the Power of Increase in the Numbers of Mankind* (London: Longman, 1820), book 1, chapter 8, citing Robertson and Voltaire.

[56] AGI, Estado, 48, 1794, quoted in Belzunegui Ormázabal, *Pensamiento Económico*, 210.

[57] 'Del Arroz', *Gazeta*, Vol. 8, no. 369 (22 October 1804), 477.

[58] Mokyr, *Enlightened Economy*, 171–97.

[59] 'Poblacion', *Gazeta*, Vol. 6, no. 286 (22 November 1802), 301–3.

in part by Society members such as the former bishop of Chiapas, Goicoechea. Their calculation of the total population came to just under one million inhabitants: more than previous estimates had shown.[60]

The question of the labour force was therefore not simply a game of absolute numbers. Reformers now engaged more closely with the fear that potential labourers might exist, but were not currently 'useful' members of the economy. The *Gazeta* asked about the number of 'useful men' in each province as part of the chorographic reports, and was concerned with the availability of '*brazos utiles*', that is, useful men ready to work. Although some provinces which were considered depopulated, such as Chiapas, were described as being in a state of 'misery', there was now an optimistic undercurrent that promised that 'opulence' was instead possible.[61] Political economists across the empire made similar arguments. The authors of the *Semanario del Nuevo Reyno de Granada* blamed their province's economic woes on the lack of useful, healthy, and productive labourers.[62] Newly created 'useful' citizens would be the basis for a more prosperous nation. A *Gazeta* reader from Yucatán suggested that 'savage' Indians could be made into 'men' through education, increasing a province's useful population.[63] The Guatemalan Economic Society's often short-lived projects to provide schools of drawing, mathematics, and spinning and weaving to the working population were examples of this, as was their design to make Indians consumers of Spanish-style clothes and shoes. In the realm of agriculture, useful workers would combine with useful landscapes to create new agricultural and trading settlements.

Trujillo's Useful Settlers

Nowhere were 'useful men' more needed than on the Atlantic and Pacific coasts, where a productive population would be key to realising the agricultural and mercantile potential which reformers saw. Traditionally, Spanish military commanders and politicians had seen areas such as the Nicaraguan coast mainly as geopolitical linchpins, and projects of the late eighteenth century to establish colonies on the Mosquitia or 'Miskito' coast effectively represented attempts to establish Spanish influence on an indigenous region dominated by

---

[60] AGCA, A1, Leg. 6073, Exp. 54876; *Gazeta*, Vol. 6, no. 286 (22 November 1802), 301–2; *Gazeta*, Vol. 6, no. 261 (31 May 1802), 132.

[61] 'Ciudad Real de Chiapa 7 de Abril', *Gazeta*, Vol. 2, no. 59 (30 April 1798), 96.

[62] Shafer, *Economic Societies*, 81–2; Mauricio Nieto, Paola Castaño, and Diana Ojeda, 'Ilustración y orden social: el problema de la población en el *Semanario del Nuevo Reyno de Granada* (1808–1810)', *Revista de Indias* 65 (2005): 683–708, at 696–9.

[63] 'De la educación en Yucatán', *Gazeta*, Vol. 6, no. 273 (23 August 1802), 201. See also Julia Varela Fernández, 'La educación ilustrada o como fabricar sujetos dóciles y útiles', *Revista de educación* (1988): 245–74.

British trading settlements.[64] Agricultural settlers were also imagined as key to holding Spanish frontier spaces elsewhere, such as the empire's southernmost borderlands of the Araucanía.[65] However, the reformers' new projects of creating useful labourers in Central America were directed at the interior rather than exterior of the kingdom. Reformers imagined that new settlers in coastal towns would foment agriculture and form legal trading relationships with the interior. In the sense that different factions within the Audiencia feared coastal towns such as Trujillo on the Atlantic coast as hotbeds of contraband, and legal trade would prevent this, these coastal towns formed their own sort of border-lands. Traditionally, Spanish towns had been established in Central America's interior, and the temperate climate of the central highlands as well as the constant threat of British attacks on the coast had maintained that status quo for centuries. The Caribbean coasts, where British settlements at Belize and the Bay Islands provided a base for privateers and pirates, were traditionally the most exposed, but even the Pacific coasts were also not immune. There was therefore some element of defence strategy in these projects, alongside well-founded fears of merchants' precious wares being stolen. The *Gazeta*'s reports about a British ship robbing all of the Pacific port of Acajutla's indigo in a dramatic attack was testament to that.[66] In the early 1790s, before the establishment of the Economic Society, a Spanish official had made similar arguments for the establishment of coastal colonies within the framework of the Bourbon reforms, influenced by the Spanish minister José de Campillo's 1743 treatise on governance. The governor of Sonsonate, Antonio López de Peñalver y Alcalá, focused on creating 'useful vassals' in new settlements on Central America's Atlantic and Caribbean coasts, by settling them with 'gypsies', 'vagrants', and 'Indians'. This project was rejected by the Minister of the Indies and remained hypothetical.[67] Nevertheless, it may have formed a blueprint for the efforts of the Economic Society and Consulado in the 1790s to develop increasingly ambitious plans in this vein for settling both

---

[64] Martínez de Salinas, *La colonización de la costa*, 47–54 and 111–46.

[65] Natalia Gándara Chacana, 'Representaciones de un territorio. La frontera mapuche en los proyectos ilustrados del Reino de Chile en la segunda mitad del siglo XVIII', *Historia Crítica* 59 (2016): 61–80, at 74–6.

[66] José Domas y Valle to the king, 8 January 1799, AGI, Guatemala, 888. The *Gazeta* later reported a story about the daring escape of the Spaniards who had been taken prisoner in Acajutla and other ports that the ship had plundered: 'Carta escrita en Lima por sugeto conocido de esta Ciudad', *Gazeta*, Vol. 4, no. 164 (7 July 1800), 290.

[67] 'Sobre medios de poblar las costas de América', Nueva Guatemala, 1792, AGI, Estado, 48, N.1. fols. 3v–20, 58. Spanish ideas about incorporating Indians into the state in this way are detailed in David Weber, *Bárbaros: Spaniards and their Savages in the Age of Enlightenment* (New Haven, CT and London: Yale University Press, 2005), 181–6; Josefina Cintrón Tirykian, 'Campillo's Pragmatic New System: A Mercantile and Utilitarian Approach to Indian Reform in Spanish Colonies of the Eighteenth Century', *History of Political Economy* 10, no. 2 (1978): 233–57.

coasts. The reformers' projects, however, at their core focused on agriculture and benefits to internal Guatemalan trade.

Reformers were particularly interested in the Honduran port-town of Trujillo. Re-established as a new Spanish settler colony after a period of abandonment in 1787, this town regularly featured in the *Gazeta de Guatemala* as the place with the Economic Society's most active *junta de correspondencia*, and as an allegedly promising site for agriculture. However, throughout the late colonial period Trujillo was not self-sustaining, but remained dependent on food imports from Spain and even the United States via Havana. Its population alongside the land was at the heart of reformers' considerations. The population had increased significantly in the 1790s as a result of the arrival of over 2000 free black people. In addition to 1000 colonists originally from peninsular Spain and the Canary Islands, around 780 Africans and Haitians who had fought with the Haitian revolution, then exiled by the French in 1795, were brought to Central America by the Spanish. Some of them remained in Trujillo permanently. Their numbers were swelled by around 1,700 Caribs originally from the British island of St Vincent. Opinions of these new settlers in the *Gazeta* ranged from suspicion of these 'foreigners', who might be collaborating with the nearby independent Miskitu nation, to pragmatism about their potential to be useful in an otherwise under-populated town.[68] One reformer to argue against the idea of Trujillo's potential prosperity was Juan Bautista Irisarri, the powerful merchant. His arguments against Trujillo were transparently an attempt to focus resources on the Pacific trade routes he favoured because of his mercantile interests. Indeed, he argued that Trujillo should be closed and asked for permission to settle the Pacific harbour of Acajutla instead with some of Trujillo's population of black Caribs (a plan that was approved in 1797, but we can assume that it did not bear fruit, since Irisarri was still looking for new settlers for Acajutla in 1802).[69] Aside from Irisarri's self-interested scheming to move populations across the entire Audiencia, his arguments about Trujillo also clearly spell out the Bourbon reform era's vision of what 'useful' colonists might look like. Firstly, Irisarri admitted that some of Trujillo's existing colonists were 'useful' to the state, but to him they were people in the wrong place. Labourers would only count as

---

[68] Elizet Payne Iglesias, *El puerto de Truxillo*: Un viaje hacia su melancólico abandono (Tegucigalpa: Editorial Guaymuras, 2007), 99–102 and 120–1; Nancie L. González, *Sojourners of the Caribbean: Ethnogenesis and Ethnohistory of the Garífuna* (Urbana, IL: University of Illinois Press, 1988), 39–72; Juan Manuel Aguilar and Sergio Antonio Palacios, *La Ciudad de Trujillo. Guia Historico Turística* (3rd ed., Tegucigalpa: Instituto Hondureño de Antropología e Historia, 2003), 35–8. 'Nueva Guatemala 26 de junio', *Gazeta*, Vol. 1, no. 21 (26 June 1797), 164–8. The Caribs had been deported from St Vincent by the British to the island of Roatán off the Honduran coast, which the Spanish retook from the British in 1797, and 'resettled' some of this free black population to Trujillo. The 'resettlement' is described as voluntary in the *Gazeta* and by some historians, as forced by others.

[69] 'Poblacion del puerto de Acajutla', *Gazeta*, Vol. 6, no. 246 (15 February 1802), 37–8; Dym, *From Sovereign Villages*, 41.

useful in the correct geographical context. Because Trujillo was so ill-connected with the rest of Central America, nothing good could come of their presence. After all, most of the northern coast to him was empty hinterland, where the king barely had 300 vassals in an area of 6400 square leagues. Even though there were also 1000 colonists who had originally come to Trujillo from the Canary Islands, Galicia, and Asturias, these people would be more useful if they served the country back home on the peninsula. Other Catholic settlers from Europe would also not be suitable: he dismissed the idea of German or Swiss settlers because it was 'not suitable to situate them next to the sea'. Expressing suspicion of the Honduran and Nicaraguan coast's independent Miskitu, he suggested that black settlers (*los Negros*) could potentially succeed, but only 'if they felt like it [*quando les diera la gana*]'. Men, their customs, and the specific location as well as the type of place they lived in were all connected to their potential for utility.[70]

The development of reformers' agricultural reform ambitions was not a simple question of superficially adding infrastructure to the landscape in the form of a harbour or road, drawing lines on maps. Successful projects would have to consider the entire cultural landscape involving land, agriculture, geopolitics, and the motivations of settlers. In this, the reformers' local projects, ostensibly concerned with planting a few more acres of crops to make specific towns 'useful', merged with ideologies that were clearly to be exercised at a larger scale, at that of the state. Foucault's concept of 'biopower' is appropriate for describing the grand Bourbon project of marshalling the resources of the state, and seeing both land and people included in these 'resources'. Foucault's concept, of course, relies not so much on the direct exercise of state power as on the normalisation of certain behaviours, attitudes and customs within society. His contention is that in the eighteenth-century, a new physiocratic and utilitarian idea of controlling populations arose that no longer relied exclusively on regulations, but on creating a 'desire' in actors within the population to fulfil a government's priorities, in a way that they imagine is within their individual interest.[71] As Santiago Castro-Gómez explains, biopolitics in the context of Spanish Bourbon reformers meant a reconfiguration of emphasis from quantity to quality to population, which was a key issue in the writings of the economic thinkers, Bernardo Ward and the Count of Campomanes. The idea that managing the quality of the population was a question of the science of political economy was central to such strategies of governance, as one

---

[70] 'Sigue Chirimia', *Gazeta*, Vol. 5, no. 194 (9 March 1801), 412. Carlos III had established new settlements with Catholic Germans and Swiss in Andalucía: Martínez de Salinas Alonso, *Colonización de la costa*, 18.

[71] Michel Foucault, *Security, Territory, Population: Lectures at the Collège de France, 1977–1978* (New York: Picador, 2009), 69–76.

Bogotá official explained.[72] In essay contests, enlightened enthusiasts from New Granada to Guatemala invoked political-economic writings and theories. Concrete proposals, however, could be more influenced by the detailed local knowledge of land, its history, and people that reformers so valued in other contexts. It is these small-scale applications of the idea that labourers should be marshalled as 'resources' that appeared in Irisarri's arguments, but more frequently in the diverse reform projects of the Economic Society and sympathetic governors working outside of the Economic Society, like Antonio Gonzáles Mollinedo. Applying Foucault's criteria of different historical epochs' approaches to 'biopower', José Rossi y Rubí's attempts to legislate a cacao crop into existence, or individual priests and administrators overseeing Indians' harvests, discussed in Chapter 1, were part of an *ancien régime* top-down form of motivation. By contrast, Irisarri's careful weighing up of different possible population groups as having an 'interest' in the success of the coastal colonies might hint at a new eighteenth-century conception of directing not the population's actions, but their interests. These new approaches of personal motivations, mutual interest, or even *intéressement*, to use Michel Callon and Bruno Latour's term, which applies a more thoroughly negotiated mutual action, appeared in different contexts. While reformers might be able to imagine regulating the minutiae of Indian lives close to the capital, with substantial administrative and clerical presence, the new approaches of mutual interest might have been more appropriate to remote coastal places. Perhaps this was the reason that they paid special attention to potential settlers' assumed 'intrinsic' motivations towards the land.

While Irisarri's comments made it clear that many reformers saw poor black, *mestizo,* and Indian labourers as little more than pawns in reformers' agricultural and mercantile projects, the Economic Society also posited that economic development, particularly in new agricultural settlements, would be driven by self-interest. Several contributors to the *Gazeta* insisted that free settlers, rather than enslaved labourers, should be at the forefront of any projects of coastal development. For instance, one prominent author argued under the pseudonym *Don Farruco* that among the reasons that the Honduran harbour of Omoa was not more prosperous were its inhabitants: 'only' black slaves, who had no personal interest in the progress of the port's fortunes. Omoa would therefore be resigned to a continued state of 'useless wastage' which would be 'very harmful to the exchequer'.[73] It is likely that this also represented the editors' opinion, since the article occupied the title page of the *Gazeta* in 13 consecutive weeks in 1800. In

---

[72] Santiago Castro-Gómez, 'Siglo xviii: el nacimiento de la biopolítica', *Tabula Rasa* 12 (2010), 31–45, at 36.
[73] 'Continua el discurso sobre Motagua', *Gazeta*, Vol. 4, no. 164 (7 July 1800), 287.

his sentiments about personal interest and the cost to the exchequer regarding slaves, another reader drew on Adam Smith's definition of slavery as economically inefficient from the *Wealth of Nations* cited elsewhere in the *Gazeta*: 'The experience of all ages and nations, I believe, demonstrates the work done by slaves, though it appears to cost only their maintenance, is in the end the dearest of any. A person who can acquire no property, can have no other interest but to eat as much, and to labour as little as possible.'[74] As ever, not all Society members agreed with this approach to labour. In 1797, Juan Ortiz de Letona, writing from Trujillo as the director of the local branch of the Economic Society, suggested to his contacts in Guatemala City that cacao cultivation in Trujillo would need additional labour. Because labour in this remote area was expensive, he asked that the Society might arrange for black labourers [*negros*] to be subsidised and hired out to local plantation owners at a cheaper day rate than labourers commanded in order to create such cacao plantations. Although he did not specify whether he meant for the labourers to come from Trujillo's free black population, his idea was modelled on an arrangement in the military harbour of Omoa, where enslaved men owned by the Crown were hired out to do work on other projects than the fort they were there to build. The director of the Society declined the plan to organise labourers for the cacao trees because of its cost. He suggested that once the present war with the British was over, there might be more 'well-rested' men available for such works.[75] In the case of Omoa at least, the solution proposed by *Don Farruco* was to settle the port with free 'active and hard-working' residents, 'free men, I repeat, who are animated and sustained by interest, the desire to improve their fortune'.[76] Only such men could dare to envisage 'something impossible' – the impossible dream of a prosperous and healthy port-town.

Trujillo was a test case for the widely debated potential of the Caribbean and its agriculturally active inhabitants. In 1802, following the visit of a military inspector to the kingdom's coastal regions, the captain-general drew up new regulations for all military companies on the Caribbean coast and in Guatemala's northern region, that is Omoa, Trujillo, Izabal, and Petén. Concerns with the climate were written into all aspects of the arrangements concerning the troops. The document acknowledged that 'the climate does not permit tiring out the men', presumably by expecting as much physical strain from them as their counterparts in milder lands. Caribbean garrisons also

---

[74] 'Cajón de maulero', *Gazeta*, Vol. 6, no. 253 (5 April 1802), 82, citing (in Spanish translation) Adam Smith, *An Inquiry Into the Nature and Causes of the Wealth of Nations*, Volume I (2nd ed., London: Strahan and Cadell, 1778), 471; see also Thomas Hopkins, 'Adam Smith on American Economic Development and the Future of the European Atlantic Empires', in Reinert and Røge, *Political Economy*, 53–75, at 63–4 on Smith and slaves in colonies.

[75] 'Calificacion de la calidad del cacao nuevamente plantado en Trujillo', 9 February–20 June 1797, AGCA, A1.6, Leg. 2006, Exp. 13811.

[76] 'Continua el discurso sobre Motagua', *Gazeta*, Vol. 4, no. 164 (7 July 1800), 287.

received better salaries and pensions than battalions in other parts of the region, since it would 'make their stay in such unhealthy climates more tolerable'. But with the acknowledgement that these were hardship postings also came an obligation. The favourable conditions were also an incentive 'that they should dedicate themselves to agriculture, and that they may therefore take care of their plantains, and the fields on the lands which should be distributed amongst them ... In this way, these men will be doubly useful to the state'.[77] The regulations for these battalions therefore effectively declared that soldiers sent to the coasts to man the fortifications should behave as agricultural settlers. Like most military matters, these rules were formally signed off by the king, but the text itself made reference to the captain-general in Guatemala City laying out these rules. This locally informed influence was reflected in the detailed stipulations for how the soldiers should behave in the specific environment of the Caribbean, and that the regulations repeated concerns about the climate long held by regional administrators and reformers. The captain-general who created these regulations was Antonio Gonzáles Mollinedo, who would later insist on the Economic Society's re-establishment. Unsurprisingly, this military document oozed with reformers' agricultural aims. The regulations of 1802 were followed up in 1803 by detailed explanations (again dictated by Gonzáles) on the distribution of lands to the members of said military companies, suggesting that this was a project that commanders at least tried to implement. Soldiers could choose their preferred plot near the fort, and could keep it as long as they cultivated it. If a soldier were to neglect his agricultural duties, his land would be taken away 'in order to give it to one of his comrades who is more industrious'. Soldiers would be able to ask their commander for help if they needed seeds to get this project off the ground, 'for instance yams, potatoes, rice, and so on'. In addition to agriculture, soldiers should also tend to chickens and cows 'in their houses and barracks'.[78] This wording eliminated any distinction between military and agricultural spaces, confirming the dual mission of these Spanish fortifications. All these detailed regulations followed on from concerns that reformers of the Consulado and Economic Society had listed as theoretical obstacles to the development of coastal settlements. This time, they proposed concrete ways to overcome these difficulties through careful planning and the manpower of the military.

[77] 'Reglamento de las Compañías fijas de Omoa, el Golfo, Trugillo, fuerte de S Carlos, y Peten, según el Plan formado por el Sr Subinspector general de la tropas de este reyno Brigadier D Roque Abarca, y aprobado por SM en Real Orden de 5 de Octubre de 1802', AGCA, A1.1, Leg. 6091, Exp. 55306.

[78] 'Reglas dictadas por esta capitania general en 21 de Mayo de 1803', AGCA, A1.1 Leg. 6091, Exp. 55306. The Spanish colonists of Louisiana in the 1770s also had a dual military and civilian settler role, but their agricultural role was envisaged as following their military activity: Martínez de Salinas, *Colonización de la costa*, 20.

These high-flying plans ultimately ended in failure. The salaries and careful regulations do not appear to have done much to attract settler-soldiers to the coast, and the reputation of these coastal settlements did not improve significantly. The original regulations had specified that the garrisons should be filled according to the imperial state's hierarchy of *castas*. Commanders of the garrisons should be 'whites' (*blancos*), or, if positions could not be filled, commanders should at least be '*gente de buen color*' – 'of good colour', from as high a *casta* as possible. Regular soldiers at Omoa and Trujillo could be drawn from the *mulato* and free black residents of those towns, but if there were not sufficient men, 'vagrants' (*vagos mal entretenidos*) would also do, as would 'delinquents' convicted for minor crimes. Echoing the vision of Bourbon reformers and Economic Society members who hoped to convert 'vagrants' into 'useful' settlers, men were often rounded up specifically to be co-opted as paid and unpaid workers. With the justification of reducing crime and disorder in the new city of Guatemala, the government ordered in 1799 that unemployed and vagrant city-dwellers should be imprisoned for eight days for being lazy. If they 're-offended' after being released, depending on the district judge's estimation of their character, they could be handed over directly to landholders or city residents to be put to work.[79] It appears that the garrisons continued to be understaffed, partly because of a high mortality rate in the tropical climate that reformers' optimistic rhetoric could not wish away. In 1804, two years after the first set of regulations was printed, Gonzáles Mollinedo wrote to a provincial governor asking him to communicate a request to local judges.[80] It was very important, he noted, that judges enter in the record when a prisoner was incarcerated only for minor crimes. Since the letter came with a copy of the instructions for coastal garrisons attached, we can assume that the captain-general considered these prisoners who were incarcerated for 'vagrancy, or a small crime' good candidates for working in coastal fortifications. It indicated that the manpower destined for the coastal settlements continued to be a challenge, hampering the Economic Society and Guatemalan government's optimistic plans. As late as 1815, three Indian men from Chichicastenango received four or five year sentences for rebelling against the governor of Totonicapán, the maximum sentences for these crimes, to be served in Omoa and Trujillo.[81]

Neither the government nor the Economic Society gave up on the potential of Trujillo to become a healthy agricultural paradise. Juan Ortiz de Letona, the Economic Society's correspondent in that town who oversaw the

---

[79] 'El dia 17 se publicó el Bando de tenor siguiente', *Gazeta*, Vol. 3, no. 123 (23 September 1799), 110.
[80] AGCA, A1.1, Leg. 6091, Exp. 55306, fol. 18.
[81] Coralia Gutiérrez Álvarez, 'Racismo y sociedad en la crisis del Imperio español: El caso de los pueblos del Altiplano Occidental de Guatemala', in Herrera and Webre, *La época colonial*, 265.

acclimatisation projects of Ramírez's Sumatran plants, made sure that his colleagues in Guatemala City were kept well informed of the province's potential. Yet the unwavering optimism of the Economic Society and its newspaper was questionable. The success stories of the Society finding wild cacao and the cultivation of Ramírez's Sumatran rice nearby, which the newspaper touted as a minor agricultural triumph in 1803, were certainly the exception rather than the rule. It remained dependent on food imports. Its colonists complained until at least the early 1800s about the quality of the land they had been given, and many were still waiting for the farms that they had been promised. To some extent, this was probably due to colonists from the Canary Islands and Spain trying to grow plants that were simply not suited to the tropical climate. Some of the free black settlers who had arrived from British and French Caribbean islands in the 1790s had more luck as fishermen.[82] Yuca and other roots also grew well: by 1803, José Rossi y Rubi contributed specific information to the *Gazeta* concerning the type of yuca that settlers on the coast cultivated.[83] Given the role of enslaved Africans in spreading knowledge about rice-growing technologies in North America (specifically the Carolinas), it is also worth considering that the knowledge of settlers of African descent may have been a factor in making rice cultivation in Trujillo a success alongside their expertise in cultivating yuca.[84] After all, the *Gazeta* specifically commented that when it came to rice production, Honduran cultivators would do well to take Carolina as an example. The editors believed that the mills used in Valencia (as suggested as a model by their Spanish correspondents) would be inadequate for American rice, and looked to 'La Carolina' for models instead.[85] It seems that with some understanding of the local climate, settlers were in a position to make agricultural progress, and given the success stories around rice and yuca, the efforts of Juan Ortiz de Letona to acclimatise plants and encourage agriculture seem worthwhile. However, his attempts to create a flourishing agricultural colony look quixotic in light of the lengths he went to protect his attempts at creating a botanical garden. In 1802, two female colonists of Trujillo started legal proceedings against Ortiz de Letona, accusing him of killing their cows that were grazing on common land. The treasury official clearly considered himself above the law here. Even so, his defence of his actions was remarkable: he argued that the cows simply had to go because they disrupted and threatened his agricultural projects.[86] To him, the means justified the enlightened end: a blunt expression of the argument used by

---

[82] Payne Iglesias, *Puerto de Truxillo*, 214–21.
[83] 'Cultivo Beneficio de la Yuca Dulce', *Gazeta*, Vol. 7, no. 290 (14 February 1803), 17–19.
[84] Judith Carney, *Black Rice: The African Origins of Rice Cultivation in the Americas* (Cambridge, MA: Harvard University Press, 2001).
[85] 'Del cultivo y exportacion de arroz', *Gazeta*, Vol. 8, no. 367 (8 October 1804), 462.
[86] Payne Iglesias, *Puerto de Truxillo*, 116.

colonisers across the early modern world, that a superior management of a landscape through 'improvement' could justify violence and theft, bringing claims about reformers' superior management of landscapes to their perhaps inevitable conclusion. The violent struggles involved in planting crops in this frontier town rather betrayed the optimistic rhetoric about Trujillo's potential.

Improving Climates

Rather than merely having the goal of settlers eking out an existence of subsistence agriculture, reformers now pursued the more ambitious goal of a positive transformation of landscapes taking place in wider areas, beyond roads and harbours. When provincial merchants as well as the *Gazeta de Guatemala* held up the Caribbean harbour of Trujillo as areas of future progress, they emphasised the supposed potential of their agricultural hinterland as much as their mercantile possibilities. Imbued with optimism about the potential of the land, they sought to transform parts of the country that were previously considered peripheral or even desolate. Even aside from the question of who should populate these new settlements, however, the offhand tone in which reformers promised that untold riches would pour forth from the soil hid severe concerns with the climate of Guatemala's coasts. One of Juan Bautista Irisarri's more forceful arguments against maintaining settlements on the Caribbean coast concerned climate. He conceded that Trujillo's natural resources might appear abundant, but doubted that its 'putrid fevers' and 'deadly climate' [*climas mortiferos*] could be overcome in the long term, a situation which he contrasted with the southern coasts' 'healthy' climate.[87] Irisarri's scepticism about northern harbours was plainly informed by his trading interests, but in this case he voiced an uncomfortable truth.

Trujillo was widely considered an unhealthy place where tropical fevers raged, and even proponents of plans to increase the harbour's use for maritime trade acknowledged this. Settlers would be in mortal danger. At one point, the editors of the *Gazeta* tried to argue with this perception on the basis that the mortality rate in the hospital of Trujillo was actually lower than that of hospitals in Guatemala City or Madrid, but even so, they had to concede that Trujillo was an unhealthy environment. 'But so are all places on the coast of America', they shrugged in the harbour's defence.[88] The most positive comment they could make about Trujillo was apparently that death anywhere in the Caribbean, not just in Trujillo in particular, was inevitable. As the *Gazeta* drily remarked, the most common remedy against the tropical fevers that these environments

---

[87] 'Chirimia por el Sur', *Gazeta*, Vol. 5, no. 193 (2 March 1801), 408–20. Chirimia was Irisarri's alias in the *Gazeta*.
[88] 'Nota', *Gazeta*, Vol. 1, no. 35 (2 October 1797), 278–9.

provoked was 'the excessive use of liquor'. Rum was indeed the only 'medicine' listed on a summary of items sent to Trujillo in March 1793.[89] Reformers nevertheless sought to find solutions against the destructive *temperamento*, the climate or particular nature, of the Honduran coast. Since the land and its cultivation by humans were considered the basis for economic wealth, reformers sought out the potential of humans to improve landscapes. Climate had strong implications for humans, but human actions were also considered to influence climate. Climate needed to be controlled if the 'useful' settlers envisaged by reformers were to survive, and agriculture and trade were to thrive.

Deforestation was seen as the best solution for getting rid of the overly humid climate. The governor of Sonsonate, who proposed to 'settle the coasts of America' in 1792, for instance, was convinced that once the surplus of trees was cut down, the sun's rays would help to 'evaporate the humidity of the endless shade, and the rotting leaves, so that the air may penetrate, and purify'. Then 'beans, wheat, maize, rice, potato, yucca, vegetables, and cotton' could be planted, and only six months later, he assured the reader, 'plenty of crops will be obtained, capable of sustaining [the settlers], and of extracting the surplus to the islands and port-towns of these Kingdoms'. He listed flax as a potential crop, to produce linen that would challenge the hegemony of textiles produced by 'foreigners', that is, the British. He estimated that it would take about half a century for Guatemala to become self-sufficient in its linen consumption, 'and not need foreigners'. The simple act of establishing an agricultural colony by clearing the dense tropical forest would therefore have advantages for the kingdom's position more generally, and supply materials for consumption within the country as well as for export.[90]

Another swampy, forested place that reformers fixated on was the harbour of Omoa on the Caribbean shore of Honduras. Omoa was generally acknowledged to be an unhealthy place. President Domás y Valle pointed to specific problems with its geographical location: swampy mangroves in a windward position that produced 'putrefaction', as well as a sheltered position which meant that the prevailing winds could not 'ventilate' the harbour, making it 'one of the most deadly of the world'.[91] Problematically, it was also 'distant from the oversight of the government, so separated from travel routes and communication with the capital'. Yet, this distance and unhealthy situation could be overcome. The *Gazeta* argued that 'the root cause' of the harbour's problems was 'not, as some persuade themselves, the unhealthiness of its climate', because this state

---

[89] 'Continua el discurso sobre Motagua', *Gazeta*, Vol. 4, no. 162 (23 June 1800), 272. Sugarcane *aguardiente* sent to Trujillo: AGCA, A.3, Leg. 1012, Exp. 18614.

[90] 'Sobre medios de poblar las costas de América', Nueva Guatemala de la Asunción, 1792, AGI, Estado, 48, N.1. fols. 2–9.

[91] Domás y Valle, Guatemala, 3 November 1799, AGI, Guatemala, 706, fol. 3.

could be partially resolved with fewer trees which would lead to more ventilation. Since 'the quality [of unhealthiness] is reduced in proportion with the deforestation which is done in its environs', there was yet hope for the harbour given some simple interventions.[92] Omoa, Trujillo, and other key locations of the Caribbean could be made liveable.

The warehouses that stored import and export goods at the inland harbour of Izabal, too, were subject to considerations about the relationship between ventilation, forests, and health. In 1805, a royal order approved the transfer of the warehouses to a new site and the construction of a new road to them, as well as plans for a settlement of fifty people. While the old warehouses were by all accounts in need of renovation, their removal to a new site was also an act of controlling the landscape.[93] The decision to traverse the countryside and establish a transport route had merely been the first step: to make it permanent and let a Spanish settlement take root, a deeper intervention in the landscape was necessary. The Consulado spent the next few years trying to improve the initially provisional new storehouses. The engraved map accompanying an 1808 report showed a plan of the new settlement and the new road leading to it (Figure 5.1). It touted the advantages of the new location over the old warehouses, and projected a convincing image of nature tamed by architecture. The document was created by the engineer Juan Bautista Jauregui, and printed by the engraver Pedro Garci Aguirre, who had also made Rossi y Rubí's road through the mountains into a dignified formal document. The explanation printed underneath the plan spoke of the fortuitous location of the village and storehouses: 'They are located on a ventilated peak; to their east there is a great forested plain which is advantageous for the purpose of constructing them as formal buildings that are regular, comfortable, and beautiful.' The neat plan, an Enlightenment dream of a well-ventilated, grid-patterned settlement, was as of yet unfortunately pure fiction. It was published in order to promote the enterprise among merchants and officials, and raise funds, a masterpiece in wishful thinking. The Consulado seemed convinced that their project would be successful, but the small print underneath the map made it clear it was merely a 'proposal'.

Notably, the Consulado also assumed that the new, healthier, and more orderly location would have no trouble attracting settlers, simply because of the 'advantages that the place [terreno] offered' in terms of its agricultural potential, and the possibilities for cattle-holding. They wanted to mirror the move of the storehouses to a better location with an improvement in personnel. Bernardo Sanchez, the current administrator of the storehouses, had been convicted twice already of smuggling contraband. His appointment eleven years previously was supposed

[92] 'Continua el discurso sobre Motagua', *Gazeta*, Vol. 4, no. 164 (7 July 1800), 287.
[93] 'Cartas del Presidente Antonio Gonzales', 1809, AGI, Guatemala, 625.

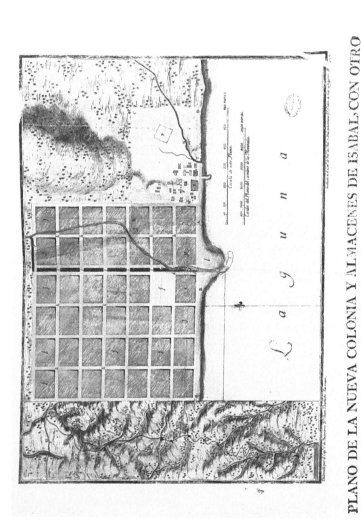

PLANO DE LA NUEVA COLONIA Y ALMACENES DE ISABAL, CON OTRO

de su camino en la montaña hasta el rancho del Mico, levantados de orden del M. Y. S. Presidente D. Antonio Gon-
zalez, à solicitud del Real Consulado de Guatemala para manifestar en ellos el estado en que se hallan, è indicar las me-
joras de que son susceptibles con lo demás que se ha estimado conducente para perficcionar è ilustrar el Proyecto de
dicho Cuerpo.

Figure 5.1 Juan Bautista Jauregui, 'Plano de la nueva colonia y almacenes de Ysabal con
otro de su camino en la montaña hasta el rancho del Mico (golfo de Honduras). Delineado
por el Cap.n de Yngenieros Juan Baptista Jauregui', 1807. Ministerio de Cultura y Deporte.
Archivo General de Indias. MP-Guatemala, 306(1).

to be temporary, but it seemed that no replacement had been found. The merchants were sure that the 'lethal climate of the old storehouses' [*el mortífero clima*] had driven the appointment of such an unsuitable 'common man'.[94] Once again, the specific attributes of a geographical location were bound up with the health as well as moral qualities of the people tasked with its maintenance. The unhealthy reputation of the old storehouses had repercussions for the economic life of the country, preventing the proper administration of the harbour, but the Consulado was working to change this. After all, as late as 1786, the harbour at Izabal had been an administrative no-man's-land, still at the beginning of being integrated into the Audiencia's formal and legal trading routes. A chapel run by Guatemala City's Dominican order, established on inherited land near the harbour, had become the main place to unload and store all of Guatemala's commercial wares as they arrived or departed from Spain or Cuba. When the merchants complained about the Dominicans overcharging them for landing and storage, the king intervened and ordered the establishment of a customs-house on the coast in 1786, with the Crown's administrator finally appointed in 1807. New storehouses in a healthy place were expected to solve this problem of inadequate oversight instantly, allowing the imperial administration to appoint 'a suitable subject with good qualifications' to the post. They did not entirely solve the problem, since imperial administrators, merchants, and Dominicans all continued to wrangle over the right to charge for the storage of goods at Izabal.[95] In addition, the location remained vulnerable to attacks: one panicked commentator feared that it would only take a few invading British traders to turn the Motagua 'into the Thames'.[96] Nevertheless, the physical move and transformation of the storage site to become more healthy was part of a conscious effort (albeit by different factions with their own financial interests) to connect the harbour more closely with the interior. Remoteness conditioned by unhealthy surroundings was therefore tied up in differing political visions of how the Atlantic coast should be integrated into the political economy of the Audiencia.

By 1812, the imaginary had at least partly become reality. The new location did not please everyone, and the new road was not completed to a satisfactory standard, as a later Consulado report noted. However, one party to the lengthy dispute was happy to hear that even one of the plan's detractors thought that 'the climate had improved somewhat' with the move. He admitted that 'it got better in the sense that since the move one does not hear about people dying

[94] Real Consulado, *Informe aprobado por la Junta*, 36.
[95] 'Expediente instruido por el Consulado de Guatemala', Antonio Gonzales, 17 December 1809, AGI, Guatemala, 625.
[96] 'Informe, que por acuerdo de la Junta de Gobierno del Real Consulado dieron, su Consiliario Dn José Munilla, y Tesorero Dn Rafael Trulló, sobre la navegacion del Rio Motagua en 14 de Febrero de 1817', 4, AGI, Guatemala, 892.

daily as they did at the previous [store houses]. Instead, they die in the natural proportion with respect to the population in which people die everywhere.'[97] Dying at regular rather than alarming frequencies was, in this case, a real compliment for Izabal. The superiority of the new site could be explained by reference to its airs: the new site was 'more ventilated and bathed in sunlight than the old one', which had been 'surrounded by heavy mountains and swamps'.[98] The carefully researched intervention in the merchants' cultural landscape had been a success. The Consulado and Economic Society also worried about the climate of the surrounding region. Swamps and waters, of course, were closely linked to disease, but the nearby Motagua river that merchants would partially travel along on the way to the new warehouses had fortunately been anecdotally confirmed in its healthiness. The fishermen who lived in the village of Gualán and spent two months a year in close contact with its waters allegedly did not 'experience diseases'. The twenty-nine men who participated in the 1792 reconnaissance with engineer Porta y Costas, too, stayed healthy after the expedition, despite being exposed to rains.[99] Such empirical evidence was as good as any for determining a place's climatic properties. After the 1773 earthquake, too, engineers prospecting for a new site had been told to look for 'old men' in towns – if they had survived, it was probably a healthy site. These promising signs meant that if the old dream of making the whole Motagua navigable could be made reality, it would be a much healthier alternative to dealing with any of the land routes leading to the storehouses: a healthy river would lead to a ventilated town, paving the way for a renaissance of Central American maritime trade.

This understanding of the interactions between humans, trees, and climate voiced by reformers built on a long tradition of different theories regarding the effect of forests on the surrounding air and its humidity. Scholars and officials were worried about forests because the shade from the trees prevented ventilation and they were consequently associated with humidity and swampy land. This in turn created a link to human health. A causal connection between forested, damp land and unhealthy environments went back to at least the sixteenth century, while neo-Hippocratic theories of the relationship between man, air, water, and place were popular from Spain to Scotland.[100] Colonial officials in British India likewise equated the drainage of marshes, the clearing of wasteland and jungle with creating a disease-free environment, although, as

---

[97] 'Informe sobre la navegacion del Rio Motagua en 14 de Febrero de 1817', 36–7, AGI, Guatemala, 892.

[98] Real Consulado, *Informe aprobado por la Jun*, 37.

[99] Porta y Costas, *Relacion del reconocimiento que ... practico el ingeniero ordinario D Antonio Porta*, 16.

[100] Grove, *Green Imperialism*, 154–5; Fernando Rodrigo, 'The climate of Granada (southern Spain) during the first third of the 18th century (1706–1730) according to documentary sources', Draft paper for Open Access Journal *Climate Past. Discussions* (2018), 1–2.

Richard Grove has shown, there were some colonial contexts that associated deforestation with decline.[101] Proposals which linked trees to the quality of the air and indeed the entire 'economy of nature' had been made by practitioners of pneumatic chemistry in 1770s Europe, based on the work of Stephen Hales and John Priestley. The development of eudiometry, that is measuring the 'virtue' or 'breathability' of air, was particularly closely linked to attempts to regulate natural environments such as swamps and forests in order to improve the health of surrounding populations.[102] In the *Gazeta*'s Great Plantain Debate of 1798, one contributor well-versed in current debates around 'fixed air' had already raised the possibility that trees might in fact be healthy. Another contributor to the *Gazeta* from Mexico also insisted in 1801 that planting trees was beneficial for economic and practical reasons (providing fruits, timber, firewood, shade), but also aesthetic ones (leafiness). He dismissed arguments that trees would prevent healthy airs from circulating as irrelevant, since, as he pointed out, they need not be planted closely together. He even invoked some environmental and weather-related benefits: trees would give 'humidity to the atmosphere, mitigate the heat, calm the winds, moderate the cold'.[103] However, most Central American officials and merchants continued to believe that trees and their associated humidity were generally detrimental to human health.

Such theories fed into enlightened concerns with sources of clean air and public hygiene. Alexander von Humboldt, during his Mexican travels, was also concerned with similar matters of the effect of agriculture and human involvement on the natural environment. Environmental historian Richard Grove observes that Humboldt examined Spanish historical documentation of Lake Valencia in Venezuela in order 'to demonstrate that human population shifts and ensuing deforestation had been directly responsible for the changes in lake level', an activity which Gregory Cushman explains in modern terms as 'the discovery of human-caused climate change in South America'.[104] For the eighteenth century, however, these worries about Trujillo's climate can also be related to concerns within Spanish American society about the interplay of

---

[101] Arnold, *Tropics and the Travelling Gaze*, 48; Jonsson, *Enlightenment's Frontier*, 80; Grove, *Green Imperialism*, especially 158–67 and 301–8.

[102] Kenneth Thompson, 'Trees as a Theme in Medical Geography and Public Health', *Bulletin of the New York Academy of Medicine* 54, no. 5 (1978): 517–31; Simon Schaffer, 'Measuring Virtue: Eudiometry, Enlightenment, and Pneumatic Medicine', in Andrew Cunningham and Roger French (eds.), *The Medical Enlightenment of the Eighteenth Century* (Cambridge University Press, 1990), 281–318, at 300–18; Emma Spary, *Utopia's Garden. French Natural History from Old Regime to Revolution* (Chicago, IL and London: University of Chicago Press, 2000), 138–46.

[103] 'Del plantío de árboles. México, 27 Mayo 1801, Lic. Mateo Zorrilla', *Gazeta*, Vol. 5, no. 224 (17 September 1801), 566–7 and 227 (28 September 1801), 578–9.

[104] Grove, *Green Imperialism*, 364–8; Gregory Cushman, 'Humboldtian Science, Creole Meteorology, and the Discovery of Human-Caused Climate Change in South America', *Osiris* 26, no. 1 (2011): 16–44, at 32.

vegetation, air, and a region's specific natural 'character', debates which developed even without recourse to precision instruments of the sort that Humboldt or his host in Bogotá, Francisco Caldas, commanded. Humboldt and the Guatemalans, of course, differed in their opinions on whether a drier climate was a better or worse one. Among Central American reformers (with perhaps the exception of its Mexican contributor cited above) there was little evidence of the conservationism of trees either on environmental grounds or even in the interest of resource management that scholars have observed in some other early modern contexts. The Central American government was not blind to resource management, for instance determining a closed fishing season, and issuing a prohibition against fishing with the aid of the *barbasco* plant poison, because this would kill young fish.[105] On the matter of trees, however, there was a broad consensus that their removal would signify a drier climate, and therefore improvement.

By the late 1790s, the *Gazeta* brought another aspect to the debate about Trujillo's climate. The editors unearthed promising historical information going even further back that raised the possibility that despite its much-maligned environment, Trujillo's unhealthiness was not innate. They took it on good authority that Trujillo had once been 'marvellous in its fertility and freshness', that it had been remarkably healthy before being settled by Spaniards, who committed 'some notable errors in the planting of this new town in the last third of the [eighteenth] century'.[106] The wrong kind of agriculture and human action had led to its deterioration. Here they probably referred to fruitless attempts by Spaniards to grow crops such as maize and wheat in the tropical climate, and on stony soil which was completely unsuited to such plants.[107] Alongside the diseases that befell farmers and soldiers in this hot and humid climate, the initial failure of agriculture contributed to an understanding of Trujillo, Omoa, and other Caribbean ports as 'unhealthy'. The *Gazeta's* editor, however, made clear that he considered the errors committed by humans, rather than the place at fault here: 'What had been the fault of men was attributed to the climate.'[108] José Celestino Mutis in New Granada and José Alzate in New Spain believed that any cause of decline and decadence in America since the conquest had been the 'error' of men.[109] Those, however, were mismanaged landscapes on a grand scale. On the more manageable scale

[105] 'Doña Maria Josefa Soliz sobre que se le permita poder pescar en el Rio de Gualan', 1803, AGCA, A1, Leg. 177, Exp. 3622.

[106] 'Noticias historicas de la antigua Truxillo', *Gazeta*, Vol. 6, no. 284 (8 November 1802), 285–6.

[107] Manuel Rubio Sánchez, *Historia del Puerto de Trujillo*, Vol. II (Tegucigalpa: Banco Central de Honduras, 2000), 319; Payne Iglesias, *Puerto de Truxillo*, 121. For the significance of the long history of failed Spanish attempts to grow European crops, see Earle, *Body of the Conquistador*, 67–78.

[108] 'Noticias historicas de la antigua Truxillo', Gazeta, Vol. 6, no. 284 (8 November 1802), 285–6.

[109] Peset, 'Ciencia e independencia en la América española', 213.

of just one Honduran town, the editor tried to come up with a concrete plan for putting the newspaper's theoretical musings about the relationship between history and geography into action. Expanding its remit from records of specific natural disasters and harvest to more lofty predictions on climate, the editor (likely Ignacio Beteta) was convinced that he had stumbled upon Trujillo's true identity in historical documents. If human intervention could destroy a healthy climate, it could also restore it:

May that colony [Trujillo] return to the state in which it was when it was destroyed by the Dutch [in the seventeenth century]: let the mountains be made transitable, where no human foot has penetrated since that time: let the evergreen woods be removed, whose tops are always covered by dense vapours, with which the atmosphere is suffused, and which will produce more or less palpable effects on the bodies exposed to contact with the airs, for reasons of their quantity, humidity, and heat; finally Trujillo should be made a settlement in the manner, as it should be, open, well ventilated, with a regular police, and a traffic such as it had in the two previous centuries, and then its climate will be without a doubt more healthy than those of Havana or Vera Cruz.[110]

In this analysis, while the treetops and their 'vapours' contributed to the unhealthiness of the region, cutting down these woods was only part of the solution. The trees' vapour would disappear, with the implication that the new drier town would be a healthier one. Alongside being a 'well ventilated' town, hosting as much commercial traffic as in the previous centuries would also contribute to its being a healthy place. The introduction of commercial traffic would presumably play a role because, as Rossi y Rubí had also learnt, only well-used roads stayed clear of trees. Climate could therefore be altered in a general sense by deforestation and agriculture, as British observers in Scotland and North America also agreed.[111] Historical and botanical evidence therefore pointed to Trujillo's climate having been caused by a mismanaged landscape, rather than being sickly by nature. Crucially, Central American reformers considered these processes reversible. If it had become dangerous through 'the fault of men', it could be remedied through the work of men. Beteta here subscribed to a current of thought common to many environmental thinkers of the era. Ravi Rajan interprets the idea that 'human environmental misdeeds could be undone' as an attempt to ensure the 'continued generation of wealth through cultivation', a symbiotic philosophy of resource management that resonates with the work of Guatemala's reformers.[112]

The 'impossible dreams' of creating thriving trading towns on the Caribbean coast may have been ambitious, but editors and contributors to the *Gazeta* certainly took them seriously. Echoes of the French Enlightenment's concern

---

[110] 'Trujillo 4 de Septiembre', *Gazeta*, Vol. 1, no. 35 (2 October 1797), 279.
[111] Jonsson, *Enlightenment's Frontier*, 48–9, 73–81.
[112] S. Ravi Rajan, *Modernizing Nature: Forestry and Imperial Eco-Development 1800–1950* (Oxford University Press, 2006), 34.

with the interactions of climate and men were apparent in various *Gazeta* articles in which contributors suggested improvements to Central American nature. One article entitled 'alterations of the climate' in particular referenced Montesquieu's take on the importance of man's interventions in nature, citing Egypt and China as shining examples of areas which man made habitable and useful through irrigation and drainage.

> Climate [*el clima*] is a compound of powers and influences, which plants and animals, alongside all things that breathe, contribute to fostering in its corresponding seasons: and man is situated in it, as sovereign of the earth, to change it through artifice. Ever since he stole fire from the heavens, and made steel obedient to his hand: since he subjected not just animals, but also other men to his will, and these as well as plants he changed for his purpose; he has contributed in various ways to the changing of climates [*temperamentos*]. – In former times Europe was thickly forested, almost impenetrable, as were other regions which are now cultivated. Exposed to the rays of the sun, its habitants have also varied with the climate. The surface of Egypt would have always remained covered by the waters and silt of the Nile, if not for the art and industry of man, who has gained that country from the waters, acclimatising ... to this type of artificial surroundings [*accomodandose á esta especie de clima artificial*]. ... Time will let us see the extent of his prowess in this land; and whether we can, which would not be a great miracle ... convert Omoa into a leisure garden and Trujillo into a place of convalescence for the elderly, which it is said it was in ancient times.[113]

Omoa and Trujillo's fame 'in ancient times', drawn from a variety of sources including Beteta's archival documents, would be the foundation of their environmental transformation in the future. Ideas about historical change of climates through cultivation had also led European observers to assume that Canada's climate in the eighteenth century was equivalent to that of Germany or France in ancient times. Scottish observers used this assumption to argue that climate could be improved by 'the historical force of civilised settlement'. They also argued about the causes of climate variability within a country.[114] Guatemalan reformers, too, believed in the improvability of climate through settlement. Rather than getting bogged down in questions about the variability of climes across a whole continent, their vision of an improved climatic future was rather more specific. As the *Gazeta's* own epistemology demanded, they used local archives to understand the site's past, and sought to understand the particular micro-geography of a settlement to see where previous attempts at transformation had gone wrong. The French writer Jean-Baptiste Moheau had declared in 1778 that 'it is up to the government to change the air temperature and to improve the climate', laying out a new vision of the relationship between

---

[113] For Montesquieu, see Clarence Glacken, *Traces on the Rhodian Shore: Nature and Culture in Western Thought from Ancient Times to the End of the Eighteenth Century* (Berkeley, CA: University of California Press, 1976), 577.

[114] Jonsson, *Enlightenment's Frontier*, 74–7.

governance, society, and climate.[115] While the reformers were not 'a government', their comments about the future settlers of places they wanted to exert environmental control over show that their agricultural actions and internal colonisation projects were, however, an important vision of governance.

Drawing on this idea of the improvability and reversability of landscape and climate interventions, reformers justified a whole array of projects by claiming that the transformation of Guatemalan landscapes was not simply an improvement of the current situation, but would actually restitute a previous, more prosperous state of affairs, a semi-mythical golden age. Rossi y Rubí, the governor who built the ill-fated road through the mountains of his province in 1800, declared that he had been 'fortunate to find fragments of older roads which followed this same route, when along it there existed innumerable villages which have gradually got lost'.[116] For Rossi, this was a clear sign that through his work, older trade routes could be restored. Material remains within the landscape helped to confirm future possibilities, as they had done at Palenque, alongside archival documents. When the merchant Irisarri surveyed the disused Iztapa harbour as part of his efforts to open up trade on the Audiencia's Pacific coast, he was also following historical models. He thought that his actions would 'discover the old harbour of Guatemala', since Iztapa was famous for being the site where the sixteenth-century *conquistador* Pedro de Alvarado had constructed a small fleet to travel to Peru in the hope of gaining a slice of its famed riches.[117] Further trading expeditions from this harbour by the Economic Society also drew the arc from Alvarado to contemporary reformers.[118] Irisarri was clearly conscious of the harbour's significance in conquest history when he judged Iztapa as being capable of hosting great trading expeditions. He expressed surprise that the natural harbour (now largely silted up) formed by the Michatoya river's mouth had been abandoned, since he judged it superior even to the important Acajutla or Sonsonate harbour. He suggested that perhaps this was because of an 'accidental introduction [*introducción accidental*]' of the waters of the Los Esclavos river into the Michatoya river.[119] This once again suggested that an 'accident', an unintended situation, had led to the abandonment of a natural harbour in the past. It might even have been caused by men, and therefore might be reversed, for instance by building a canal. Ultimately, Irisarri changed his mind and favoured Acajutla in the Guatemalan province of Sonsonate over Iztapa, but his blithe reaction to the

---

[115] Cited in Migletti and Morgan, 'Introduction', 7–8.
[116] José Rossi y Rubí, Malzatenango, 10 May 1800, AGCA, A1.21, Leg. 207, Exp. 4171, fol. 5v, 45.
[117] 'Comercio del Sur', 1798, AGCA, A1, Leg. 169, Exp. 3421, fol. 1.
[118] Real Sociedad, *Octava junta pública*, 25. Manuel Rubio Sánchez, 'Puerto de Iztapa o de la Independencia. primera parte', *Antropología e Historia de Guatemala* 8, no. 2 (1956): 24–49, at 28–30.
[119] 'Comercio del Sur', 1798, AGCA, A1, Leg. 169, Exp. 3421, fols. 6v–7.

sandbanks that he encountered at the latter (enough to make local mariners doubt the feasibility of the project) was typical of the reformers' optimistic belief in the improvability and perfectability of places, buoyed by historical precedent.

Conclusion

The productions generated by the soil, and how to reap profit from them, took centre stage in the late eighteenth century. The Economic Society and the Consulado focused especially on plans to transform 'wilderness' into economically useful spaces. Such notions of 'wilderness' were culturally defined and included landscapes such as the Motagua valley and mountainous regions, which were inhabited by indigenous people rather than being the blank canvas that reformers saw. Beyond building roads, reformers now pursued a deeper transformation of sites. Ideas about humans' power to transform and manage landscapes and their airs went hand in hand with the Economic Society's search for the bountiful qualities of the land, with concrete suggestions of how an inhospitable environment could be altered and improved to generate profit for the colony and its merchants. Although Spanish colonial officials had taken the idea of human settlements as contingent upon natural disasters such as earthquakes into account for a long time, the reformers' new ambitions to make use of broader swathes of the countryside means they were confronted with new questions. Their relationship with nature and landscape was an active one, and marked by a temporal as well as a spatial and intellectual dimension. Reformers and merchants considered the effects which settlers or agriculture had had on the landscape in the past, and the potential of humans to impact and change the landscape to their economic advantage in the future. Reading archives was an essential component of placing the current landscape within a historical continuum, a narrative that could be one of progress, but was not necessarily linear.

The 'climate' of a place could be influenced by human actors, from Economic Society members to soldiers stationed on the coast and mandated to become part-time farmers. If lands had been mismanaged by previous generations, measures such as deforestation or drainage might improve them, and allow the building of new harbours and roads in places hitherto considered uninhabitable. New settlers would make places liveable, healthy, and productive. When reformers considered the 'problem' of bringing people into the landscapes to be reformed, they expected these labourers to carry out the physical work of transformation through managing vegetation along roads, agriculture, and trade. Rossi y Rubí's daring road through the mountains, narrated initially as a feat of engineering, turned out rather to be a project of managing Indian labour from different villages that Rossi emphasised his leadership in. The Economic Society, Consulado and government's designs for the management of people grew increasingly sophisticated, and connected to the particular attributes of places and environments.

The *Gazeta* and Economic Society's reform designs nevertheless tended to emphasise questions of landscapes' and climates' potential over questions of its inhabitants. Perhaps as a direct consequence of land disassociated as far as possible from questions of labour and politics, eighteenth-century reformers' ideas proved remarkably resilient to the political changes of the nineteenth, as Chapter 6 will show. The only occasionally clearly defined and localised *patria* that reformers imagined in the eighteenth century would, of course, find its uses in the political ideologies of independence, as would the optimistic idea that future prosperity was built into the new nation's geography. For instance, this belief would resonate in the statesman José del Valle's expectation that Guatemala already possessed 'all principles of prosperity to an eminent degree': the country already contained its own promise of future prosperity.[120] However, the connection between the colonial and independent period was also far more practical, relying on the banal, yet powerful connections between landscapes past and present through specific projects of road building, agriculture, and most importantly geography.

---

[120] José Cecilio del Valle, *Discurso del presidente del Poder Executivo a la apertura del Congreso Federal de Guatemala. En 25 de Febrero de 1825* (Guatemala City: Imprenta Nueva de Juan José de Arévalo, 1825), 31.

6   Independence and Useful Nature

Independence from Spain in 1821 did not signify the end of 'enlightened reform' as it had been interpreted in Central America for the previous three decades. The idea of bettering landscapes, rather than merely believing in their bountiful fertility, had become the narrative of the *patria*. Working for such betterment was what defined self-anointed patriots both before and after independence, even though the political context had been transformed. Both the colonial and the independent period accorded a central status to the management of landscapes, with a new, more international set of actors involved in debates in the independent era. The 'useful' landscape of enlightened reformers had become a part of Central American identity, and contributed to the geographical imagination that formed the foundation of the independent nation-states of Central America, but also its interactions with foreign governments. The legacies of colonial debates around landscapes were particularly evident in Guatemala, now an autonomous state and part of the Federal Republic of Central America, proclaimed in 1824 after a short-lived union with Agustín Iturbide's Mexican empire.[1] Led by a re-established Economic Society under its director José Cecilio del Valle, and supported by a sympathetic government under Mariano Galvez (1831–8), political discourses also continued to revolve around some of the topics that Bourbon-era reformers had emphasised: the country's natural wealth, its privileged geopolitical position between two oceans, and the use of scientific knowledge, especially geographical knowledge, in governance and nation building.

Independence-era statesmen drew only 'selectively' on Bourbon ideologies, as Jordana Dym has stressed. After all, the political context was now profoundly different.[2] However, independence was not a clean break with the colonial past either. A number of scholars have discussed ways in which the new republics of Latin America, including Central America, were shaped by the political and social structures of the Bourbon period, and explored the

---

[1] This new political entity included Guatemala, El Salvador, Honduras, Nicaragua, and Costa Rica, but not Chiapas, which had declared independence from Guatemala in 1823 and joined Mexico in 1824.

[2] Dym, *From Sovereign Villages*, 260.

significant intellectual continuities which existed in the first half of the nineteenth century.[3] In Guatemala City, the former capital of the Audiencia de
Guatemala and an important centre of power within the independent federation, the relatively peaceful transition between the colonial and republican
periods helped reformers to maintain some institutional and personal continuities in the realm of scientific knowledge in particular. The declaration of
independence on 15 September 1821 was merely the beginning of a lengthy
process of answering questions about the future of the region.[4] Historians
have rightly emphasised the political turbulence of the independence period.
While the declaration itself happened without any bloodshed in Guatemala
City and there were no confrontations with Spanish forces, it triggered
a period of civil war and political upheaval which was not fully settled until
the middle of the nineteenth century. Nevertheless, in the seventeen years
between Central America's declaration of independence from Spain in 1821
and the disintegration of the Federation of Central American Republics in
1838, there were periods of peace and relative stability in which the governments of the federation and individual nations set about the business of
creating the structures of stable nation-states.

Central America did, to paraphrase Jorge Cañizares-Esguerra, develop
a scientific self-assuredness which pre-dated political autonomy in the
Bourbon period.[5] Beyond such broad sentiments, the concrete ways in which
the scientific imagination of the late Bourbon period influenced the construction of new nations are less well understood. Stuart McCook has pointed out
that the first institutions to sponsor national scientific surveys in the nineteenth
century were 'leftovers' from the colonial period: Economic Societies.[6]
However, the continued role of the Economic Society in Guatemala at least
constituted a conscious choice by subsequent associations of scholars and
governments to re-establish an institution that they saw as particularly suited
to solving the republic's problems, and one that was mostly exempt from the

---

[3] Matthew Brown, 'Enlightened Reform after Independence: Simón Bolívar's Bolivian
Constitution', in Paquette, *Enlightened Reform*; Dym, *From Sovereign Villages*;
Jaime Rodríguez, *"We Are Now the True Spaniards." Sovereignty, Revolution, Independence,
and the Emergence of the Federal Republic of Mexico, 1808–1824* (Stanford University Press,
2012); Del Castillo, *Crafting A Republic*.

[4] See for instance: Rodríguez, *Cádiz Experiment*, 147–237; Ralph Lee Woodward, 'The Aftermath
of Independence, 1821–c.1870' in Leslie Bethell (ed.), *Central America since Independence*
(Cambridge University Press, 1991), 1–36; Luján Muñoz and Zilbermann de Luján, *Historia
general de Guatemala* Vol. III, 419–52; Julio César Pinto Sória, *Centroamérica, de la colonia al
estado nacional: 1800–1840* (Guatemala: Editorial Universitaria de Guatemala, 1989);
Andrés Townsend Ezcurra, *Las provincias unidas de Centroamérica: fundación de la
república* (San José: Editorial Costa Rica, 1973).

[5] Cañizares-Esguerra, *Nature, Empire and Nation*, 127.

[6] Stuart McCook, *States of Nature. Science, Agriculture and Environment in the Spanish
Caribbean, 1760–1940* (Austin, TX: University of Texas Press, 2002), 16–17.

criticism levelled at other institutions of the colonial state. In this, the independent country's reformers did not just draw on vague ideologies of the Bourbon period, but used its archives, communication networks, and project blueprints in practical ways.

Projects around the management of landscapes established both in colonial and independent times now travelled beyond Central America's borders. The Economic Society's colonial-era insistence on framing Central America as a place of transnational connections, open to foreign scientific expertise even as it was translated for local circumstances, was a rhetorical stance that was equally applicable to the newly international context of the independent period. Guatemala's foreign relations saw a dramatic change with emancipation from Spain. Independence brought foreign capital, but also an influx of people – foreign investors, speculators, and adventurers – who contributed to redefining Central America's relationship with other countries: through foreign (mainly British) loans and European colonisation projects, but also in scientific contexts. Although this meant that on the one hand natural-historical and geographical projects were shaped by foreign influence, on the other Spanish colonial documents as well as new surveys commissioned by Guatemala's independent government came to play a key role in new transatlantic exchanges of information. Rather ironically given the disdain British representatives heaped on knowledge produced in a Spanish American context, or that of a successor state, they profited from cartographic material in the Spanish government's old archives now. New political contexts meant that the established use of the sciences and scientific networks to frame Guatemala's place within Central America and against the wider world gathered new significance.

Inheriting Enlightenment

Like independent governments in other parts of the Americas, Central Americans set up a stark rhetorical contrast between themselves and the supposed 'dark ages' of the colonial period. This resulted in the paradox of statesmen renouncing in the strongest terms not just the politics of the Spanish state, but also its scientific and social policies, despite often building on them. Although the belief that Enlightenment and learning formed the basis of economic progress had roots in the ideologies of the Bourbon period, Guatemalans now reframed the idea of enlightened reform as a departure from colonial times. The lawyer and statesman José Cecilio del Valle argued shortly after the establishment of the Guatemalan Republic, in 1824, that the Spanish government had felt threatened by the idea of an enlightened America. The Spanish government had 'feared' American competition on an economic and intellectual level, he explained, because 'enlightenment and wealth would give [Americans] knowledge of their rights and power to maintain them'.[7] Valle, one of the intellectual authors of Central American independence, more than any other individual embodied the belief in the power of enlightened learning and science to bring progress and prosperity to a country. Just as late eighteenth-century reformers had done, he lamented the idea of a badly educated population and declared his intention to remedy the situation, swearing to 'procure the Enlightenment' of the people. He even proclaimed that he would worship the sciences as deities if this were not heresy.[8] The Gálvez government of the early 1830s also lamented that the revolution had destroyed many of the country's educational establishments, and put the lack of intellectual life down to one major adversary: the Spanish government, 'enemy of enlightenment and of the New World's prosperity'. By contrast, the government now proposed the creation of a republican Academy of Sciences: 'the first monument to the sciences in this beautiful country'.[9]

In many ways, Valle and the Gálvez government levelled arguments against a straw-man here. As Lina del Castillo has pointed out in the case of independent Colombia, the protestations against the intellectual backwardness of the former Spanish government often served contemporary political aims rather than reflecting the context in which colonial scientists like José de Caldas in New Granada or José Felipe Flores in Guatemala had worked.[10] In Guatemala, after all, the Academy replaced the University of San Carlos, an illustrious institution whose scholars and students had engaged at least to some extent with the works of European enlightenment philosophers and scientists. (Incidentally, the Academy reverted back to the name of San Carlos in 1840.) Indeed, Guatemala's leader Mariano Gálvez himself was a graduate of San Carlos. Biographers have further linked Valle's scientific interests to the intellectual climate in which he received his education in Guatemala City, at San Carlos, from the enlightened thinker Liendo y Goicoechea among others. Jacobo de Villaurrutia and Antonio García Redondo, key members of the Society, were his friends and mentors, and he had acted as one of the *Gazeta de Guatemala*'s censors from 1805.[11] The 1820s and 1830s certainly

---

[7] José Cecilio del Valle, *Discurso del gobierno supremo de Guatemala sobre la renta de tabacos. Leido en la Asamblea el día 11 de Octubre de 1824* (Guatemala City: Arévalo, 1824), 3. See also Alejandro Gómez, *José del Valle: el político de la independencia centroamericana* (Guatemala: Universidad Francisco Marroquín, 2011).

[8] José Luis Reyes, *Apuntes para una monografía de la Sociedad Económica de Amigos del País* (Guatemala City: Editorial José de Pineda Ibarra, 1964), 194–5; José Cecilio del Valle, *Memoria sobre la educación* (Guatemala City: Imprenta de la Union, 1829), Foreword.

[9] Gobierno de Guatemala, *Boletín Extraordinario: Decreto Septiembre 30 de 1832* (Guatemala City: Imprenta de la Union, 1832), unpag.

[10] Del Castillo, *Crafting a Republic*, 50–1.

[11] José Edgardo Cal Montoya, 'El discurso historiográfico de la Sociedad Económica de Amigos del Estado de Guatemala en la primera mitad del siglo XIX', *Anuario de Estudios Centroamericanos* 30, no. 1 (2004): 87–117, at 91, 97; Louis Bumgartner, *José del Valle of Central America* (Durham, NC: Duke University Press, 1963), 22; Jorge Mario García

brought new policies, for instance plans to roll out education projects for 'all classes'.[12]

José Mariano Gonzáles, the director of the new Academy of Sciences, seemed well aware that there might be an exception to the tirades against the backwardness of the Spanish period. In a speech, he listed the scholars whom he did consider important within Guatemala's learned history, including Antonio Liendo y Goicoechea, 'the censor of our first Economic Society'. He also singled out the colonial Economic Society for having 'illustrious prelates' as its correspondents, and quoted a letter to the editors of the 1799 *Gazeta de Guatemala* ('nuestra gazeta') verbatim because its description of patriotism appealed to him.[13] The focus on learned friars and bishops was perhaps more acceptable than any other aspects of the Spanish colonial state. The Declaration of Central American Independence had specifically pointed to 'the usefulness of the clergy in maintaining harmony within the nation'.[14] The Academy of Sciences also modelled the monthly periodical which it briefly published in 1835–6, the *Mensual de conocimientos utiles*, on both the colonial *Gazeta de Guatemala*'s own mission statement regarding the need to spread enlightened knowledge to an un-enlightened population.[15] Even more clearly, with its distinctly scientific focus, it revived the format of the Economic Society's short-lived *Periódico de la Sociedad Económica de Guatemala* (1815–16), which had represented a top-down attempt at the dissemination of useful information. The *Mensual*, with its earnest articles on topics from population statistics to geography, continued a tradition of newspapers for the dissemination of useful and enlightened knowledge, with the stated aim of exploring the country's 'sources of wealth'.[16]

In addition, the independent state's liberal reformers were also apparently happy to bracket out the Economic Society and its legacy of institutions and publications from the general disdain in which they held the colonial period. Despite the stark rhetoric disavowing the pre-independence era, they made room for key ideas from the Economic Society's ambit. This was not as paradoxical as it might seem. As Jordana Dym explained, processes of transformation that had started in the Bourbon era were still in motion in the 1820s and 1830s.[17] After all, the Economic Society had styled itself as a more progressive entity than the broader Spanish government at key points in its history. It was commonplace even during the colonial era to denounce

Laguardia, *José del Valle, ilustración y liberalismo en Centroamérica* (Tegucigalpa: Departamento Editorial de la UNAH, 1982).

[12] Gobierno, *Boletín Extraordinario*.   [13] *Ibid.*   [14] Rodríguez, *Cádiz Experiment*, 149.

[15] Blake Pattridge, *Institution Building and State Formation in Nineteenth-Century Latin America*: The University of San Carlos, Guatemala (New York: Peter Lang, 2004), 128–30.

[16] Sociedad Económica, *Prospecto del Mensual de la Sociedad Economica*, 3.

[17] Dym, *From Sovereign Villages*, 260.

the past as less enlightened. In the case of the late eighteenth century, this meant contrasting the current, 'enlightened' Bourbon period under Charles III and IV with the supposedly backward past of the seventeenth and early eighteenth centuries. An early editorial in the *Gazeta*'s first year, for instance, painted an unflattering picture of Spain under the later Habsburgs, but added that since Spain's Bourbon government had dismantled some of the old barriers to trade, it was now America's moment to make use of its potential.[18] After the *Cortes* period in particular, which led to a reassessment of the relationship between the Americas and the Crown, these attacks on Spain's intellectual policies continued, but no longer exempted the Bourbon period from criticism. One member of the Economic Society grumbled after 1811 that specific members of the Spanish court who 'could not bear American Enlightenment' had been responsible for the Economic Society's 1800 suspension.[19] Such declarations of change and renewal even within the colonial Society may have made it easier for its heirs to square its rhetoric with the priorities of the independent state.

Institutions such as the Economic Society and Consulado were swiftly re-established and explicitly drew on the work of their colonial predecessors, even while they also consciously put the stamp of a new political era on their projects. Immediately after independence in 1821, a group of educated men including members of the state's legislative and judiciary bodies founded a patriotic society (*tertulia patriótica*) in Guatemala City to discuss Central American economic and intellectual development. Their proposal to re-establish the Economic Society with José Cecilio del Valle as its director was put into practice in 1829.[20] Members were quick to draw a line under narrow comparisons between the old and the new Society. Its statutes certainly would have to be re-written, they argued, because 'the old ones approved on 21 October 1795 are absolutely not adaptable to our circumstances'.[21] A comparison of the 1795 and 1830 statutes, however, reveals more similarities than differences. Differences, of course, included new patronage and the ultimate purpose of the wealth they imagined to be creating: the old Society's aim to contribute to the 'prosperity of the monarchy' (*bien de la monarquía*) became the new society's 'desire for the general prosperity of the State, and the people [*pueblos*] that it consists of'. The colonial institution had its ambitions of prosperity carefully prescribed: it would act within those matters and trades

---

[18] 'Comercio', *Gazeta*, Vol. 1, no. 9 (10 April 1797), 67.   [19] Shafer, *Economic Societies*, 217.
[20] *Ibid.*, 360; Manuel Rubio Sánchez, *Historia de la Sociedad Económica de Amigos del País* (Guatemala City: Editorial Académica Centroamericana, 1981), 17–19.
[21] Bernardo Escobar to the General Secretary, 13 December 1825, AGCA, B92.4, Leg. 1394, Exp 32288, fol. 4. Localised *tertulias patrioticas* existed in various Central American cities through-out the 1820s, their main aim apparently to raise money for city and army coffers: AGCA, B92.4, Leg. 1394.

[*ramos*] that were 'compatible with the Metropolis', detailing in another document that it was most interested in those agricultural productions that Spain did not have in its own soil.[22] The independent Society had no such restrictions. However, this suggests a reframing of the political landmarks that gave the Society legitimacy rather than its core aims. Through the Economic Society's projects, Guatemalan reformers of the independence period doubled down on the idea that knowledge about the geography and natural productions of the land could cement the political and economic fortunes of their country.

The new Economic Society was established with the motto '*El zelo unido produce la abundancia*' (united zeal produces abundance). Abundance was represented on its seal through a cornucopia, but the idea of 'zeal' was just as important as any providential imagination of the national space for its connotations of patriotism. The idea of united zeal could already be found in the colonial Economic Society's insistence that a patriot would always work for 'the common good' and stood against the 'egotism' and 'monopolies' of the few, and the frequent description of men working for the patria as possessing *zelo*.[23] This core value of personal application for the common good was paricularly visible in the writings of José Cecilio del Valle, the philosopher and statesman of Guatemalan independence. He conceived of the concept of citizenship partly as an active choice which seemed to draw directly on Bourbon-era enlightened reformers' patriotism expressed through actions and useful interventions rather than birthplace. To Valle, birthplace was important in the narrative of making men who they were, for instance, in the education and loyalties that it imbued in them. However, a 'love' for the fatherland and a willingness to apply oneself were even more crucial to being a citizen, as Marta Elena Casaús Arzú has argued.[24] Valle's ideal citizen of the new republic was, in his essential attributes, not unlike the patriot of the 1790s who applied himself to changing the region's economy through agricultural or manufacturing projects. Even Valle's writings on the role of indigenous people in this new nation seemed to reflect these attitudes to some extent, emphasising the contribution of Indian labour to creating wealth from the earth (through mining, for instance).[25] We might interpret this stance on Indian citizenship as being defined by, but perhaps even contingent on, this capacity to 'apply himself' to the cause of Guatemalan prosperity, although the status of the indigenous

---

[22] Real Sociedad, *Estatutos de la Real Sociedad Económica*; Sociedad Económica de Amigos del Estado de Guatemala, *Estatutos de la Sociedad Economica de Amigos del Estado de Guatemala* (Guatemala City: Imprenta de la Union, 1830); 'Discurso sobre las utilidades que puede producir una Sociedad Económica en Guatemala', 1795, AGI, Estado, 48, N.7, 3v–4.

[23] E.g. in Real Sociedad, *Quinta junta pública*.

[24] Marta Elena Casaús Arzú, *Las redes intelectuales centroamericanas: un siglo de imaginarios nacionales (1820–1920)* (Guatemala: F&G Editores, 2005), 23.

[25] *Ibid.*, 33–5. Compare political definitions of citizenship and the special application process for men of African origin in 1812 in Dym, *From Sovereign Villages*, 117.

citizen remained separate from that of the *ladino*.[26] Reflecting the complex roles accorded to foreigners in this new country, a 'chosen' homeland further opened a door for naturalised citizens to act in the benefit of this new homeland, and become 'creators of wealth' who worked towards the common cause. A re-established Economic Society welcomed proposals from 'residents' as well as 'citizens', again suggesting that place of birth could be disregarded in favour of a person's intentions. Foreigners could even become the association's director, as long as they could command two-thirds of the vote (Guatemalans only needed a simple majority).[27] The figure of the enlightened patriot, reminiscent of the *Gazeta*'s ideal contributor who travelled with his eyes open rather than with a 'hollow head', seemed to continue to exist in the figure of Valle's ideal citizen, contributing to the national imaginary.

The reinvention of a colonial institution in a new independent era was no mean feat, helped not just by the social positions of its former and current members, but also its self-presentation as standing for a 'community of citizens'. This emphasis on a 'united' group of patriots working for the common good in the newly re-established Economic Society is reminiscent of the self-fashioning of French naturalists at the *jardin du roi* described by Emma Spary in the aftermath of the French Revolution, where the institution's survival depended on not being associated with the *ancien régime*'s courtly patronage networks, but was now presented as a republican resource of technological knowledge.[28] However, the Guatemalan Economic Society's independent incarnation was able to remain closely connected to its colonial roots, not just through the idea of actions for the common good embodying patriotism, but also through the colonial Society's repeated insistence that what they were producing was not abstract knowledge, but practical advice for the betterment of the country. The *Gazeta de Guatemala*'s participatory networks and ability to draw in experts from cochineal growers to readers of navigation manuals also had already formed what might be called a communal repository of technical knowledge. The new Economic Society's aims and participants were fundamentally familiar. The new society would work to 'foment education, agriculture, industry and commerce', while the old one had encouraged 'agriculture, industry, commerce, arts, and trades'. Both the new and old statutes suggested the automatic enrolment of members of government as well as the university directorate as members (*socios natos*), even if that government now looked different. Both encouraged the participation of priests.

---

[26] Arturo Taracena Arriola, *Etnicidad, estado y nación en Guatemala, 1808–1944*, Vol. I (Antigua Guatemala: CIRMA, 2002), 43–5.

[27] Jordana Dym, 'Citizens of Which Republic? Foreigners and the Construction of National Citizenship in Central America, 1823–1845', *The Americas* 64, no. 4 (April 2008): 477–510. Sociedad Económica, *Estatutos de la Sociedad*, 14.

[28] Spary, *Utopia's Garden*, chapters 4 and 6, especially 157–8.

The personal attributes they demanded of its members and directors were also similar. Both councillors of the new Society and the director of the old were described as men who had to be *amantes del bien publico*, or 'lovers of the public good', although the director of the new Society was also given the additional task of having demonstrated 'patriotism' throughout his career.[29] The original 1829 project to re-establish the Society even suggested the continued use of its colonial name (*Sociedad Económica de Amantes de la Patria*), but this had been changed by 1830 to *Sociedad Económica de Amigos del Estado de Guatemala*, a minor change in the sense that it was still generally referred to as Sociedad Económica, but signalling a reframing towards a new political entity, a state rather than a more vaguely defined *patria*. Nevertheless, echoes of the *patria* of the colonial period, which at least technically encompassed the whole kingdom, remained in the new Society's statutes: the mission statement of the newly re-established Economic Society 'of the State of Guatemala' no doubt raised eyebrows in Central America's other capitals when it stressed that its members were equally interested in the prosperity of El Salvador, Honduras, Nicaragua, and Costa Rica.[30] In addition to being termed a 're-establishment', contemporaries also saw the Economic Society of the independence era as the heir to the colonial one, and a continuation of the colonial project. This was apparent in the preservation of a continuous archive of the various epochs of the Society, which included documents from 1798 as well as 1831. It was also considered the same legal entity as its colonial predecessor. In a dispute about the payment of interest on the mortgage for the Society's headquarters, the lender (the administrator of a Guatemala City convent) listed the history of the Economic Society as an almost seamless narrative, from the 1790s until the 1840s. In each period that the Society lay dormant, the administrator stressed how a small but determined group of members had preserved the Society's interests and its headquarters. This narrative was a necessary part of the legal argument supporting the debt payment, but it was also a narrative of an uninterrupted existence of the same Economic Society conceded by the government and the Society itself.[31]

In addition to such formal continuities, the new Economic Society's work on agricultural knowledge and natural history also echoed its predecessor. It

---

[29] Sociedad Económica, *Estatutos de la Sociedad*; Real Sociedad, *Estatutos de la Real Sociedad*; Bernardo Escobar to the General Secretary, 13 December 1825, AGCA, B92.4, Leg. 1394, Exp 32288, fol. 4.

[30] Sociedad Económica, *Prospecto del Mensual de la Sociedad Económica de Amigos del Estado de Guatemala* (Guatemala City: Imprenta de la Unión, 1830). For Guatemala City as a regional metropolis that continued to be resented by other countries, see Dym, *From Sovereign Villages*, especially chapter 6.

[31] 'Inventario del archivo de la Sociedad Economica, 1847', AGCA, B. Leg. 1390, Exp. 32099; 'Administrador del Beaterio de Santa Rosa al Ministro del Dpto de Hacienda', 18 January 1843, AGCA, B92.1, Leg. 1390, Exp. 32088.

encouraged the propagation of specific crops through contests and correspondence between members, and encouraged the collection of 'useful' specimens. By 1831, the Society had imported olive tree grafts from Mexico and offered a silver medal 'to the farmer who planted 500 grape vines'. Various members had reported on potential imports which might be cultivated in Guatemala, while others offered land for planting them.[32] One such project was the potential cultivation of silkworms. A member of the Society, Gregorio Rosales, corresponded with the government of Guatemala on this question. He noted that all known instances of silkworms in Guatemala had unfortunately disappeared in 1826, but there were colonial precedents for their cultivation. He himself had cultivated them back in the colonial period, and he referenced the Economic Society's 1790s efforts to import the worms from Oaxaca.[33] Rosales was of the opinion that this industry should be encouraged among 'poor people, and if it should be possible indigenous people'. Unlike his colonial predecessors, he referred to indigenous people as *los indígenas*, but he maintained the idea that they would have to be slowly and carefully enlightened by Spanish or urban leaders. The text he recommended as a handbook on this matter was Rozier's *Dictionary* (specifically, a passage from Volume 8), the same book published by the Madrid Economic Society that colonial-era reformers had repeatedly sought out for its agricultural guidance. He added two caveats, which also closely resembled eighteenth-century reformers' attempts to modify European scholarship to suit the 'circumstances of the country': Rozier's text itself would need 'some additions adjusted to our climate [*clima*]', and at any rate the French text might be too long and therefore not suitable 'for the class of citizens among whom it should be noted'. With a small amount of adjustment to this knowledge, silkworms would be just one of the aspects of agriculture and industry that would flourish in Guatemala.[34] In addition, again building on colonial antecedents, José del Valle, in his function as director of the Economic Society, in 1831 proposed the 'establishment of a museum or cabinet of natural history'. As its 1796 model did, it would rely both on a secular administrative network, with district governors being asked to send in 'the items most deserving of attention among the stones, metals, half-metals,

---

[32] Reyes, *Apuntes para una monografía,* 110–1; Robert Smith, 'Financing the Central American Federation, 1821–1838', *The Hispanic American Historical Review* 43, no. 4 (1963): 483–510, at 504–5. See also Elise Piazza, 'La Sociedad Económica de Amigos del País de Guatemala, 1821–1854' (Tulane University BA Honours thesis, 1975).

[33] Colonial precedents: Concurrencia General, 25 June and 22 July 1795, HSA, HC 418/563. The Economic Society had distributed silkworms to its members in 1797: AGCA, A 1.1, Leg. 2007 Exp. 4347, fol. 25.

[34] Josef Gregorio Rosales al Secretario General del Gefe de Estado, 15 August 1833, AGCA, B95.1, Leg. 3618, Exp. 84593. Although Rosales did not explicitly link his work on silkworms to the Economic Society in his 1833 letter, he promised to send a treatise on them 'very soon'. He had presented a similar treatise (about the utility of potatoes) to the Society in 1833: AGCA, B. Leg. 1390, Exp. 32099.

bitumen, volcanic products, seeds, plants, and dissected animals', and on a network of members across the region to contribute potentially 'useful' natural productions for study in the capital. One such member was the Irish-Guatemalan Juan Galindo, a government official stationed in Guatemala's northern region, Petén. In 1831, the Society's leaders specifically asked him to send 'the cotton seeds which he had offered, and all that he might find useful and worthy of placing in a museum', conflating economically useful natural productions with 'museum-worthy' objects.[35] If Bourbon-era reformers had imagined Central America as an unexplored land full of potential, a sort of *tabula rasa* upon which a new, useful economy could be built, the new republican context encouraged even stronger dreams of finally developing the land's unfulfilled potential.[36] In doing so, they leaned on colonial antecedents.

Past and Present Infrastructure

Knowledge of the land in the form of geographical information stood at the heart of independent governments' continued concern with 'useful nature' and scientific methods of studying it. Geography acquired a particularly important status as a science that was both outward- and inward-facing. Looking beyond Guatemala and beyond Central America, independent governments were well aware of the necessity to represent the nation internationally and later developed ambitious mapping projects that sought to define its borders. The new Economic Society's newspaper under Valle published a serialised narrative geography that proceeded department by department and included population statistics, reminiscent of the colonial *Gazeta de Guatemala*'s 'Description of Guatemala'.[37] However, there were no apologies to readers who might get bored with the dry narrative: chorographical descriptions of the kind that the *Gazeta* had championed were now a foundation of the new state's governance. As Valle explained: 'It is very important to be a geographer in order to be a legislator. There are misunderstandings when one does not know the ground that one treads on, nor the men who live there.'[38] However, the geographical projects that the new Central America states most frequently sponsored were initially limited to narrative geographical descriptions or schematic route-maps for the maintenance of harbours, land surveying, and road-building. Road-building projects were one of the most pressing tasks for independent governments. The infrastructural

[35] 'Junta ordinaria de la Sociedad Económica de 1o de agosto de 1831', in Reyes, *Apuntes para una monografía,* 161–2, 167.
[36] McCreery, *Rural Guatemala*, 22.
[37] Sociedad Económica, *Prospecto del Mensual de la Sociedad Económica,* 6; Sociedad Económica, *Mensual de la Sociedad Económica de Amigos del Estado de Guatemala*, no. 1 (Guatemala City: Imprenta de la Union, April 1830), 5.
[38] Sociedad Económica, *Mensual de la Sociedad Económica de Amigos del Estado de Guatemala*, no. 2 (Guatemala City: Imprenta de la Union, May 1830), 26.

elements of state power, as Miguel Centeno and Agustin Ferraro have argued, have been underrepresented in the literature of Latin American state-building of the nineteenth century in comparison to its Weberian territorial and economic aspects, yet they were an immediate priority.[39]

Valle also saw a lack of suitable infrastructure as hindering Guatemala's progress. 'We do not have roads', he declared.[40] In 1825, a government minister warned that maintaining harbours on the basis of specific knowledge of their 'climate, their population and their needs' was essential to prevent foreign invasions, building on the colonial Society's efforts to draw these spaces into the mercantile networks of the interior, and modify their climate to make them more liveable.[41] Such complaints had, through the Economic Society and Consulado, already formed part of the construction of notions of a colonial *patria*, presaging this nineteenth-century importance of a nation not just built on national borders, but on improved internal connections. Independence-era projects to improve communications explicitly resurrected and referenced colonial-era projects. Specific locations featured in the new as well as the old geographical order. A project to build a bridge over the river Salamá near the city of Quetzaltenango, which had become mired in disputes about labour, funding, and uncertainties about the terrain when José Rossi y Rubí proposed it in 1803, was now on the agenda of a private company authorised by the Guatemalan government to build a bridge and charge a toll for it.[42] In the early 1830s, the government commissioned several surveys of the harbour of Iztapa and approached an engineer to create a plan for 'flattening' the road between Gualán and Izabal that had so occupied the Consulado thirty years earlier.[43] Many of the road-building projects remained unrealised due to financing problems, including the fact that Indian labour levies were now outlawed (at least in theory – some towns did not abolish the *requerimiento* despite a government mandate to do so after independence) and aside from prisoners and military forces, the labour force would have to be paid through 'patriotic subscriptions'.[44]

---

[39] Miguel Centeno and Agustin Ferraro, 'Republics of the Possible: State-Building in Latin America and Spain', in Centeno and Ferraro (eds.), *State and Nation Making in Latin America and Spain. Republics of the Possible* (Cambridge University Press, 2013), 3–24, at 13.
[40] Del Valle, *Discurso del presidente*, 31.
[41] Orden 356, Guerra (17 December 1825), TNA, FO 254/1, 235.
[42] 'Proyecto presentado por el alcalde mayor de Suchitepequez, sobre construir un puente sobre el río Samalá' (1803) AGCA, A1.22.33, Leg. 207, Exp. 4178; '... para evitar que los indígenas ... obstaculicen dicho trabajo', 1811, AGCA, A1.1 Leg. 6113, Exp. 56242; 'Orden no. 10', *Boletín Oficial* Vol. 34 (21 March 1834), 300–1.
[43] 'Orden no. 15', *Boletín Oficial* Vol. 25 (10 April 1833), 229.
[44] 'Al C. Cayetano Resinos'. AGCA, B. Leg. 1398 Exp. 3216. McCreery, *Rural Guatemala*, 111–12. Woodward, *Class Privilege*, 57–8, also notes a new labour levy for all men to work on roads three days every year, but this seems to have been put in place after 1838 in Guatemala.

To aid with the road-building projects, surveyors and governments enthusiastically used colonial geographical documents. Where officers of the Spanish Crown in the eighteenth century had often complained of insufficient geographical knowledge, the leaders of independent Central America welcomed even the knowledge of that archive, which after all contained centuries' worth of geographical questionnaires, narratives, and plans. Despite the anti-Spanish rhetoric employed by independent governments, individual politicians seemed to hold the information amassed by the colonial state in great regard. Alejandro Marure, an influential Guatemalan politician, historian, and lawyer, blamed a lack of available archival material from the colonial period for what he considered a terrible state of disinformation. Looking back at the 1820s in 1845, he blamed the turbulence of the independence period and particularly the years 1821 to 1823, when Central America was part of Agustín Iturbide's Mexican empire, for the current lack of geographical information:

the antecedents which existed in the archives of the former Captaincy-General suffered the most scandalous plunder: some had been taken away privately, some were sent to Mexico, where they never arrived; so that, when the National Assembly wanted to have before it those [papers] which referred to the business in question, it was informed that there was not a single one left of those which had been preserved in the former Geographical Depository of the Kingdom.[45]

The removal of colonial geographical documents impeded the work of the new country's parliament, Marure complained. When surveyors did have access to specific survey documents from the colonial period, they created conscious links between past and present infrastructure projects. In 1841, a re-established *Consulado de Comercio* picked up the project of connecting the port of Santo Tomás and the Motagua river by building a shipping channel or a bypass road, quoting a survey on the same topic which had been conducted by a previous incarnation of the Consulado more than thirty years earlier: 'In the year 1792 the engineer Antonio Porta indicated to the Spanish government a project of communication between the port of Santo Tomás, and the Motagua river, by way of a road, or better, by way of a canal.'[46] This document quoted directly from the 1792 survey, as well as follow-up surveys that engineers had performed on behalf of the government or Consulado in 1836 and 1837. It should not be surprising that specific faults in the road network identified by the Economic Society and Consulado in the 1790s should still be relevant. The insistence of Bourbon-era reformers to design geographical projects from a point of view that would be useful to Central America before any larger

---

[45] Alejandro Marure, *Memoria historica sobre el canal de Nicaragua* (Guatemala: Imprenta de la Paz, 1845), 12 (partly quoting an 1826 report of J. F. Sosa).

[46] 'En 1792 el Ing P presentó un proyecto de comunicación entre el puerto de Santo Tomás y el río Motagua.' (Consulado papers, 1841), AGCA, B92.2., Leg. 3612, Exp. 84389.

geopolitical considerations that Madrid might have been interested in meant that independence-era governments were more easily able to use these blueprints for their own projects. Independent governments appeared to accept the Consulado and Economic Society as organisations concerned with Guatemala, rather than an unwelcome remainder of the colony.

Although the use rather than creation of archival documents is difficult to trace, references to the use of colonial surveys demonstrate that where independence-era leaders had access to Spanish geographical archives, they very much put them to use in a pragmatic manner. Sylvia Sellers-García has argued that in the independence period, the colonial archive was transformed from a 'living archive' into a mere repository of old documents.[47] This repository nevertheless occupied a significant role. In the era of boundary commissions and international treaties, one of the most obvious functions that tied archives to the present in post-colonial Latin American states was to preserve colonial boundaries to define the limits of new nations, disputed though such claims often were.[48] Yet archives were much more frequently used for internal matters. Geographical documents were even sent around the country. Evidence for the use of such documents comes from a government secretary's letter to the military officer and administrator Juan Galindo, somewhat impatiently asking him to return some documents from the colonial period: 'The government knows that the plan of Lake Izabal, and the history written by *padre* Citizen Francisco García Peláez, are in your possession; and since both things are urgently needed, the chief of state has requested me to ask you for them, asking you to send them as quickly as possible.'[49] The plan in question was most likely a colonial-era map of Izabal created by the Consulado: the Economic Society had itself previously requested such a document from the archives in 1831.[50]

Historical precedents also acquired an additional dimension in the independence-era road commissions. Archival material could help to support suppositions that a road might have existed in a particular place in the past, but that it had been forgotten through mismanagement or deliberately obscured. José del Valle ominously noted that contraband featured in the Guatemalan economy simply because of Central Americans' secret knowledge of the countryside that was hidden to the state: they 'know the roads and paths that lead to [contraband]'.[51] This tied in with older perceptions of indigenous people

[47] Sellers-García, *Documents and Distance*, 11–16, 161–2.
[48] Tamar Herzog, *Frontiers of Possession: Spain and Portugal in Europe and the Americas* (Cambridge, MA and London: Harvard University Press, 2015), 265–6.
[49] 'Palacio del Superior Gobierno', Guatemala, 13 August 1834, AGCA, B95.1, Leg. 1398, Exp. 32614.
[50] 'Borrador de la nota dirigida por la Secretaria General del Gobierno al Jefe Politico', Guatemala, 11 June 1831, AGCA, B119.4. Leg. 2556 Exp. 60097, fol. 8. In 1831, this document was in the Consulado archives, which by 1834 had been consolidated with the government's.
[51] José del Valle, *Boletín Oficial*, Vol. 7 (1 July 1831), 94.

having different access to roads and land, and in particular being able to move around the mountains more freely. The road-building commissions in the 1830s often seemed intended as a way to formalise such 'secret' routes and connect them to the 'official' road network by declaring them existent. In addition to employing local guides, the commissions relied on historical precedent from archives for discovering such paths. An 1841 commission that looked back at previous reports on a proposed route between the Golfo Dulce and the interior noted a description of an alternative path given by 'an old mule driver' in 1836, who reported on a contraband route through the Mico mountains, which at the time had apparently not been followed up. The commission noted that 'the words of the mule driver alone should not entirely be credited; but there is no harm in [considering] this notice in case another reconnaissance [of that region] is made'.[52] Road commissioners described such hidden routes as if they were not engineering projects, but ideal forms or hidden archaeological objects. One engineer spoke of his delight in 'finding' a new road to the sea from Izabal, quite possibly the old muleteer's path through the Mico mountains.[53] With this idea of uncovering hidden advantages, road-builders of the early independent period also harked back to the imagination of colonial reformers who appealed to the transformative power of infrastructure. Just as the colonial Economic Society had imagined that road-building and forest management could 'restore' a supposedly more prosperous past, independent road-builders also imagined that their efforts would help to recreate past glories. The military officer and politician Juan Galindo praised the possibilities that a good road to the Caribbean ports might bring, describing its possibilities in almost mythical terms:

Suppose there were a good cart-road from here to Refugio. . . . The journey will easily be made on this good road from here to the sea in a day, and [the city of] Guatemala will itself become like a port city: its productions of grain, sugar, leather etc will be shipped without excessive costs . . . Guatemala will thus legitimately recover its old commercial supremacy [*recobrando asi Guatemala con legitimidad su antigua supremacia comercial*].[54]

Efforts to base nineteenth-century roads on historical precedents therefore echoed the colonial past not just through researches in the colonial archives, but also in the way they embraced the rhetoric of the recovery of a lost past and perfectibility. Through uncovering, describing, and fixing the bounty that Guatemalan nature provided, as well as re-establishing roads that appeared

---

[52] Guatemala, 5 January 1841, AGCA, B92.2, Leg. 3612, Exp. 84389.
[53] 'Plano del reconocimiento del terreno por parte de don Cayetano Recinos', 1834, AGCA, B92.2, Leg. 3612, Exp. 84354, fol. 7.
[54] Sociedad Económica de Amigos del Estado de Guatemala, 'Guatemala, 23 October 1830', *Memorias de la Sociedad Economica de Amantes de Guatemala* (Guatemala City: Imprenta de la Union, 1831), 22–3.

lost, they would transform the country into a prosperous one. Engineers and statesmen of the independent era used colonial reformers' rhetoric, but also their material legacy of reconnaissance documents to resurrect their projects, which they saw as congruent with their own priorities.

Transatlantic Designs and Travelling Archives

Research in colonial archives by road surveyors formed just a small part of the reframing of knowledge in new contexts that characterised scientific information about Central America in the 1820s and early 1830s. It was a sign of changed times that the British consul saw the project of building a road between Gualán and Santo Tomás (the colonial Consulado's own pet project now being recreated by the Guatemalan government on the basis of colonial surveys) as mutually beneficial for Belize and Guatemala, and urged his countrymen to buy shares in the road-building company.[55] Colonial documents were now being used in changed political circumstances, and, put in the context of global politics, rather strange ones. After all, instances of the British state benefiting from the colonial archives of the Spanish state ran counter to the pervasive narrative of Britain's own agents, who insisted that both imperial Spain and its successor states had been incompetent guardians of the American continent's natural wealth, and needed British naturalists to 'rescue' it instead by, for instance, taking cinchona bark (quinine) to India.[56] Outside of the most obviously political exchanges, too, the opening of Guatemala to foreigners after independence constituted a dramatic expansion of possibilities for the creation and exchange of scientific knowledge in general and geographical knowledge in particular, both by Guatemalans and by foreigners. In the colonial period, Central American naturalists had prided themselves on their international connections. Now, independence meant that it was easier than ever to send their findings to foreign scientific bodies, or see themselves as members of more than one national centre of science. In turn, European knowledge about Central America in the early 1830s, reproduced in the journals of geographical or scientific societies and collected by naturalists, was not necessarily the result of European scientists and travellers touring the region in search of facts and knowledge and reporting home. Instead, it often drew on the knowledge of Central Americans or long-term foreign residents. Central American archival knowledge, new road surveys commissioned by Guatemalan governments, and individuals' transatlantic correspondence all formed part of new networks of knowledge-making.

---

[55] Mario Rodríguez, *A Palmerstonian Diplomat in Central America: Frederick Chatfield, Esq.* (Tucson, AZ: The University of Arizona Press, 1964), 103.
[56] Philip, *Civilizing Natures*, 259–61; Drayton, *Nature's Government*, 50–55, 229–38.

Foreigners and foreign-born Central Americans now played an increasingly important role in the exchange of geographical and scientific information. By the 1840s, 'scientific travellers' from North America and Europe would start to tour Central America. In the immediate aftermath of independence, by contrast, the foreigners who poured into the country were more likely to arrive with commercial or diplomatic aims rather than with the express aim of scientific investigation or collecting. However, many of them established residence in Central America, took on roles that led them to collaborate with locals, or even became naturalised citizens.[57] They used their international background to act as intermediaries and correspondents for scientific societies and individual collectors within Guatemala and elsewhere. One such intermediary was George Ure Skinner, one of the partners of the mercantile firm Klee, Skinner and Company. He was described by a fellow Briton as 'the owner of extensive estates in Guatemala', an amateur collector of plants and antiquities. He contributed many specimens to British orchid collections and shot rare Guatemalan birds, including a quetzal, for inclusion in the collections of the Manchester Natural History Society around 1836 and 1837.[58] However, scientific information also travelled across the Atlantic in both directions. In 1837, Skinner facilitated a mutually beneficial exchange of scientific information between the Guatemalan Academy of Sciences and Sir William Jackson Hooker, then Regius Professor of Botany at the University of Glasgow. Hooker, via his contact Skinner, in 1837 requested a swap of items: the herbarium of Dr Pérez of Guatemala for his collected publications on botany. When the Guatemalan government forwarded this request, it noted appreciatively that Hooker's works 'consists of many volumes and is worth 3000 pesos'.[59] The Glaswegian botanical library could be added to the collection of the Guatemalan Academy of Sciences. Given that the Academy's library was in such a bad condition and received so little funding that its librarian resigned in protest in 1834, this was a good exchange for both.[60] This was not a case of a European scientist going on a tour of Central America and collecting plants himself; it was a specific botanical collection that the Glasgow professor was interested in, and at Guatemala's Academy Hooker found a counterpart in Pérez, whose work and collection he valued. The experience of such scientific contacts 'on the ground' clearly ran counter to the rhetoric of openly imperialist botanists who engaged, as Philip described for the case of Clements Markham,

---

[57] Dym, 'Citizens of Which Republic?'

[58] James Bateman, *The Orchidaceae of Mexico and Guatemala* (London: Ridgway & Sons, 1843), Tables XII–XIII. A contemporary described Skinner as being from Manchester, while other accounts describe him as being from Newcastle or Scotland.

[59] Government to the Academy of Sciences, Guatemala, 4 May 1837, AGCA, B, Leg. 1075, Exp. 22879.

[60] 'José Mariano Gonzales renuncia officio', 25 September 1834, AGCA, B, Leg. 1075, Exp. 22742. See also Pattridge, *Institution Building*, 132–3.

in 'biopiracy' to steal cinchona plants with the justification of Bolivian incompetence.[61] Many of the recorded scientific interactions between foreigners and Central Americans, such as George Ure Skinner's collecting, seemed to rely on personal connections and were perhaps established fortuitously through commercial contacts. Skinner was certainly not significant within international scientific circles save for his unique position as a geographically well-placed enthusiast. However, this particular request for the botanical collection was made through the Central American government offices, suggesting that the Gálvez government's ministers saw such international exchanges as an important component of their remit.

Another example of a key scientific correspondent was Juan (John) Galindo, a colourful addition to the Central American elite. An Irish-born naturalised citizen of the Central American Federation, he had come to Guatemala in 1827, had a brief but successful career in the Liberal army under Morazán, and was then appointed governor of Guatemala's northern region, Petén. He also represented Guatemala as a diplomat in negotiations with Britain.[62] Like Bourbon bureaucrats before him, he was interested in Guatemala's northern tropical regions in connection with land and river surveys, and in the region's flora. However, the frame of reference for the results of his investigations was new: he sent the results of his investigations to the Royal Geographical Society, Royal Horticultural Society, the Medico-Botanical Society in London, and the *Société de Géographie* in Paris, as well as the Guatemala City government.[63] Even when the Guatemalan Economic Society specifically requested that he contribute specimens to Guatemala City's new natural history museum, he simultaneously sent pieces of rocks and petrified wood to Charles Lyell at the Geological Society in London.[64] There was no doubt about Galindo's status as a Guatemalan: Galindo feuded for years with the British consul Frederick Chatfield, who considered the Irish-born Galindo a traitor for representing the Central American Federation against the country of his birth. Galindo in turn wrote irate letters to Foreign Secretary Palmerston in London, complaining that Chatfield was thwarting his Central American career by trying to publicly discredit him as a turncoat and British subject.[65] His correspondence

---

[61] Philip, *Civilising Natures*, 259.

[62] Ian Graham, 'Juan Galindo, Enthusiast', Estudios de Cultura Maya 3 (1963): 11–35; William Griffith, 'Juan Galindo, Central American Chauvinist', *The Hispanic American Historical Review* 40, no. 1 (1960): 25–52; Robert Aguirre, *Informal Empire*. Mexico and Central America in Victorian Culture (Minneapolis, MN and London: University of Minnesota Press, 2005), 72–4.

[63] Galindo, 'On Central America', 119–35; Juan Galindo, 'Medico-Botanical Society. Communication from Coronel Galindo', *The Literary Gazette: A Weekly Journal of Literature, Science, and the Fine Arts* 1010 (28 May 1836): 344; Graham, 'Juan Galindo', 27–8.

[64] Juan Galindo to Charles Lyell, 28 October 1831, Correspondence relating to untraced collections, Mineralogy Library, Natural History Museum, London.

[65] Mario Rodríguez, *Palmerstonian Diplomat*, 132, 144.

with European scientific societies was likely intended as a mixture of self-promotion (he was eagerly awaiting his 'diploma as an honorary member of the R.G.S.' in 1839 and received a medal from the French *Société de Géographie*), sociability with British acquaintances (he had social connections with British representatives who were not Chatfield), or the promotion of Central America.[66]

The colonial geographical documents which Guatemalan governments found so useful in identifying infrastructure projects were also part of these exchanges. Given the long history of consulting antecedents from the archives for new geographical projects, it was not a surprise that the Consulado had looked back at older surveys of the Motagua region when it designed a route for the road between Gualán and Santo Tomás. It was perhaps less obvious how interested British geographers would be in this archive, too. A sketch-map of the Motagua river, drawn up by a British engineer who surveyed the region in 1837 for the Guatemalan government (one Henry Gardiner), ended up in the archives of the Royal Geographical Society (RGS) in London. The map included a reference to the 'intended canal', which the 1792 and 1796 surveys of the river commissioned by the Consulado had briefly discussed, but which had been dismissed as too expensive and impractical since.[67] An engineer working for the Guatemalan government, Carlos Meany (a native of Trinidad and naturalised Guatemalan), sent this sketch-map to London, alongside an even more detailed map of the river which was also inscribed with the Consulado's specific activities. This survey map, consisting of 28 separate sheets, contained detailed soundings of the depth of the Motagua river. According to its title, it was the record of an 1812 Consulado expedition (by the canon Dighero), copied by Carlos Meany in 1829. Although it was recorded at the RGS as being from Porta y Costas's 1792 notes, it retained annotations in Spanish such as '2a dormida 5 de Abril 1812' (second overnight camp), or 'Siembras de Maiz' (maize field), rendering Dighero's 1812 journey of exploration and places along the route.[68] The practical use of these maps would have been limited: after all, later Consulado commissions had raised questions about the river's navigability, and by 1817 they had declared navigation attempts a folly. Without the weighty *expedientes* which usually related successive explorations to each other within the archive, it was unclear to what extent this context was transmitted alongside the maps. In their new context of

---

[66] Juan Galindo to the RGS, Guatemala, 15 January 1839, RGS, Corr. Block 1834–40, Galendo; Graham, 'Juan Galindo', 16.
[67] Henry Gardiner, 'River Motagua Central America', 1837, RGS, Guatemala S/S 10.
[68] RGS, Guatemala S/S 10. Francis Herbert, 'Guatemala, Colonisation, and the Royal Geographical Society in the 1830s: Some Contemporary Manuscript and Printed Documents in the RGS's Collections', *Journal of the International Map Collectors' Society* 107 (2006): 6–12, at 10, mentions these documents, but they do not appear to be related to the maps of colonisation projects that the RGS also holds of this region.

the RGS archive, however, they demonstrated British geographers' ability to be informed of the topographies of the world, and a tangible representation of the country in which British commercial interests were continuously expanding.

From railways to the Panama Canal, the influence of foreign capital and geopolitical ambitions in Central America was clear, to the point that infrastructure projects may appear as something that 'happened to' Central America in the nineteenth and twentieth centuries. However, the importance of colonial Spanish archival records, even in the context of British speculation about a possible trans-isthmian canal through Nicaragua, is underlined by a note from the editor of the journal of the RGS. He explained that 'a Spanish MS. existing in the archives at Guatemala, and copied by [the British diplomat] Mr. Thompson, which states that the engineer Don Manuel Galisteo executed a survey, in 1781' was the only information on Lake Nicaragua's water levels available at that time.[69] As late as 1845, the politician and historian Alejandro Marure made a point of reclaiming the putative Nicaraguan canal project as an authentically Central American one. He pointed out that this was not a newfangled idea, the Spaniard Gil Gonzales Dávila having been the first person to survey the Pacific coast with this idea in mind in 1522.[70] That is not to say that all infrastructure projects were home-grown: of course they acquired ever more transnational dimensions in this newly international era, and the need for raising capital meant that the involvement of foreign banking houses was almost inevitable. However, transnational exchanges of information were not limited to the imposition of 'foreign' ideas upon Central America. In 1836, Juan José Aycinena, a scion of the once all-powerful Aycinena merchant family, was living the United States at the height of that country's canal-building era. Inspired by the Erie Canal, Aycinena drew up a project for a Central American interoceanic canal that he sent to the president of the Federal Republic, General Morazán, from New York. Morazán in turn commissioned the Anglo-Irish engineer Juan (John) Baily to investigate the possibility of such a canal.[71] Although the Central American Congress authorised an exploration of the possibility of a Nicaraguan canal and Baily started to map the region, the political situation rapidly deteriorated in 1837, and the Central

---

[69] Galindo, 'On Central America', 120. An electronic search of Thompson's texts suggests that he did not include this source in his published account: George Thompson, *Narrative of an Official Visit to Guatemala from Mexico* (London: John Murray, 1829). A different communication by the Guatemalan government thought that it was instead a 1783 exploration by the engineers Joaquín de Isasi and Agustín Cramer that recorded Lake Nicaragua in the colonial period.

[70] Marure, *Memoria histórica*, 3.

[71] *Ibid.*, 25–8. David Chandler, 'Peace Through Disunion: Father Juan José de Aycinena and the Fall of the Central American Federation', *The Americas* 46, no. 2 (1989): 137–57, at 140–5; Thomas Leonard, 'Central America and the United States: Overlooked Foreign Policy Objectives', *The Americas* 50, no. 1 (1993): 1–30, at 8.

American Federation disintegrated in 1838 before any concrete steps could be taken towards completing the dream of the canal.

This transatlantic co-construction of geographical knowledge was not confined to the use of colonial precedents. The work of John Bailey is another example of the way in which the complicated allegiances of foreigners working within an increasingly international Central America shaped knowledge about the region. Born in Britain, John Bailey came to Guatemala to participate in loan negotiations on behalf of a British banking house.[72] Although he settled in Guatemala (he died in Guatemala City in 1852), he represented both British commercial and Guatemalan official interests at various times. His familiarity with the intellectual elite of Guatemala is evident in his role as translator of Domingo Juarros's *History of Guatemala* into English.[73] He was an engineer and was engaged in cartographic projects on behalf of the Guatemalan and Nicaraguan governments throughout the 1830s and 1840s.[74] One of his most widely distributed works was a comprehensive map of Central America, printed in London in 1850, at the height of the canal frenzy.[75] Yet this map was not originally intended to be a British project at all: the surveys for it came from within the Guatemalan government. In 1837, General Morazán, the president of the federation, had commissioned Baily to explore the possibility of an interoceanic canal across Nicaragua. The proposed route would connect the harbour of San Juan del Sur with the Caribbean coast via Lake Nicaragua. The political crises that eventually led to the disintegration of the federation in 1838 meant that Baily never finished his survey, but he successfully mapped most of Lake Nicaragua as well as the San Juan and Tipitapa rivers, and evidently had enough data to form the basis of a new map. Baily's correspondence states that he initially deliberately kept the results of the survey to himself. After all, he had been in the employ of the 'Federal Government on a service supposed to be of advantage to the whole country', and the map was conceived of as a Guatemalan government project.[76]

However, the implosion of the government in 1838 meant that the engineer was only paid a small part of the fee he was due. Therefore, in 1842, he contacted the RGS in London instead in the hope of availing himself 'of any

---

[72] Smith, 'Financing the Central American Federation', 487.

[73] 'Baily, John', in Rojas Lima, *Diccionario histórico-biográfico*, 156; Domingo Juarros, *A Statistical and Commercial History of the Kingdom of Guatemala*, transl. John Baily (London: J. Hearne, 1823).

[74] AGCA, B95.1 Exp. 32600 Leg. 1398 fol. 8; AGCA, B95.1 Leg. 1398 Exp. 32689; Marure, *Memoria historica*, 28.

[75] 'Map of Central America Including the States of Guatemala, Salvador, Honduras, Nicaragua and Costa Rica, the Territories of Belise and Mosquito, with Parts of Mexico, Yucatan and New Granada. Engraved from the Original Drawing of John Baily' (London: Trelawney Saunders), 1850.

[76] John Baily to the RGS, Guatemala, 9 September 1842, RGS/CB3/38.

other mode of compensation'. He stated that he was encouraged to do so by the British consul, Frederick Chatfield, as well as correspondence with one Captain Barnett of the HMS Thunder (a British survey ship), and a meeting with Barnett's assistant surveyor, Lieutenant Lawrence, in Granada, who 'on inspecting the Maps he strongly recommended, as well as Capt Barnett, that I should make them known to the Geographical Society'.[77] The small community of Britons living in 1830s Central America appear to have been well connected, whether they formally worked together or not. It seems that it was these connections as much as Britain's geopolitical interest in the region that led to the eventual publication of Baily's map of Central America, including all 'proposed routes between the Atlantic & Pacific Oceans', in London in 1850. Nevertheless, the gathering of the geographical information which eventually appeared in the pages of the RGS's journal and the published map had originally been devised and commissioned by the Central American government.

Even aside from these connections and interests, however, it would not have been news to the RGS that a new survey of Central America was taking place. Regardless of the government's financial position, it seems that the survey results would have made it to London upon their completion. While Baily was still surveying, Juan Galindo was writing to the RGS himself, announcing that he was waiting anxiously for the results of the survey so he could bring the Society Baily's results. He framed his position as one of a disinterested scientific bystander, waiting for the 'completion of the labours of a corps of engineers employed by the Central-American government, to examine the facilities for opening a ship canal by the waters of Nicaragua, to complete my description of that commonwealth, in continuance of the series of the different states composing this federation, promised by me to the Royal Geographical Society'.[78] While Baily, a Briton, saw himself as a loyal employee of the Guatemalan government while he was fulfilling his contracted commission, Galindo, a naturalised Guatemalan, here appeared more conscious of his position as an intermediary between the Guatemalan government and Britain's learned societies. Galindo had previously sent numerous letters to the RGS, containing, for instance, details about Lake Nicaragua or French maps of the Salvadoran coast. He earnestly framed this information as simply scientific, sending the position of islands off the Costan Rican coast simply because they had 'never before correctly laid down' and in order to 'correct' erroneous maps.[79] Juan Galindo saw British geography as the international standard to aspire to. Writing to the RGS, he was effusive in his praise of the work that British survey ships were carrying out on the

---

[77] *Ibid*. HMS Thunder survey maps are at TNA in Kew.
[78] Juan Galindo, Guatemala, 15 January 1839, RGS/CB2/210.
[79] Juan Galindo, Various letters 1836–1839, RGS/CB2/210.

Central American coast: 'The nautical surveys of our coasts – that on the North side to be completed by Lieut. Barnett, brig Lark – & that on the Pacific side by the surveying squadron under capt.n Belcher, will finally place our geography on a respectable footing: we are much indebted to England.'[80] Opposed as Galindo was to Britain's Central American geopolitical ambitions in general, he apparently saw the RGS and scientific correspondence as quite distinct from any British imperial aims.

Representing the New Country

Galindo's position might be characterised as either self-serving (as informant of two governments at once), or naïve. After all, most foreign involvement in Central America at this time had little to do with mere scientific detachment. Foreigners played an important role as investors, as well as founders and representatives of colonisation companies. The disproportionate influence of Britain in nineteenth-century Latin America has traditionally been described as 'economic' or 'informal' imperialism (not to mention, of course, the direct colonisation of British Honduras). Yet Latin American countries were also in desperate need of political recognition and foreign investment, and proactively courted it. London, of course, was a centre of Spanish American exiles throughout the early 1800s, playing host to Latin American revolutionary intellectuals such as Francisco de Miranda. Latin Americans residing in London in the 1820s wrote articles with the aim of gaining political support for the new republics from Britain, and promoted their countries as lucrative investment opportunities.[81] The government of Mariano Gálvez supported British and other European colonisation enterprises in Guatemala, even though many of them failed or, in the case of Gregor MacGregor's scheme to sell shares in a colony in the fictional Central American location of Poyais, turned out to be entirely fraudulent. They hoped to create a thriving economy and infrastructure in the jungles of Guatemala's northern region, Petén, and the east coast, which were sparsely populated relative to Guatemala's central and western regions.[82] Central American Enlightenment reformers had stressed the potential of these coasts for years, and had grappled with the question of how best to make agriculture possible in those humid climes and persuade settlers to live there throughout the 1790s and early 1800s. In 1825, one of the

---

[80] Juan Galindo, Guatemala, 15 January 1839, RGS/CB2/210.
[81] Matthew Brown (ed.), *Informal Empire in Latin America: Culture, Commerce and Capital* (Oxford: Blackwell/Society for Latin American Studies, 2008); Aguirre, *Informal Empire.*
[82] William Griffith, *Empires in the Wilderness: Foreign Colonization and Development in Guatemala, 1834–1844* (Chapel Hill, NC: University of North Carolina Press, 1965); Karl Offen, 'British Logwood Extraction from the Mosquitia: The Origin of a Myth', *The Hispanic American Historical Review* 80, no. 1 (2000): 113–35, at 133–4.

many projects José del Valle was engaged in was 'a plan for the settling with foreigners the territory bordering upon the port and river of San Juan in Nicaragua'.[83] From a Central American perspective, foreign capital was to fund these settlement projects decades in the making, once and for all integrating the elusive Izabal region into the economic life of the new nation.

Other aspects of Central America's relationship with Britain during the first decades of independence were more strained. By the 1830s, Guatemalan politicians were sufficiently wary of the political power that British economic speculation and colonisation projects along the ill-defined Belizean-Guatemalan border represented that they implored the United States' help against Britain on the basis of the Monroe doctrine. (Although Honduras, Nicaragua, and Costa Rica followed suit in the 1840s, the US did not regard the matter pressing enough to intervene).[84] In the 1820s, Guatemalan and British diplomats also clashed over the issue of slavery. While Britain was a bastion of liberal inspiration to many reformers, slavery was still legal in much of the British empire (much as Colombians, as Lina del Castillo has observed, at the time argued for greater racial equality than their sometime model, the young United States, could provide).[85] Although forced labour practices continued in some parts of the country, Guatemala had formally abolished slavery with independence in 1821. This turned out to have practical consequences for residents of neighbouring British Honduras (Belize), who complained in 1826 that slaves in their territory would escape over the border to Guatemala, where they would be free. The Guatemalan government refused to return the formerly enslaved people to Belize, a source of tension between the countries throughout the late 1820s.[86] Guatemala was almost unique in the former Spanish American lands for having a long-established land border with a British territory, a situation that intensified the investment that Guatemala courted, but also the political interference it did not.

Regardless of diplomatic disagreements, foreign capital was tremendously important to the new Central American countries. Central American government bonds were floated on the British market alongside those from other Latin American countries in 1825, but with little success.[87] Conscious of the need to borrow money, statesmen like José del Valle believed that investors would inevitably be attracted to Guatemala once they had been informed of the natural resources and other circumstances of the country. Valle urged one of his correspondents in London to publish his outline of the politics and governance

[83] Thompson, *Narrative*, 209.   [84] Leonard, 'Central America and the United States', 5.
[85] Del Castillo, *Crafting a Republic*, 81–5.
[86] Reports of the Governor of British Honduras, TNA, FO 15/9, 21, 101–8; Marcial Zebadúa, *Manifestación pública del ciudadano Marcial Zebadua sobre su mision diplomatica cerca de Su Magestad Británica* (Guatemala City: Imprenta de la Union, 1832), 9.
[87] Smith, 'Financing the Central American Federation', 486–8.

of Central America (previously published in Valle's own *Redactor General*) in 'the most reputable newspapers' of London. He imagined that knowledge of Central American geography would encourage further investment and scientific interest in the region, and ultimately facilitate the recognition of its independence.[88] Here, Valle built on a long tradition of seeing correct geographical representation abroad as going hand in hand with an internal recognition and exploitation of natural resources. Valle wrote about Central America's obvious geopolitical importance in 1826: 'It is enough to see the map of the continent and its divisions to understand this after just a few moments of reflexion. It is necessary that the Republic of Central America should exist in order to conserve the equilibrium of the New World.'[89] An 'impartial' history of Guatemala, designed to override previous biased chronicles and histories (by Spaniards and foreigners), would also help to 'secure public opinion in Europe and America'.[90]

Valle's attempts at promoting Guatemala's history, law, and government showed a multi-faceted picture of the country. Above all, however, it would be natural wealth coupled with its advantageous location that would lure investors. Unlike in Mexico, where intellectuals, as Jorge Cañizares-Esguerra has asserted, 'fought against the type of landscape aesthetics first introduced by Humboldt', Central American politicians were not opposed to presenting a one-dimensional view of their homeland as a land of nature where they perceived it to be in their best economic interest.[91] This characterisation of Guatemala as a place of natural wealth was reflected in the writings of men like Juan Galindo, who emphasised the richness of Guatemalan nature in their dealings with foreign scientific societies.[92] While the British engineer John Baily's report on Central America came with its share of dismissive comments about inefficient government, he may have been reflecting Guatemalan propaganda rather than just an 'orientalising' British view to some extent when he touted the 'remarkable capacity for colonisation' in its very title and praised the 'natural richness of the soil'.[93] Indeed, if the idea of unquestionably rich American landscapes had been cemented in the Spanish American imagination through the influence of Jesuits exiled to Italy in the 1760s, it could be said that the idea of American bountiful nature (stripped of its 'patriotic' connotations and now widely circulating in Central America as simple scientific fact) was

[88] Valle to Rocaforte, 10 July 1825, in *Cartas de José Cecilio del Valle*, 38–9.
[89] Valle, *Redactor General*, no. 27 (28 September 1826).
[90] 'Historia', *Mensual de la Sociedad Económica*, no. 3, 69.
[91] Cañizares-Esguerra, *Nature, Nation and Empire*, 154–5.
[92] Galindo, 'On Central America', 124.
[93] John Baily, *Central America: describing each of the states of Guatemala, Honduras, Salvador, Nicaragua, and Costa Rica; their natural features, products, population, and remarkable capacity for colonization* (London: T. Saunder, 1850), ix.

now being re-exported across the Atlantic by Hispanic Americans peddling an investment opportunity.

In fact, statesmen presented the natural world of Central America as one of the chief reasons that foreigners might be persuaded to invest in the region. Nowhere was the insistence on the transformative power of natural resources (and science as a tool for identifying them) more strongly expressed than in José del Valle's proposal to the Guatemalan Congress to invite a 'scientific expedition, composed of an astronomer, a botanist, a geologist, a physico-mineralogist, a physician and corresponding draughtsmen and artists'. This expedition would not operate for the benefit of a narrow audience, as some of the Spanish expeditions had done: the scientists' purpose would be to 'study Central America and make it known abroad'. Valle thought that it should not be an autonomously Guatemalan undertaking, but that it should be formed by an 'Anglo-Guatemalan company', expanding upon the already existing commercial and scientific relationships between the two countries. He hoped that the 'spirit of speculation', which he saw reigning in Britain, would take this opportunity by the horns.[94] It may of course be that it was Central America's paucity of resources that prompted Valle to frame this project as one of foreign investment and speculation rather than a homegrown scientific commission in the mould of the 1820s chorographic commission in Colombia.[95] However, he also appeared to see any ability to attract international scientific attention as a way for Guatemala to present itself on the world stage – an undertaking that would give the country both scientific and political legitimacy.

Like other Central American reformers, Valle thought that Spain's imperial knowledge of these lands had been at best incomplete. He framed this not just in terms of Spain's failure (Guatemala having been unjustly overlooked by most Spanish botanists), but in terms of the fabulous natural wealth of America, which could not be fully explored even in a century of expeditions. Even in South America, the best-explored part of Spain's former possessions, there were 'immense spaces which still have not been seen by the botanists' observing eyes'.[96] In 1829, Valle also wrote to Alexander von Humboldt asking him to come to Central America. They had previously exchanged letters in 1825, in which Humboldt encouraged Valle to study Central America's volcanoes.[97] Valle repeated the foundation myth of Bourbon scientific interest: that America

[94] 'El Lic José Cecilio del Valle, presenta al gobierno del Estado, un plan para hacer venir una expedición científica, integrada por un astrónomo, un botánico, un geólogo, un físico-mineralogista, un medico y los dibujantes y artistas correspondientes para que estudien a Centro América y lo de a conocer en el extranjero' (5 October 1825), AGCA, B85.1 Exp. 82664 Leg. 3599.

[95] Applebaum, 'Reading the Past'; Del Castillo, *Crafting a Republic*.

[96] Del Valle, *El Amigo de la Patria*, Vol. 2, no. 11 (26 July 1821).

[97] Humboldt to Valle, 30 November 1825; Valle to Humboldt, 29 October 1829, in Del Valle, *Cartas de José Cecilio del Valle*, 48–51.

was a vast and unexplored land. The country's fabulous plant wealth was no accident: it could be understood by putting it in the context of global climatic zones, as he explained. Quoting the *Encyclopédie*'s natural history articles and Conrad Malte-Brun's *Universal Geography*, he explained that Guatemala was located in the earth's 'torrid zone', which was marked by the happy coincidence of uniting 'the beautiful and the useful'. Central America therefore deserved to be studied more thoroughly. After all, the continent of America in general had furthered the careers of the greatest European scholars, from Hernández to Ruiz and Pavón, Humboldt and Bonpland, and expanding their work to Central America would drive naturalists' careers as well as the country forward in a mutually beneficial arrangement.[98] Valle pragmatically stated that further research would demonstrate the 'value of our lands', referring specifically to the export value of natural productions. For him, scientific botany did not exist without economic botany. Guatemala held much in store for students of both. Valle did not see a contradiction in asserting Guatemala's status as a modern and progressive place, and resting its chief claim to global fame on its natural wealth: filling in the gaps of received colonial knowledge, natural history, and foreign investment would support each other.

Not all Britons interested in Guatemala reduced it to a 'land of landscapes', of course. The polymath José del Valle in particular quickly established an international reputation as a scholar. The British emissary George Alexander Thompson's description of his visit to Valle's study, overflowing with books and manuscripts, and his visit to the mint, convey respect for the men he encountered in Guatemala.[99] Valle corresponded with the English political theorist and utilitarian philosopher Jeremy Bentham throughout the 1820s, sending him copies of his writings in the newspapers *El Amigo de la Patria* and *El Redactor General*. Valle was particularly inspired by Bentham's designs for a civil penal code. Bentham in turn seemed to see a kindred spirit in Del Valle, praising him as the 'most estimable man that late Spanish America has produced' to two of his other correspondents, and naming him as one of the recipients of twenty-six mourning rings to be distributed after his death. In another instance, Bentham compared Valle and the Bengali reformer Ram Mohan Roy, seeing them both as 'kindred souls', part of an international elite of reformers bringing enlightened improvement to their countries. Bentham also helped to amend and augment a list of books that Valle proposed to buy in Europe and have them shipped to Guatemala, on legal matters as well as on

---

[98] Del Valle, *El Amigo de la Patria*, Vol. 2, no. 11 (26 July 1821). Malte-Brun in his *Geographie universelle* added that the torrid zone produced men who were brutes alongside these large vegetables, which Valle chose to ignore.

[99] Jordana Dym, 'La reconciliación de la historia y la modernidad: George Thompson, Henry Dunn y Frederick Crowe, tres viajeros británicos en Centroamérica, 1825–1845', *Mesoamérica* 40 (2000): 142–81.

science and natural history.[100] Bentham, hoping that one of the new Latin American nations might adopt his Codes of Civil and Penal Law, also saw an opportunity for the expansion of his ideas in Guatemala and courted Valle as a follower, hoping that he would be elected Guatemala's president (he narrowly lost to Manuel Arce in 1825). A correspondent's suggestion that José del Valle would be 'the most efficient as well as the most willing instrument you can employ' when it came to his relations with Guatemala suggests that Bentham saw himself as a sort of puppet master, pulling philosophical strings through like-minded people across the world.[101]

In addition to select intellectuals maintaining such correspondence, expert foreigners were also welcome in Guatemala. In the 1820s, the new nations made conscious use of the people and knowledge contained in the newly expanded global networks they had access to. If Spain was considered to have held back scientific progress, Western Europe was the model for rapid development and scientific expertise. Latin American countries recruited foreigners in the wake of independence for their technological expertise or their ability to conduct research. Colombia and Mexico, for instance, 'imported' foreign experts to set up museums of natural history, as Helen Cowie has shown.[102] In 1825, the government empowered its envoys to promise free passage, land, and funding for the initial cost of cultivation to anyone with expertise in growing vines or olives, while in 1835, a Guatemalan merchant was tasked with recruiting a civil engineer while travelling to Havana, London, and Paris.[103] There was no suggestion that this would undermine national autonomy: it was a pragmatic recruitment decision. The principle of this phenomenon was not entirely new. The colonial Economic Society and *Gazeta* had previously expressed the idea that non-Hispanic foreigners might offer useful knowledge for the development of Central America, but in the colonial period it had primarily been texts and plants, not people, who made their way across the Atlantic. Closer to home, Valle's Economic Society also revived a specific failed plan from the Society's *Cortes* era to request a visit to Guatemala from Andrés del Río, a Spanish mineralogist working in Mexico. In 1824, Valle corresponded at length with Del Río, hoping that his German associate might make the journey to Guatemala. The plan faltered, both because of an initial lack of funds and because the geologist eventually

---

[100] List of books, Bentham MSS, UCL Archives, London, UK, Box 12/348. Letter 3300 to Jean Baptiste Say (18 January 1827) and Letter 3393 to Rammohun Roy (1827–8), in Jeremy Bentham, *The Collected Works of Jeremy Bentham: The Correspondence of Jeremy Bentham*, Vol. XII, ed. J. H. Burns and H. L. A. Hart (Oxford Scholarly Editions Online, 2015).

[101] Letter 3289 from Sarah Austin (18 December 1826), Bentham, *Collected Works* Vol. XII; James Schofield, 'Jeremy Bentham: Legislator of the World', *Current Legal Problems* 51 (1998): 115–47.

[102] Cowie, *Conquering Nature*, 152–3.

[103] Del Valle, *Discurso del presidente*, 29; AGCA, B85.1, Leg. 3599, Exp. 82664, fol. 21.

travelled to Colombia instead.[104] Valle remained confident in his plan, however, and promoted it as late as 1830. He thought that 'useful men' coming from Europe to Central America would help the progress of 'Enlightenment and wealth'.[105] While European countries' practical and economic aims were sometimes seen with suspicion, Guatemalan intellectuals of the independence period often embraced these countries as scientific partners.

Independent Cartographies

In the 1820s and 1830s, the Guatemalan government, the Economic Society, the Consulado, and even foreign speculators broadly agreed that gathering geographical information, from current surveys as well as colonial archival material, was key to unlocking the natural wealth of Guatemala and represent the country to the rest of the world. However, cartographic representations were in short supply, as Guatemalans and foreigners were well aware. In the 1820s, existing maps of Guatemala gained a rather frosty international reception. As Jordana Dym has noted, a traveller to Central America complained that he had trouble finding any maps that were not 'more calculated to mislead than inform'.[106] A member of Britain's boundary commission in Belize made his opinion clear when he spoke of a map he had been given to him by the Guatemalan government. He expected it to be 'undoubtedly full of Errors, arising from the total want of proper instruments at the line of surveying the Country'.[107] Tellingly, Alexander von Humboldt reported that in response to a written request for a map – any map – of the interior Guatemala, perhaps one from the archives, he merely received 'a small, very strange plan of the plateau between New Guatemala and Lake Atitlán, a copper engraving printed in the country itself in 1800': Rossi y Rubí's map of his 'enlightened road'.[108] It was probably Valle who sent this to him. British observers also continued to complain about the state of geographical disinformation while still using Spanish cartographic material. Although the diplomat George A. Thompson professed his disappointment at the official maps available to him during his visit to Guatemala, the title page of his published *Narrative* was illustrated with a map based on archival material that was circulating

---

[104] Del Valle, *Discurso del presidente*, 28; Del Valle, 'Carta geografica', *Mensual de la Sociedad Económica*, no. 3, 56–7; Andrés del Río to Valle, 11 April 1827, in Del Valle, *Cartas de José Cecilio del Valle*, 62–3.
[105] Valle to Rocaforte, 10 July 1825, in Del Valle, *Cartas de José Cecilio del Valle*, 38–9.
[106] Jordana Dym, '"More Calculated to Mislead Than Inform": Travel Writers and the Mapping of Central America, 1821–1945', *Journal of Historical Geography* 30 (2004): 340–63.
[107] 'Guatemala, supplementary. Boundary Commission' (1827), TNA, FO 15/9, 130v.
[108] Alexander von Humboldt, 'Ueber den neuesten Zustand des Freistaats von Centro-Amerika oder Guatemala', in Héctor Pérez Brignoli, *Zentralamerika/Centroamérica* (San José: Editorial de la Universidad de Costa Rica, 2011), 39–109, at 40.

among Guatemala's scholarly elite: copies of different plans of Iztapa that he brought back to Britain were marked as 'copy of a survey made by the scientific & venerable Canon Dighero' (who had been responsible for many of the Consulado's harbour explorations alongside Irisarri) and 'lent to me by Don José del Valle'. Another map had come from the military commander Manuel Arzú.[109] In addition, Valle may have helped George A. Thompson to pencil in new information about independent Central America's territorial divisions on an older map when he visited his office.[110] Nobody was in a position to ignore what existing maps there were of Guatemala. Indeed, it is possible that it was Central America's particularly under-funded public purse that made independent governments and British speculators and diplomats alike particularly dependent on its colonial archives.

Guatemalan statesmen agreed that there was a lamentable paucity of cartographic depictions. José del Valle complained that 'we still do not have an exact map of Central America' in 1830. Much like Galindo did through his RGS correspondence, Valle also sent a manuscript map to London with a young Guatemalan traveller to reduce the cartographic misinformation that circulated.[111] As early as 1824, José de Barrundia's liberal government, on the advice of Valle, had resolved to procure a 'map that should describe, with the best possible exactitude, the territory of the state with the current division of counties'.[112] It did not become a reality until the early 1830s under the auspices of Mariano Gálvez's government. In 1831, it noted that the Economic Society had already started to collect data relevant to the compilation of a map of independent Guatemala, and granted it 100 pesos for the work of engraving and printing it.[113] Shortly after, the engineer Miguel Rivera Maestre set to work on a series of prints that would depict the state of Guatemala within the Federation. The collection of eight maps was completed in 1832. It consisted of one map of Guatemala as a whole (Figure 6.1), and one map of each of its seven departments. The new map would supersede older colonial maps, even in a literal and material way. Ever short of materials, the Economic Society pointed out that in the archive of the former Consulado de Comercio, a printing plate existed on which the Consulado's 1808 map of Izabal was engraved. The government now asked the citizen in charge of this archive to pass the plate on to the Economic Society, so it would be able to re-use it for the purpose of the new map. The old map of Izabal was 'of absolutely no use' now that the new 'geographical map of

---

[109] TNA, MPK 1/54; Thompson, *Narrative*, 202.
[110] Thompson, *Narrative*, 321. Valle disputed this later: Dym, 'Democratizing the Map', 166.
[111] Del Valle, 'Carta geografica', *Mensual de la Sociedad Económica*, no. 3, 59–60.
[112] Antigua Guatemala, 15 October 1824, AGCA, B, Leg. 193, Exp. 4215.
[113] 'Se concede a la Sociedad Economica la cantidad de 100 pesos para abrir la lamina', 15 April 1831, AGCA, B. Leg. 1390, Exp. 32077.

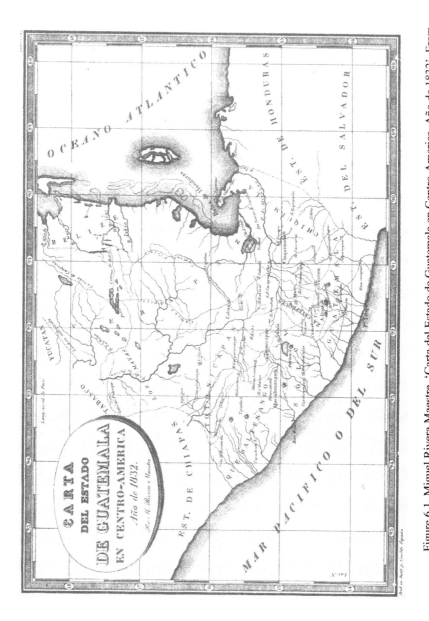

Figure 6.1 Miguel Rivera Maestre, 'Carta del Estado de Guatemala en Centro-America. Año de 1832'. From *Atlas guatemalteco*, 1832. Map reproduced by permission of Ministerio de Relaciones Exteriores, Guatemala, from *Atlas guatemalteco. Año de 1832. Edición facsimilar* (Guatemala City, 2001).

this State' had been drawn up. The new atlas would replace the cartography of the colonial period literally as well as intellectually.[114]

And yet, this first map of independent Guatemala made by Guatemalans, like many of its predecessors, purported to be proudly unscientific. Just as José Rossi y Rubí's map of Suchitepéquez province relied on estimated distances, and the *Gazeta* had declared all mathematical geography 'futile' in 1797, the makers of the *Atlas Guatemalteco* acknowledged that it lacked mathematical precision, but chose not to apologise for this. Scientific exactitude might be something to aspire to at a later point, but the Economic Society and Congress in 1831 advocated a step-by-step approach in which Guatemala should be geographically described differently from 'more advanced' countries. The government noted that fixing any latitude or longitude for the majority of places on the map would require 'excessive' resources and time. As a consequence, they explained that it was 'at the present still not possible' to create a map 'with mathematical exactitude, because the operations and necessary calculations for determining geographical positions have still not been carried out'.[115] However, their attitude was not one of resignation. Instead, they calmly explained that the map they were producing was exactly what could be expected of a map of Guatemala at this point in time:

One cannot arrive at exactitude in just one big jump. [Even] the more advanced nations are arriving there step by step. It is not possible to start with mathematically exact maps. At least in the beginning we are satisfied with those who approximate exactitude, and this is what was proposed to the Society.[116]

Approaching, but not reaching, exactitude was enough for Congress. Although the *Guatemalan Atlas* was born from a desire to represent the country's borders internationally, it did so on its own terms. Of course, this was probably at least in part a rhetorical distraction to explain away the lack of funding for this project, but there was also an undertone of 'patriotic' science in these descriptions of exactitude 'appropriate' to Guatemala. Patriotic considerations also appeared in discussions around which meridian to use for the map: José del Valle favoured the American meridian demarcated by the 1742 geodesic mission to the Andes as a point of hemispheric pride, although practicality won the day and the map used the Paris meridian instead.[117] The simplicity of the new map of Guatemala, at least in its makers' rhetorical defence, was a return to the values of reformers of the previous decades, defined by an application of

---

[114] 'La Sociedad Económica solicitó la lámina en que está grabado "el punto de Izabal", lámina que debe estar en el archivo del extinguido Consulado de Comercio', AGCA, B119.4. Leg. 2556, Exp. 60097, fol. 8.

[115] 'Se concede a la Sociedad Economica la cantidad de 100 pesos para abrir la lamina', 15 April 1831, AGCA, B. Leg. 1390, Exp. 32077.

[116] *Ibid.*   [117] Del Valle, 'Carta geografica', *Mensual de la Sociedad Económica*, no. 3, 63.

scientific knowledge in so far as it was entirely appropriate to the circumstances of, and state of knowledge in, the country.

The emphasis of the new atlas was not solely on boundary measurements, but it inevitably made political statements in the representation of the borders of the new state of Guatemala. Some of these boundaries were conditioned by political disputes (for instance, the extent of British claims to Belize was downplayed), but others relied on a long-established link between history and geography, a further sign that this map built on decades of Guatemalan geographical discussions and reflected internal priorities.[118] The northern boundary of the 1832 map, as Jens Bornholt has pointed out, 'carefully curves around' the ruins of Palenque in order to situate them in Guatemalan territory.[119] This seemingly well-defined boundary was wishful thinking on the Guatemalan government's part. After all, Palenque in the Spanish colonial period had been considered part of Chiapas, and it was only now that Chiapas had broken away from Guatemala and joined Mexico that a space was carved out of the border to make it seem otherwise. The new state also attempted to claim ancient ruins on its southern border, by surveying the archaeological site of Copán (near the Honduran–Guatemalan border) for inclusion in the maps and descriptions of the atlas. In a letter to his Honduran counterpart, the leader of Guatemala's government, Mariano Gálvez, conceded that in his effort to map the 'precious ancient monuments which the territory of Guatemala contains', the surveyor would probably have to enter Honduran territory, too.[120] Although he formulated it as a polite letter of recommendation for the engineer who was about to carry out his work in Honduras, in the context of the ruins of Palenque being claimed by the same project, it is clear that the overall intention was to 'usurp' these monuments, technically just outside its borders, to the new Guatemalan state's territory. In the end, Copán and Palenque were not included in the final set of maps. However, versions of the *Atlas* printed after 1834 contained an additional three maps by Rivera Maestre: the Maya sites of Utatlán, Quiché, and Iximché, which were indubitably contained in the territory Guatemala (see Figure 1.2 for Quiché). For Galvez, mapping the ancient ruins was a 'patriotic' project, designed to 'ennoble the state'. Valle, too,

---

[118] See also Dym, 'Democratizing the Map', 164–70; Arturo Taracena Arriola, Juan Pablo Pira, and Celia Marcos, 'La construcción nacional del territorio de Guatemala', *Revista Historia* 45 (2002), 9–33.

[119] Jens Bornholt, *Cuatro siglos de expresiones geográficas del istmo centroamericano, 1500–1900* (Guatemala City: Universidad Francisco Marroquín, 2007), 140.

[120] 'Manuel Jonama es comisionado para reconocer las ruinas de Copán/Trascribase al C Juan Galindo nombrado en lugar del C Jonama', 1834, AGCA, B95.1, Leg. 1398, Exp. 32575. The engineer Jonama, who also oversaw several road reconnaissance projects, was initially commissioned to carry out the task, but refused the commission. Juan Galindo proceeded to map Copán instead. His papers were later published by Sylvanus Morley. See Graham, 'Juan Galindo', 26–9.

claimed indigenous history as central to the new state's history, even while doubting the ancient builders' 'degree of enlightenment'.[121] Geographical interest in archaeological ruins projected the narrative of a new nation even while its potential social ramifications for the status of indigenous people continued to be ignored.

Despite these cartographic decisions, which revolved around long-standing internal debates about the significance of indigenous pasts and ruined cities for Guatemala, Central Americans also recognised the increasing importance of conforming with internationally relevant ideas of space and measurement. Even while road-building commissions explicitly valued the information of 'an old mule driver', José del Valle's concept of useful knowledge specifically excluded such knowledge as a mule-driver might possess. The mule-driver's experiential knowledge was not 'scientific': it was 'without knowledge, observations, or experiments', and no longer fitted into the framework of international science that Valle imagined Guatemala should aspire to.[122] By 1837, the engineer Rivera Maestre, author of the *Atlas* maps, submitted a report to the Guatemalan government in which he proposed to create a better system for accurately converting Central American units of measurement, *varas*, into European metres.[123] Rivera Maestre's national map, and Valle's certainty that accurate geographical information about Central America would result in the republic's recognition by other governments, show that international standards, rather than local applicability, were increasingly the measure of the usefulness of a geographical strategy. While the many different definitions of a 'league' in Spanish America were not a concern as long as the local units of measurement were agreed on, a more universal system of measurement was becoming necessary in a world where maps were used to represent Central American spaces to other countries and foreign investors. Much, perhaps, as Rossi y Rubí's 1801 attempts at translating local medicinal knowledge to Europe had failed, parochial and global ways of doing geography clashed here. The evolving use of maps among local and foreign actors is perhaps best symbolised by an 1834 Honduran map. While the governor of the department of Gracias in Honduras drew up a map of his department that focused on the territory's governability (giving the distances between towns), and accompanied it with a 'statistical, chorographic and historical summary' of the department, it was re-used in 1853 by the North American Squier expedition. The English-language annotations that now appeared on the map formed

---

[121] AGCA, B, Leg. 1398, Exp. 32619–22; on Valle, the indigenous population, and this map, see Dym, 'Democratizing the Map', 167–8; Del Valle, 'Carta geografica', *Mensual de la Sociedad Económica*, no. 3, 61–2; 'Historia', 64.

[122] Del Valle, *Mensual de la Sociedad Económica*, no. 2, 25.

[123] Robert Claxton, 'Miguel Rivera Maestre: Guatemalan Scientist-Engineer', *Technology and Culture* 14 (1973): 384–403, at 395.

a calculation of the area of the department, and the price of land per square mile or *caballería*.[124] While, perhaps simply in a rhetorical justification built on necessity, the Guatemalan Congress defended practical and empirical knowledge over mathematically based geography, the complex international connections of Guatemala City meant that statements arguing for a uniquely Guatemalan view of geography were becoming increasingly irrelevant. However, this was not an alien, European-style positivism imposed on Guatemalan mapmakers from abroad, but a decades-long negotiation between Guatemala's own traditions about exploiting and recording its natural resources, its persistent narrative of its own peripherality, and the country's increasingly international ambit.

Conclusion

The geographical and scientific imaginations of landscapes developed in the intellectual and administrative world of Central America during the late Bourbon period became a critical part of the national narrative of Central America's new republics, in particular Guatemala, after 1821. While working in a radically different political context, independent governments pinned their hopes on the revival of some of the projects that the Economic Society and Consulado had championed. Their forceful arguments for the power of Central American nature, and the emphasis on its hitherto incomplete use, rested on the same logic that Bourbon-era reformers had employed. It represented a desire to break with the past, and to inscribe a new independent symbolism on Central American landscapes. However, even here there were echoes of the colonial Economic Society's self-confident declarations that their work formed a new era of locally focused, widely communicated, and useful science specifically for Central America. The vague boundaries of the colonial *patria* did not translate to the independent one in a meaningful geographical way, but the processes of acting on the countryside through infrastructure and attracting the interest of men who would exploit the country's natural riches through geography and natural history resonated, not just in internal representations of Guatemala, but also in its dealings with the rest of the world.

This influence was of a practical nature as much as an ideological one. Independent governments and the re-established Economic Society and Consulado eagerly made use of colonial archives. In particular, they re-examined specific plans for building roads and navigating rivers that nation-builders of the nineteenth century and reformers of the eighteenth alike had seen as crucial for the progress of Guatemalan internal or external commerce.

---

[124] José M. Cacho, 'Mapa itinerario del Departamento de Gracias: formado en 1834, por el ciudadano José M. Cacho' (1834), Bancroft Library, University of California at Berkeley.

Foreigners, too, showed themselves highly interested in these geographical antecedents, with the British consul at Belize being particularly interested in the revival of a road-building project initially designed by the Consulado in the 1790s. The 28-page Consulado route-map of the Motagua river that appeared in the archive of the Royal Geographical Society, and José Rossi y Rubí's 1800 map of his mountain road that Alexander von Humboldt found puzzling, are emblematic of these travels of maps through space and time. In addition to such colonial antecedents, new surveys commissioned by the independent Guatemalan government and geological specimens collected on its behalf also travelled across the Atlantic through new networks of scientific exchange and ambiguous political loyalties. All these approaches to recording Guatemalan geographies built on Bourbon-era traditions of defining society's relationship with the landscape and building up knowledge that would allow governments to exploit the natural resources found therein. Both consciously and unwittingly, independent governments of Guatemala, a new generation of statesmen influenced by ideas of enlightened reform, as well as British investors and statesmen, profited from the legacies of this knowledge.

Conclusions

Looking at the country they sought to reform, the self-proclaimed enlightened enthusiasts of Central America saw all sorts of ruins: archaeological sites, the remains of the old city of Antigua as a background against which the new capital of Nueva Guatemala was built, fields destroyed by locusts, the decline of indigo exports, and roads covered by mudslides. The eruption of a volcano might emit sufficient fire to resemble 'the mouth of hell', while rivers were so difficult to navigate that European standards of engineering did not apply. Yet enlightened approaches to science and 'useful knowledge' promised to bring change to Guatemala, in the sense of material improvement as well as in the paternalistic sense of an ordered, 'moral' society that started with a controlled landscape. A new focus on nature arose simultaneously from the Bourbon Reforms' emphasis on natural resources, and the emergence of new patriotic discourses about land in Spanish America. The rhetoric of fulfilling the region's potential and of discovering dormant opportunities was particularly salient in the projects of the Economic Society and the Guatemala City merchant association, the Consulado de Comercio. Balanced against a pervasive view that Central America was particularly remote from the rest of the world and this sense of ruin that underpinned reformers' actions, the self-proclaimed enlighted patriots believed that their practical interventions in the landscape could transform the region's fate.

By 1838, Central American landscapes had not been transformed as the Bourbon-era reformers had hoped. Lack of money and political turmoil all too often put a stop to their grand projects. Nevertheless, their optimistic reimagination of Central America had taken hold and informed the ways in which nineteenth-century statesmen constructed their own vision of an independent Guatemala. Central America was no longer seen as peripheral: it asserted its promise as a region at the centre of the American continent, between two oceans. It could draw on cultivated and 'wild' landscapes that contained a multitude of climates and natural resources which had borne fruit, or, rather, had produced specific 'useful' crops. Its countryside might look harsh, but experience showed that it could be surveyed and managed. Even the famously tropical and humid coasts were not an obstacle: they had perhaps been mismanaged, but were not necessarily unhealthy. Key to all of this was a thorough knowledge of landscapes and their particular properties, not in a generic sense of 'American' or even 'Guatemalan' landscapes, but increasingly a focus on local and regional geographies, picking out individual places to connect to a landscape made useful through patriotic actions: the harbours of Iztapa or Izabal, or the allegedly promising agricultural zone of Trujillo. Crucially, building superficial infrastructure would not be enough. Reformers attempted more permanent interventions in the landscape, trying to improve the health of a place and its airs by attempting to intervene in its vegetation and ventilation. Independence-era scholars, none more than José del Valle, agreed that plenty more could be done to further the study and exploitation of these landscapes, but the groundwork had been laid.

Some critics regarded this as blind optimism. Cynicism or, one might say, realism made some appearances in public discourses about nature. Occasionally, archival documents or newspaper articles indirectly recorded the voices of farmers (or even *hacienda* owners) who complained that new-fangled crops and agricultural strategies were being imposed on them by superiors or city-dwellers who really did not know best. One commentator made fun of those Economic Society members who believed in the superiority of the Guatemalan soil even though Venezuelan soil could also provide good indigo. In a similar vein, the repeated attempts to navigate the Motagua by 'addicts of navigation', despite the apparent impossibility of the project, were taken by a nineteenth-century observer as 'a new proof that in this Kingdom there is an immense distance between speculation and practice'. Having a difficult landscape to deal with was no excuse: this lack of successful projects persisted 'even in those things whose execution is considered trivial in other places'.[1] This unusually harsh criticism came from an outsider, a Spaniard, though long-term resident of Guatemala. Most reformers, by contrast, blamed any setbacks on the chronic lack of funding and the overly ambitious nature of some of their projects. They believed in the fundamental bounty of Central American nature, but above all their own practical actions that were transform-ing the landscapes around them, at least in small ways.

A variety of projects drawing on broadly 'enlightened' ideals as well as responses to specific local grievances created new relationships between inland cities and the coasts, and between urban elites and the surrounding countryside, and even urban nature such as trees or small botanical gardens. Scholars and administrators engaged in a constructive negotiation of different intellectual traditions, taking into account local, lived experiences and views of nature as well as the broader ideals of Enlightenment. The distinctly localised approaches that reformers insisted on led to carefully defined descriptions of

---

[1] 'Informe de Dn José Munilla, y Tesorero Dn Rafael Trullé', AGI, Guatemala, 892, fol. 6.

particular disciplines of knowledge, including geography, history, and botany. Throughout, a stubborn insistence on practicality over theory led to a distinctly Guatemalan interpretation of the purpose of science and Enlightenment. It was a practical Enlightenment that, obsessed with tracing the effects of and Economic Society initiatives through the countryside, exchanged plant material, and negotiated its own perspective on the applicability of 'useful knowledge'. In botany and medicine the *Gazeta*'s editors stressed that theoretical knowledge was at best of limited use: it was no replacement for directly observing or experiencing phenomena. To some extent, this attitude was motivated by the core mission of the newspaper serving a wide public, who might not be familiar with learned language, as the editors never ceased to point out. It nevertheless resulted in the ruthlessly impatient creation of a new kind of enlightened knowledge: one that would be 'useful' to Guatemalans specifically. Even examples of principles of political economy or the industrial application of scientific knowledge were only of use if they had a proven application in Guatemala: they needed to be adapted to the 'circumstances of the country'. When it came to writing history, this rejection of established but unproven authority led Economic Society members to defend the use of 'original manuscripts' from archives to base historical narratives on. The most impassioned tirades in favour of empirical and local knowledge, however, were reserved for geography. Parish priests and other *Gazeta* readers contributed their own observations to the newspaper in an attempt to form a new body of knowledge, directly relevant to Guatemala, countering the outdated publications of European geographers. Individual reformers also felt emboldened to conduct their own practical projects of reconnaissance and mapping.

Since learned treatises were not the ultimate answer to finding useful science, empirical knowledge must be found within Central America. In identifying guidelines for collecting knowledge from the countryside, reformers turned to established traditions of knowledge-gathering. Some, like the idea that Indians possessed knowledge of the countryside and its plants invisible to Spaniards, or that routes through the mountains or up and down rivers must exist if Indians or contrabandists used them, rested on an understanding of the distinct social and geographical spaces inhabited by a divided population. Others drew on bureaucratic practices. While reformers saw no use in a closed bureaucratic circuit in which reports were read by a tiny circle of men, or a bureaucracy whose red tape crossed the Atlantic and added years to infrastructure projects, they appreciated the usefulness of its established modes of information-gathering, and the ability of the Church hierarchy to reach the most far-flung corners of the kingdom through its parish priests. In the case of botany as well as geography and archaeology, they emphasised the Society member, merchant, or farmer's physical position within the landscape. Information networks were conceived of as two-way connections, fundamental not just for gathering useful knowledge, but for disseminating it. The publication of the Economic Society's work (and to some extent, the Consulado's work) was a key goal of their enlightened aims. Progress and Enlightenment could be said to have taken foothold only when it had reached 'the public'. It was an Enlightenment of Guatemala, but more specifically of the 'patriots' and 'useful men' who would apply themselves in its service: scientific knowledge was worth nothing as a disembodied (European, possibly written in Latin) entity. It was its application by a range of people to their everyday tasks which would bring a truly prosperous and successful country.

Despite this insistence on local relevance, reformers did not believe that they were constructing a parochially Guatemalan phenomenon. Instead, they also valued the transnational aspects of enlightened learning and saw themselves as participants in a global world of learned communication. Guatemala was peripheral in the sense that it was less wealthy than large viceroyalties, and counted fewer harbours and trade routes. However, it also formed the centre of its own networks of scientific communication. By simultaneously emphasising the image of an isolated Central America with few outside connections, and appearing to offer solutions to these problems through science and Enlightenment, the Economic Society and the *Gazeta de Guatemala* ensconced Enlightenment ideals of science, progress, and optimism as the central tenets of Guatemalan society's approach to nature. Independence-era statesmen repeated and complicated these rhetorical patterns, turning to internal infrastructure and science projects, but also negotiating matters of geography, infrastructure, and development with foreign financiers and political powers. British governments dismissed colonial Spanish and independent Guatemalan maps even while they were reliant on them. The publication of two parallel articles on Sumatran plants in a Guatemalan and a British journal only months apart hints that supposedly 'peripheral' locations may have been much less cut off from global historical phenomena than one might suppose and forms a model for the engagement of other 'peripheral' places with global intellectual processes.

First Bourbon reformers, then independence-era statesmen, saw their work in applying scientific knowledge to the countryside as a radical departure from a less enlightened past. They saw themselves as enlightened standard-bearers, confronted with an obscurantist past. The public sphere and correspondence network represented by the *Gazeta de Guatemala* formed a genuinely novel way of distributing information and questioning established truths than previous models of correspondence, although, of course, the Economic Society's members happily accepted support from the Audiencia government in the form of labour tribute, or Consulado funds. The rhetorical position of novelty, change, and bringing light to a dark place of ignorance was a formative part of Guatemalan enlightened rhetoric. However, this rhetoric also obscured continuities which would have thrown doubt on claims of singularity. The structures of political governance and commercial practices had long incorporated their practical experience of living with natural risk and scheming to make a terrain often perceived as defined by volcanoes, mudslides, and mountain roads more prosperous and governable. In their supposedly groundbreaking 'Description of Guatemala', the *Gazeta de Guatemala* closely followed administrative formats of recording space, and drew on ecclesiastical census information. In negotiations on foreign loans and fixing national borders, both Guatemalan statesmen and foreign investors relied on archival material from the colonial period.

In its optimism, this Enlightenment could be expected to be resolutely oriented towards the future. However, the historical and mythical past also played a large role in legitimising reformers' claims that their future plans were founded on more than empty rhetoric. The landscape formed a strong connection between the past, present, and future – it became historical and historicised. Specific landscapes and places could be simultaneously constant in their promise of fertility and abundance, but also deeply malleable in conjunction with human's actions and management of the environment. Their economic potential could be measured by the accounts of eyewitnesses, historical memory, or archival documents relating to economic activity. Captain-General Estachería made a similar assumption in the 1780s when he thought that from the physical evidence of Palenque's ruins, he might be able to gain concrete evidence for the economic activities of past settlers, and the present potential of the Chiapas terrain. In agriculture, as in road-building, reformers repeated this sentiment throughout the period: they often looked to excavate a semi-mythical, more prosperous past from the current landscape. Past richness of the land served as proof of the future, and a convenient way of escaping a less promising present state.

To Bourbon reformers, the more ambitious aspects of enlightened optimism did not count as illusions as long as they were carefully tied to practical actions and small results in the cultural landscape, quantifiable observations such as the number of cacao trees, or an improved number of healthy residents of a town. When the Economic Society promised that enlightened approaches would bring projects to fruition which could hitherto only have been seen as 'fantastical' or 'impossible', they established a rhetoric of promise and transformation through learning and action which also mapped well onto the new worlds of the possible that the independent nations symbolised. New cultural landscapes bore the marks of eighteenth-century visions of useful nature.

Glossary of Colonial Administrative Terms

This glossary provides simple definitions of the sometimes bewildering array of administrative terms that denoted the Spanish government's bureaucracy and structures of power in Guatemala. For more thorough and technical descriptions of Spanish governmental structures, see Hall and Pérez Brignoli, *Historical Atlas of Central America*; Rojas Lima, *Diccionario histórico-biográfico de Guatemala*; Tarver and Slape, *The Spanish Empire: A Historical Encyclopedia*. See also diagram in Burkholder, *Biographical Dictionary*, Appendix 2.

Alcalde mayor or *corregidor*
The head of the *cabildo* or *ayuntamiento*, chief magistrate and therefore the highest representative at a municipal level. Since the jurisdiction of a *cabildo* often included a large province that extended to the boundaries of the next province, these could be powerful officers.

Audiencia
The *Audiencia de Guatemala* was the highest court in the land, made up of individuals known as *oidores*. Because it interpreted Spanish law and oversaw appeals, it effectively held executive power at a regional level. The President of the Audiencia was also the Governor of the Kingdom of Guatemala, the highest representative of the king in Central America. Because the jurisdiction of the Audiencia covered the whole administrative region of Spanish Central America, this territory was often known as the Audiencia (see also: Kingdom of Guatemala).

Captain-general (*capitán general*)
See also: President. The highest office in the Captaincy-General of Guatemala. This was a military position, but in conjunction with the civilian offices of President of the Audiencia and Governor of the Captaincy-General or Kingdom of Guatemala (all held concurrently by the same person), effectively the highest representative of all branches of government.

Governor (*gobernador*)
Gobernador could refer to either the Governor of the entire Kingdom of

Guatemala (that is, the Captain-General and President of the Audiencia), or the highest Spanish official in an intendancy, who was usually referred to as the *'gobernador intendente'*. Where there is potential for confusion, I use the English word 'governor' as a general term for regional administrators such as *alcaldes mayores* and *intendentes*, and refer to the incumbent of the office of the 'Governor of Guatemala' as President (of the *Audiencia*) or Captain-General.

Intendancy (*intendencia*)

As part of the Bourbon Reforms, Intendancies were established as new units of government in Spanish America. In 1786–7, León (Nicaragua), Ciudad Real (Chiapas), Comayagua (Honduras), and San Salvador (El Salvador) became *intendencias*, replacing *alcaldías mayores*. They were governed by an intendant (*intendente/gobernador intendente*). Only *españoles peninsulares*, Spaniards born in Spain, could hold this office. Since the intendancy reforms did not cover the whole kingdom, the offices of intendant and *alcalde mayor* existed in different parts of Central America at the same time.

Kingdom of Guatemala

Spanish administrative unit covering Central America, also known as the Captaincy-General or Audiencia of Guatemala. Technically part of the Viceroyalty of New Spain, in practice administered almost exclusively from the capital in Guatemala (Santiago de Guatemala until 1773, then Nueva Guatemala de la Asunción, as Guatemala City was formally known). In 1779, the captain-general acquired additional powers of taxation, meaning that in the period covered in this book, the Kingdom of Guatemala was a territorial subdivision with independent military powers and a government, judiciary, and treasury that was effectively independent from the Viceroyalty of New Spain.

President

President of the Audiencia, the highest representative of the king in Central America. An office concurrently held with that of Captain-General of Guatemala and the title of Governor of the Kingdom of Guatemala, uniting military, judicial, and executive power in one person. Colonial presidents of the period covered by this book: Martín de Mayorga (1773–1779), Matías de Galvez (1779–1783), José de Estachería (1783–1789), Bernardo Troncoso Martínez (1789–1794), José Domás y Valle (1794–1801), Antonio Gonzáles Mollinedo y Saravia (1801–1811), José de Bustamante y Guerra (1811–1818), Carlos de Urrutia y Montoya (1818–1821). To avoid confusion, I refer to the President of the Audiencia as President or Captain-General, while the president of the Economic Society is denoted 'director of the Economic Society'.

Province (*provincia*)

An informal term for smaller administrative units within the kingdom. *Provincia* was often used to refer to the larger hinterland under the jurisdiction of a *corregidor* or *alcalde mayor*. Used in the plural (*estas provincias*) it was a generic term for 'these lands' and could refer to several *alcaldías mayores* or the whole kingdom.

Bibliography

Manuscript Sources

Archivo General de Centro América. Guatemala City, Guatemala
  A – Colonial period
  B – Independent period
Archivo General de Indias. Seville, Spain
  Estado, Guatemala
  Indiferente General
  Mapas y Planos, Guatemala
  Mapas y Planos, Mexico
Archivo del Museo Naval. Madrid, Spain
Bancroft Library. University of California at Berkeley, USA
British Library: Western Manuscripts Collection. London, UK
Hispanic Society of America: Hiersemann Collection. New York, USA
Natural History Museum: Mineralogy Library. London, UK
The Newberry Library: Ayer Collection. Chicago, USA
New York Public Library: Map Division. New York, USA
The National Archives. London, UK
Royal Geographical Society Archives. London, UK
University College London Archives: Bentham Papers. London, UK
Wellcome Library: Archives and Manuscript Collection. London, UK

Printed Primary Sources

Anonymous. *Isagoge historico apologetico general de todas las Indias y especial de la provincia de Sn. Vicente Ferrer de Chiapa y Goathemala de el orden de predicadores*. Edited by José María Reina Barrios, Madrid, 1892.

Baily, John. *Central America: describing each of the states of Guatemala, Honduras, Salvador, Nicaragua, and Costa Rica; their natural features, products, population, and remarkable capacity for colonization*. London: T. Saunder, 1850.

Bateman, James. *The Orchidaceae of Mexico and Guatemala*. London: Ridgway & Sons, 1843.

Batres Jáuregui, Antonio, *La América Central ante la historia*. Guatemala: Sanchez y De Guise, 1920.

Bentham, Jeremy. *The Collected Works of Jeremy Bentham: The Correspondence of Jeremy Bentham*, Vol. XII: July 1824 to June 1828. Edited by J. H. Burns and H. L. A. Hart. Oxford Scholarly Editions Online, 2015.

Cacho, José. 'Extracto del resumen estadistico, corográfico histórico del Departamento de Gracias, escrito por el señor Don José Maria Cacho en el año 1834.' *Revista del Archivo y Biblioteca Nacional de Honduras* 4 (1908).

Cadena, Felipe. *Breve descripcion de la noble ciudad de Santiago de los Caballeros de Guatemala: y puntual noticia de su lamentable ruina ocasionada de un violento terremoto el dia veinte y nueve de julio de mil setecientos setenta, y tres.* Mixco, Guatemala: Oficina de don Antonio Sanchez Cubillas, 1774.

Córdova, Matías de. *Utilidades de que los Indios se visten y calzen a la española, y medios de conseguirlo sin violencia, coaccion, ni mando.* Guatemala City: Ignacio Beteta, 1798.

Corona Española. *Recopilación de leyes de los reynos de las Indias. Tomo Segundo.* 3rd ed., Madrid, 1774.

Cortés y Larraz, Pedro. *Descripción geográfico-moral de la Diócesis de Goathemala. 1770.* Edited by Julio Martín Blasco and Jesús María García Añoveros. Madrid: Consejo Superior de Investigaciones Científicas, 2001.

Del Valle, José Cecilio. *Cartas de José Cecilio del Valle.* Edited by Rafael Heliodoro Valle. Tegucigalpa: Universidad Nacional Autónoma de Honduras, 1963.

Discurso del gobierno supremo de Guatemala sobre la renta de tabacos. Leido en la Asamblea el día 11 de Octubre de 1824. Guatemala City: Imprenta Nueva de Juan José Arévalo, 1824.

Discurso del presidente del Poder Executivo a la apertura del Congreso Federal de Guatemala. En 25 de Febrero de 1825. Guatemala City: Imprenta Nueva de Juan José Arévalo, 1825.

Memoria sobre la educación. Guatemala City: Imprenta de la Union, 1829.

Espinosa y Tello, José. *Relacion del viage hecho por las goletas Sutil y Mexicana en el año de 1792 para reconocer el Estrecho de Fuca: con una introducción en que se da noticia de las expediciones executadas anteriormente por los españoles en busca del paso del noroeste de la América.* Madrid: Imprenta Real, 1802.

Flores, José Felipe. *Especifico nuevamente descubierto en el Reyno de Goatemala para la curación radical del horrible mal de cancro.* Guatemala City, 1781 and Mexico City, 1782.

Fuentes y Guzmán, Francisco Antonio. *Recordación Florida: discurso historial y demostracion natural, material, militar y politica del Reyno de Guatemala.* Guatemala City: Sociedad de Geografía e Historia, 1933.

Galindo, Juan. 'On Central America.' *Journal of the Royal Geographical Society of London* 6 (1836): 119–35.

García Redondo, Antonio. *Memoria sobre el fomento de las cosechas del cacaos, y de otros ramos de la agricultura. Presentada a la Sociedad Económica por el Socio Dr. D Antonio Garcia Redondo, Canonigo Magistral de la Metropolitana de Guatemala.* Guatemala City: Ignacio Beteta, 1799.

Gobierno de Guatemala. *Boletín Extraordinario: Decreto Septiembre 30 de 1832.* Guatemala City: Imprenta de la Union, 1832.

Godwin, William. *Of Population. An Enquiry Concerning the Power of Increase in the Numbers of Mankind.* London: Longman, 1820.

Gonzáles Bustillo, Juan. *Extracto, ô Relacion methodica, y puntual de los autos de reconocimiento: practidado en virtud de comission del señor presidente de la Real Audiencia de este Reino de Guatemala. Impreso con superior permiso en la Oficina de D. Antonio Sanchez Cubillas en el pueblo de Mixco en la Casa que llaman de Comunidad de Santo Domingo, año de 1774*. Mixco, Guatemala, 1774.

Humboldt, Alexander von. 'Ueber den neuesten Zustand des Freistaats von Centro-Amerika oder Guatemala. Aus Korrespondenz-Nachrichten von Alexander von Humboldt.' In Héctor Pérez Brignoli (ed.), *Zentralamerika / Centroamérica*. San José: Editorial de la Universidad de Costa Rica, 2011, 39–109.

Isasi, José de and Juan Antonio Araujo. Consulado de Comercio. *Ereccion de la Compañia de Navegacion del Rio Motagua. Nociones Dirigidas al publico, para completar el numero de acciones en esta Capital, y sus Provincias. Nueva Guatemala Año de 1796. Impresa con superior permiso en la oficina que dirige D. Alexo Mariano Bracamonte*. Guatemala City, 1796.

Juarros, Domingo. *A Statistical and Commercial History of the Kingdom of Guatemala*. Translated by John Baily. London: J. Hearne, 1823.

　Compendio de la historia de la ciudad de Guatemala. 2 volumes. Guatemala City: Ignacio Beteta, 1808.

Liendo y Goicoechea, José Antonio de. *Memoria sobre los medios de destruir la mendicidad y de socorrer los verdaderos pobres de esta capital*. Guatemala City: Ignacio Beteta, 1797.

López, Antonio. *Instrucción para cultivar los nopales y beneficiar la grana fina, dispuesta por el R. P. Predicador General Fr. Antonio López del S. O. de Predicadores y Cura de Cubulco la da a luz la Real Sociedad Económica de Guatemala*, Guatemala City: Ignacio Beteta, 1818.

Malthus, Thomas Robert. *An Essay on the Principle of Population, as it Affects the Future Improvement of Society: With Remarks on the Speculations of Mr. Godwin, M. Condorcet, and Other Writers*. London: Johnson, 1798.

Markham, Clements. *The Chinchona species of New Granada: containing the botanical descriptions of the species examined by Drs. Mutis and Karsten; with some account of those botanists, and of the results of their labours*. London: Eyre and Spottiswoode, 1867.

Marsden, William. *The History of Sumatra: Containing an Account of the Government, Laws, Customs, and Manners of the Native Inhabitants*. London: McCreery, 1811.

Marure, Alejandro. *Memoria historica sobre el canal de Nicaragua: seguida de algunas observaciones inéditas de Mr. J. Baily sobre el mismo asunto, escrita por Alejandro Marure*. Guatemala: Imprenta de la Paz, 1845.

Matienzo, Juan. *Gobierno del Perú, 1567*. Edited by Guillermo Lohmann Villena. Paris and Lima: Institut Français d'Études Andines, 1967.

Mociño, José Mariano. *Tratado del Xiquilite y Añil de Guatemala. Dedicado a su Real Sociedad Económica por D. Jose Mariano Moziño botanico de la Real Expedicion de N.E. con Notas Puestas por el Socio mencionado Dr. Fr. Jose Antonio Goycoechea. Año de 1799*. Guatemala, 1799.

　Las "Noticias de Nutka de José Mariano Moziño", 1793. Edited by Fernando Monge and Margarita Olmo Pintado. Madrid: CSIC and Ediciones Doce Calles, 1999.

Porta y Costas, Antonio. *Relacion del reconocimiento que, de orden del ex.mo Señor Presidente, Gobernador, y Capitan General, D. Bernardo Troncoso, practico el ingeniero ordinario D Antonio Porta, en la costa comprendida desde Omoa, hasta la punta de Manabique; y desde la barra del Rio de Motagua, hasta donde se le une el de Chicosapote, a 14 leguas de la ciudad de Guatemala. Con Superior permiso en la Oficina de D Ignacio Beteta 1792.* Guatemala, 1792.

Real Academia Española. *Diccionario de la lengua castellana compuesto por la Real Academia Española. Reducido á un tomo para su mas fácil uso. Tercera edición.* Madrid: Viuda de Joaquin Ibarra, 1791.

Real Consulado de Comercio. *Informe aprobado por la Junta de Gobierno del Consulado de Guatemala sobre el Objeto y Cumplimiento de las Rles ordenes de 13 Sept y 22 oct de 1812.* Guatemala, 1814. (in AGI, Guatemala, 892)

Real Sociedad Económica de Amantes de la Patria. *Estatutos de la Real Sociedad Económica de Amantes de la Patria de Guatemala, aprobada por S.M. en real cedula fecha en S. Lorenzo á 21. de Octubre de 1795.* Guatemala City: Ignacio Beteta, 1796.

Junta Pública de la Real Sociedad Económica de Amantes de la Patria de Guatemala: celebrada en 12. de diciembre de 1796. Nueva Guatemala en la Oficina de D. Alexo Mariano Bracamonte. Guatemala, 1797.

Noticia de la publica distribucion de los premios aplicados a las mejores hilanderas al torno, enseñadas en la Escuela Patriótica de la Nueva Guatemala. Celebrada en 4. de Noviembre de 1795. En la Nueva Guatemala: En la Oficina de la Viuda de don Sebastian de Arevalo, año de 1796. Guatemala, 1796.

Noticia del establecimiento del museo de esta capital de la Nueva Guatemala. Y exercicios publicos de historia natural que han tenido en la sala de estudios de dicho museo. Los bachilleres en filosofia don Pascasio Ortiz de Letona, cursante en leyes, y don Mariano Antonio de Larrabe en medicina. Bajo la direccion de don Jose Longinos Martinez, naturalista de la real expedicion facultativa de este reyno, y Nueva España, profesor de botanica etc. Con motivo de la apertura del Gavinete de historia natural, que en celebridad de los años de nuestra augusta reyna y señora, le dedicò, ofreciò, y consagrò dicho naturalista, en su dia 9. de diciembre de 1796. Impreso en la Oficina de la Viuda de d. Sebastian de Arevalo año del 1797. Guatemala City, 1797.

Novena junta pública de la Sociedad Economica de Amantes de la Patria de Guatemala. Fundada por el Sr. D. Jacobo de Villa-Urrutia y Salcedo, del consejo de S.M. oidor de esta real audienca, etc. Segunda despues de su restablecimiento, celebrada el dia 5. de abril de 1812. Por Beteta. Impresor de la Sociedad. Guatemala City, 1812.

Octava Junta Pública de la Sociedad Económica de Amantes de la Patria de Guatemala. Primera despues du restalecimiento [sic] celebrada el dia 12. de Agosto de 1811. Por Beteta Impresor de la Sociedad. Guatemala City, 1811.

Quarta junta publica de la Real Sociedad Economica de Amantes de la Patria de Guatemala: celebrada el dia 15. de julio de 1798. Guatemala City, 1798.

Quinta Junta Pública de la Real Sociedad Economica de Amantes de la Patria de Guatemala: celebrada el dia 16. de diciembre de 1798. Nueva Guatemala. Por la Viuda de Sebastian de Arevalo año de 1799. Guatemala City, 1799.

Tercera junta pública de la Real Sociedad Economica de Amantes de la Patria de Guatemala: celebrada el dia 9 de diciembre de 1797. Guatemala City, 1798.

Real Sociedad Económica de Amigos del País. *Periódico de la Sociedad Económica.* Guatemala City, 1815–16.

Rio, Antonio del. *Description of the Ruins of an Ancient City: discovered near Palenque, in the Kingdom of Guatemala, in Spanish America. Translated from the original manuscript report of Captain Don Antonio del Rio: followed by Teatro critico americano; or, A critical investigation and research into the history of the Americans by Doctor Paul Felix Cabrera.* London: H. Berthoud, and Suttaby, Evance and Fox, 1822.

Rivera Maestre, Miguel. *Atlas guatemalteco en ocho cartas formadas y grabadas en Guatemala: de orden del gefe del estado C. Doctor Mariano Galvez.* Guatemala City, 1834. Facsimile edition: Guatemala City: Ministerio de Relaciones Exteriores, 2001.

Rozier, François. *Curso completo, ó Diccionario universal de agricultura teórica, práctica, económica, y de medicina rural y veterinaria. Traducido al castellano por don Juan Alvarez Guerra.* 8 volumes. Madrid: Imprenta Real, 1797–1801.

Sigaud de la Fond, Joseph-Aignan. *Essai Sur Différentes Especes D'Air-Fixe Ou De Gas: Pour servir de suite & de supplément aux Elémens de Physique du méme Auteur. Nouvelle édition, revue et augmentée, par M. Rouland, Professeur de Physique expérimentale & Démonstrateur en l'Université de Paris.* Paris: P. Fr. Gueffier, 1785.

Smith, Adam. *An Inquiry Into the Nature and Causes of the Wealth of Nations*, Volume I. 2nd ed., London: Strahan and Cadell, 1778.

Sociedad Económica de Amigos del Estado de Guatemala. *Estatutos de la Sociedad Economica de Amigos del Estado de Guatemala.* Guatemala City: Imprenta de la Union, 1830.

Memorias de la Sociedad Economica de Amantes de Guatemala. Guatemala City: Imprenta de la Union, 1831.

Mensual de la Sociedad Económica de Amigos del Estado de Guatemala. Guatemala City: Imprenta de la Union, 1830.

Prospecto del Mensual de la Sociedad Economica de Amigos del Estado de Guatemala. Guatemala City: Imprenta de la Union, 1830.

Terreros y Pando, Esteban. *Diccionario castellano con las voces de ciencias y artes*, Vol. II. Madrid: Viuda de Ibarra, 1787.

Thompson, George Alexander. *Narrative of an Official Visit to Guatemala from Mexico.* London: John Murray, 1829.

Ward, Bernardo. *Proyecto economico, en que se proponen varias providencias, dirigidas á promover los intereses de España, con los medios y fondos necesarios para su plantificacion: escrito en el año de 1762.* 2nd ed., Madrid: Ibarra, 1779.

Zebadúa, Marcial. *Manifestación pública del ciudadano Marcial Zebadúa sobre su misión diplomática cerca de Su Magestad Británica.* Guatemala City: Imprenta de la Union, 1832.

Newspapers and Periodicals

El Amigo de la Patria. Ed. José Cecilio del Valle. *Guatemala City,* 1820–22.

Annals of Agriculture and Other Useful Arts. Ed. Arthur Young, Bury St Edmunds, 1784–1796.

Correo Mercantil de España y sus Indias. Madrid: Vega, 1792–1808.

El Editor Constitucional / El Genio de La Libertad. Guatemala City: Ignacio Beteta, 1820–1821.

Gazeta de Guatemala. Guatemala: Ignacio Beteta, 1797–1807.

Gazeta de Literatura / Gacetas de literatura de México. Ed. José Antonio Alzate. Mexico City, 1788–1795.

Journal of the Royal Geographical Society of London. London: Royal Geographical Society, 1831–1880.

The Literary Gazette: A Weekly Journal of Literature, Science, and the Fine Arts. London: Colburn, 1817–1863.

Mensual de la Sociedad Económica de Amigos del Estado de Guatemala. Guatemala City: Imprenta de la Union, 1830.

Mercurio de España. Madrid: Imprenta Real, 1784–1830.

Mercurio Peruano. Lima: Imprenta Real, 1790–1795. In *El Mercurio Peruano, 1790–1795*. Vol. II: *Antología*. Edited by Jean Pierre Clément. Frankfurt: Vervuert, 1998.

Papel Periódico de Santa Fé de Bogotá. Edited by Manuel del Socorro Rodríguez. Santafé de Bogotá: Imprenta de Don Antonio Espinosa de los Monteros, 1791–1797.

Periódico de la Sociedad Económica de Guatemala. Guatemala City: Ignacio Beteta, 1815–1816.

El Redactor General. Ed. José Cecilio del Valle. Guatemala City, 1825–6.

Secondary Sources

Aceves Pastrana, Patricia, ed. *Periodismo científico en el siglo XVIII: José Antonio de Alzate y Ramírez*. Mexico City: Universidad Autónoma Metropolitana, Unidad Xochimilco and Sociedad Química de México, 2001.

Achim, Miruna. *Lagartijas medicinales: Remedios americanos y debates científicos en la Ilustración*. Mexico City: Consejo Nacional para la Cultura y las Artes, 2008.

From Idols to Antiquity: Forging the National Museum of Mexico. Lincoln, NE: University of Nebraska Press, 2017.

Adelman, Jeremy. *Sovereignty and Revolution in the Iberian Atlantic*. Princeton University Press, 2016.

Afanador Llach, María José. 'Political Economy, Geographical Imagination, and Territory in the Making and Unmaking of New Granada, 1739–1830'. PhD Dissertation, The University of Texas at Austin, 2016.

Aguilar, Juan Manuel and Sergio Antonio Palacios, *La Ciudad de Trujillo. Guia Historico Turística* (3rd ed., Tegucigalpa: Instituto Hondureño de Antropología e Historia, 2003).

Aguirre, Robert. *Informal Empire. Mexico and Central America in Victorian Culture*. Minneapolis, MN and London: University of Minnesota Press, 2005.

Albi, Peter Christopher. 'Derecho Indiano vs. the Bourbon Reforms: The Legal Philosophy of Francisco Xavier de Gamboa'. In Paquette, ed. *Enlightened Reform in Southern Europe*, 229–49.

Alcina Franch, José. *Arqueólogos o anticuarios: historia antigua de la arqueología en la América española*. Barcelona: Serbal, 1995.

Almagro Gorbea, Martín, and Jorge Maier Allende, eds. *Corona y arqueología en el siglo de las luces*. Madrid: Patrimonio Nacional, 2010.

Alvar, Jaime. 'Carlos III y la arqueología española'. In Almagro Gorbea and Maier Allende, eds. *Corona y arqueología en el siglo de las luces*, 313–23.

Álvarez Peláez, Raquel. *La conquista de la naturaleza americana*. Madrid: Consejo Superior de Investigaciones Científicas, 1993.

Anderson, Benedict. *Imagined Communities: Reflections on the Origin and Spread of Nationalism*. London and New York: Verso, 1991.

Applebaum, Nancy P. 'Reading the Past on the Mountainsides of Colombia: Mid-Nineteenth-Century Patriotic Geology, Archaeology, and Historiography', *Hispanic American Historical Review* 93, no. 3 (2013), 347–76.

Aramoni Calderón, Dolores. 'Los indios constructores de Palenque y Toniná en un documento del siglo XVIII', *Estudios de Cultura Maya* 18 (1991): 417–38.

Arnold, David. *The Tropics and the Travelling Gaze. India, Landscape and Science*. Seattle, WA: University of Washington Press, 2006.

Baker, Alan. *Geography and History: Bridging the Divide*. Cambridge University Press, 2003.

Ballesteros Gabrois, Manuel. *Nuevas noticias sobre Palenque en un manuscrito del siglo XVIII*. Mexico: UNAM, 1960.

Barrera-Osorio, Antonio. *Experiencing Nature: The Spanish American Empire and the Early Scientific Revolution*. Austin, TX: University of Texas Press, 2006.

Barrios y Barrios, Catalina. *Estudio histórico del periodismo guatemalteco: época colonial y siglo XIX*. Guatemala City: Editorial Universitaria, 2003.

Baskes, Jeremy. *Indians, Merchants, and Markets*. Stanford University Press, 2000.

Bayly, Christopher A. *Imperial Meridian: The British Empire and the World 1780–1830*. London and New York: Longman, 1989.

Belaubre, Christophe. *Diccionario de la asociación para el fomento de los estudios históricos en Centroamérica (Diccionario AFEHC)*. Biographical entries for S. Bergaño y Villegas, R. Casaus y Torres, J. B. Irisarri, A. López Peñalver y Alcalá, N. Serrano Polo, J. Villegas. Online publication: www.afehc-historia-centroamericana.org/index_action_lst_type_diccionario.html

'El traslado de la capital del Reino de Guatemala (1773–1779). Conflicto de poder y juegos sociales', *Revista De Historia* 57–8 (2008): 23–61.

'Lectura crítica de la "Memoria sobre el fomento de las cosechas del cacao" del canónigo Antonio García Redondo', *Boletín de la AFEHC n°39* (December 2008). Online publication: www.afehc-historia-centroamericana.org/?action=fi_aff&id=2106

Belzunegui Ormazábal, Bernardo. *Pensamiento económico y reforma agraria en el Reino de Guatemala, 1797–1812*. Guatemala: Comisión Interuniversitaria Guatemalteca de Conmemoración del Quinto Centenario del Descubrimiento de América, 1992.

Berggren, Lennart, and Alexander Jones. *Ptolemy's Geography*. Princeton, NJ and Oxford: Princeton University Press, 2000.

Berquist Soule, Emily. *The Bishop's Utopia: Envisioning Improvement in Colonial Peru*. Philadelphia, PA: University of Pennsylvania Press, 2014.

Bethell, Leslie, ed. *Central America since Independence*. Cambridge University Press, 1991.

Blackbourn, David. *The Conquest of Nature: Water, Landscape and the Making of Modern Germany*. London: Jonathan Cape, 2006.

Bleichmar, Daniela. *Visible Empire: Botanical Expeditions and Visual Culture in the Eighteenth-Century Hispanic World*. University of Chicago Press, 2012.

Bleichmar, Daniela, Paula De Vos, Kristin Huffine, and Kevin Sheehan, eds. *Science in the Spanish and Portuguese Empires, 1500–1800*. Stanford University Press, 2009.

Bond, Dean W. 'Plagiarists, Enthusiasts and Periodical Geography: A.F. Büsching and the Making of Geographical Print Culture in the German Enlightenment, c.1750–1800', *Transaction of the Institute of British Geographers* 42, no. 1 (2017): 58–71.

Bonilla Bonilla, Adolfo. 'The Central American Enlightenment: An Interpretation of Political Ideas and Political History'. PhD dissertation, University of Manchester, 1996.

Ideas económicas en la Centroamérica ilustrada 1793–1838. San Salvador: FLASCO Programa El Salvador, 1999.

Bornholt, Jens. *Cuatro siglos de expresiones geográficas del istmo centroamericano, 1500–1900. / Four Centuries of Geographic Expressions of the Central American Isthmus 1500–1900*. Guatemala City: Universidad Francisco Marroquín, 2007.

Bowden, Martyn J. 'The Invention of American Tradition', *Journal of Historical Geography* 18, no. 1 (1992): 3–26.

Brading, David. *The First America: the Spanish Monarchy, Creole Patriots, and the Liberal State, 1492–1867*. Cambridge and New York: Cambridge University Press, 1991.

Brockmann, Sophie. 'Retórica patriótica y redes de información científica en Centroamérica, 1790–1814', *Cuadernos de Historia Moderna*, Anejo XI (2012): 165–84.

'Sumatran Rice and "Miracle" Herbs: Local and International Natural Knowledge in Late-Colonial Guatemala', *Colonial Latin American Review* 24, no. 1 (2015): 84–106.

'Surveying Nature: The Creation and Communication of Natural-Historical Knowledge in Enlightenment Central America', PhD dissertation, University of Cambridge, 2013.

Brown, Matthew. 'Enlightened Reform after Independence: Simón Bolívar's Bolivian Constitution'. In Paquette, ed. *Enlightened Reform in Southern Europe*, 339–60.

Informal Empire in Latin America: Culture, Commerce and Capital. Oxford: Blackwell / Society for Latin American Studies, 2008.

Brown, Matthew, and Gabriel Paquette, eds. *Connections after Colonialism. Europe and Latin America in the 1820s*. Tuscaloosa, AL: University of Alabama Press, 2013.

Brown, Richmond. 'Profits, Prestige, and Persistence: Juan Fermín de Aycinena and the Spirit of Enterprise in the Kingdom of Guatemala', *The Hispanic American Historical Review* 75, no. 3 (1995), 405–40.

Browning, John. 'El despertar de la consciencia nacional en Guatemala'. In Luján Muñoz and Zilbermann de Luján, *Historia general de Guatemala*, Vol. III, 627–40.

'The Periodical Press: Voice of the Enlightenment in Spanish America', *Dieciocho* 3, no. 1 (1980): 5–17.

'Rafael Landivar's *Rusticatio Mexicana:* Natural History and Political Subversion', *Ideologies and Literature, nueva época* 1 (1985): 10–30.

Buisseret, David. 'Spanish Colonial Cartography, 1450–1700.' In Brian Harley and David Woodward, eds. *The History of Cartography*, Vol. III, Part 1: *Cartography in the European Renaissance*. Chicago, IL and London: University of Chicago Press, 2007, 1143–71.

Bumgartner, Louis. *José del Valle of Central America*. Durham, NC: Duke University Press, 1963.

Burkholder, Mark. *Biographical Dictionary of Audiencia Ministers in the Americas, 1687–1821.* Westport, CT and London: Greenwood Press, 1982.

Burkholder, Mark, and D. S. Chandler. *From Impotence to Authority: The Spanish Crown and the American Audiencias, 1687–1808.* Columbia and London: University of Missouri Press, 1977.

Burns, Kathryn. *Into the Archive: Writing and Power in Colonial Peru.* Durham, NC and London: Duke University Press, 2010.

Butterwick, Richard. 'Peripheries of the Enlightenment: an Introduction'. In Richard Butterwick, Simon Davies, and Gabriel Sánchez Espinosa, eds. *Peripheries of the Enlightenment.* Oxford: Voltaire Foundation, 2008, 1–16.

Cabello Carro, Paz. *Política investigadora de la época de Carlos III en el área maya. Según documentación de Calderón, Bernasconi, Del Río y otros.* Madrid: Ediciones de la Torre, 1992.

Cal Montoya, José Edgardo. 'El discurso historiográfico de la Sociedad Económica de Amigos del Estado de Guatemala en la primera mitad del siglo XIX', *Anuario de Estudios Centroamericanos* 30, no. 1 (2004): 87–117.

Calatayud Arinero, María. 'El real gabinete de historia natural'. In Sellés, Peset, and Lafuente, *Carlos III y la ciencia de la Ilustración,* 263–76.

Candiani, Vera S. *Dreaming of Dry Land: Environmental Transformation in Colonial Mexico City.* Stanford University Press, 2014.

Cañizares-Esguerra, Jorge. *How to Write the History of the New World: Histories, Epistemologies, and Identities in the Eighteenth-Century Atlantic World.* Stanford University Press, 2001.

  Nature, Empire and Nation: Explorations of the History of Science in the Iberian World. Stanford University Press, 2006.

Capel, Horacio. *Los ingenieros militares en España, siglo XVIII. Repertorio biográfico e inventario de su labor científica y espacial.* Barcelona: Universitat Barcelona, 1983.

Caradonna, Jeremy L. *The Enlightenment in Practice: Academic Prize Contests and Intellectual Culture in France, 1670–1794,* Ithaca, NY and London: Cornell University Press, 2012

Carlson, Anthony E. 'Vast Factories of Febrile Poison: Wetlands, Drainage, and the Fate of American Climates, 1750–1850'. In Migletti and Morgan, eds. *Governing the Environment in the Early Modern World,* 153–71.

Carney, Judith. *Black Rice: The African Origins of Rice Cultivation in the Americas,* Cambridge, MA: Harvard University Press, 2001.

Carter, Paul. *The Road to Botany Bay: An Exploration of Landscape and History.* Minneapolis, MN and London: University of Minnesota Press, 2010 [1987].

Casanova, Rosa. 'Imaginando el pasado: el mito de las ruinas de Palenque, 1784–1813', *Cuadernos de la Asociación de Historiadores Latinoamericanistas Europeos* 2 (1994): 33–90.

Casaús Arzú, Marta Elena. *Las redes intelectuales centroamericanas: un siglo de imaginarios nacionales* (1820–1920). Guatemala: F&G Editores, 2005.

Castañeda Paganini, Ricardo. *Las ruinas de Palenque: su descubrimiento y primeras exploraciones en el siglo XVIII.* Guatemala: Tipografía nacional, 1946.

Castleman, Bruce. *Building the King's Highway. Labor, Society, and Family on Mexico's Caminos Reales, 1757–1804.* Tucson, AZ: The University of Arizona Press, 2005.

Castro-Gómez, Santiago, 'Siglo xviii: el nacimiento de la biopolítica', *Tabula Rasa* 12 (2010), 31–45.

Centeno, Miguel A., and Agustin E. Ferraro. 'Republics of the Possible. State-Building in Latin America and Spain'. In Centeno and Ferraro, *State and Nation Making*, 3–24.

eds. *State and Nation Making in Latin America and Spain. Republics of the Possible.* Cambridge University Press, 2013.

Chambers, Sarah. *From Subjects to Citizens. Honor, Gender and Politics in Arequipa, Peru, 1780–1854.* University Park, PA: Pennsylvania State University Press, 1999.

Chandler, David. 'Peace Through Disunion: Father Juan José de Aycinena and the Fall of the Central American Federation', *The Americas* 46, no. 2 (1989): 137–57.

Chandler, Dewitt. 'Jacobo de Villaurrutia and the Audiencia of Guatemala, 1794–1804', *The Americas* 32, no. 3 (1976): 402–17.

Chartier, Roger. *The Cultural Origins of the French Revolution.* Durham, NC and London: Duke University Press, 2010.

Chinchilla Mazariegos, Oswaldo. 'Archaeology and Nationalism in Guatemala at the Time of Independence.' *Antiquity* 72 (1998): 376–86.

Cintrón Tirykian, Josefina. 'Campillo's Pragmatic New System: a Mercantile and Utilitarian Approach to Indian Reform in Spanish Colonies of the Eighteenth Century', *History of Political Economy* 10, no. 2 (1978): 233–57.

Clark, Fiona. 'The *Gazeta de Literatura de Mexico* (1788–1795): The Formation of a Literary-Scientific Periodical in Late-Viceregal Mexico.' *Dieciocho* 28, no. 1 (2005): 7–30.

Claxton, Robert. 'Miguel Rivera Maestre: Guatemalan Scientist-Engineer', *Technology and Culture* 14 (1973): 384–403.

Coatsworth, John. 'The Limits of Colonial Absolutism: The State in Eighteenth-Century Mexico'. In Karen Spalding, eds. *Essays in the Political, Economic, and Social History of Colonial Latin America.* Newark, DE: University of Delaware, 1982, 25–51.

Concoha Chet, Héctor Aurelio. 'El concepto de montañés entre los Kaqchikeles de San Juan Sacatepéquez, 1524–1700', in Herrera and Webre, *La época colonial en Guatemala*, 19–41.

Conrad, Sebastian. 'Enlightenment in Global History: A Historiographical Critique', *The American Historical Review* 117, no. 4 (2012): 999–1027.

Cortés Alonso, Vicente. 'La antropología de América y los archivos', *Revista Española de Antropología Americana* 6 (1971): 149–78.

Cosgrove, Denis. *Social Formation and Symbolic Landscape.* London: Croom Helm, 1984.

Cosgrove, Denis, and Stephen Daniels, eds. *The Iconography of Landscape: Essays on the Symbolic Representation, Design and Use of Past Environments.* Cambridge University Press, 1988.

Cowie, Helen. *Conquering Nature in Spain and Its Empire, 1750–1850.* Manchester University Press, 2011.

Craib, Raymond. *Cartographic Mexico: A History of State Fixations and Fugitive Landscapes.* Durham, NC: Duke University Press, 2004.

Crawford, Matthew. *The Andean Wonder Drug: Cinchona Bark and Imperial Science in the Spanish Atlantic, 1630–1800.* University of Pittsburgh Press, 2016.

Cushman, Gregory. 'Humboldtian Science, Creole Meteorology, and the Discovery of Human-Caused Climate Change in South America', *Osiris* 26, no. 1 (2011): 16–44.

Daniels, Christine, and Michael Kennedy, eds. *Negotiated Empires: Centers and Peripheries in the Americas, 1500–1820*. New York and London: Routledge, 2002.

Daniels, Stephen, Susanne Seymore, and Charles Watkins, 'Enlightenment, Improvement, and the Geographies of Horticulture in Later Georgian England'. In Livingstone and Withers, *Geography and Enlightenment*, 345–71.

Daston, Lorraine. 'The Empire of Observation, 1600-1800'. In Lorraine Daston and Elizabeth Lunbeck, eds. *Histories of Scientific Observation*. University of Chicago Press, 2011, 81–114.

De Vos, Paula. 'Natural History and the Pursuit of Empire in Eighteenth-Century Spain', *Eighteenth-Century Studies* 40 (2007): 209–39.

Del Castillo, Lina. *Crafting a Republic for the World: Scientific, Geographic, and Historiographic Inventions of Colombia*. Lincoln, NE: University of Nebraska Press, 2018.

Denevan, William M. 'The Pristine Myth: The Landscape of the Americas in 1492', *Annals of the Association of American Geographers* 82, no. 3 (1992): 369–85.

Dietz, Bettina. 'Making Natural History: Doing the Enlightenment', *Central European History* 43 (2010): 25–46.

Donato, Clorinda. 'The Enciclopedia Metódica. A Spanish Translation of the Encyclopédie Méthodique'. In Clorinda Donato and Robert Maniquis, eds. *The Encyclopédie and the Age of Revolution*. Boston, MA: G.K. Hall, 1992, 73–6.

Drayton, Richard. *Nature's Government. Science, Imperial Britain, and the 'Improvement' of the World*. New Haven, CT and London: Yale University Press, 2000.

Dym, Jordana. 'Citizens of Which Republic? Foreigners and the Construction of National Citizenship in Central America, 1823–1845', *The Americas* 64, no. 4 (April 2008): 477–510.

'Conceiving Central America: A Bourbon Republic in the Gazeta (1797–1807)'. In Paquette, ed. *Enlightened Reform in Southern Europe and its Atlantic Colonies*, 99–118.

'Democratizing the Map: The Geo-body and National Cartography in Guatemala, 1821–2010'. In James Akerman, ed. *Decolonizing the Map: Cartography from Colony to Nation*. University of Chicago Press, 2017, 160–204.

From Sovereign Villages to National States. Albuquerque, NM: University of New Mexico Press, 2006.

'La reconciliación de la historia y la modernidad: George Thompson, Henry Dunn y Frederick Crowe, tres viajeros británicos en Centroamérica, 1825–1845', *Mesoamérica* 40 (2000): 142–81.

'"More Calculated to Mislead Than Inform": Travel Writers and the Mapping of Central America, 1821–1945', *Journal of Historical Geography* 30 (2004): 340–63.

Dym, Jordana, and Christophe Belaubre, eds. *Politics, Economy and Society in Bourbon Central America*. Boulder, CO: University of Colorado Press, 2007.

Dym, Jordana, and Karl Offen, eds. *Mapping Latin America*. Paperback ed., University of Chicago Press, 2011.

Earle, Rebecca. *The Body of the Conquistador: Food, Race and the Colonial Experience in Spanish America, 1492–1700*. Cambridge University Press, 2012.

'Information and Disinformation in Late Colonial New Granada', *The Americas* 54, no. 2 (1997): 167–84.

'"Padres de la Patria" and the Ancestral Past: Commemorations of Independence in Nineteenth-Century Spanish America', *Journal of Latin American Studies* 34, no. 44 (2002): 775–805.

Elliott, John H. 'A Europe of Composite Monarchies', *Past & Present* 137 (1992): 48–71.

Engstrand, Iris. 'Of Fish and Men: Spanish Marine Science During the Late Eighteenth Century', *The Pacific Historical Review* 69, no. 1 (2000): 3–30.

Entin, Gabriel. 'El patriotismo americano en el siglo XVIII: Ambigüedades de un discurso político hispánico.' In Véronique Hébrard and Geneviéve Verdo, eds. *Las independencias hispanoamericanas*. Madrid: Collection de la Casa de Velásquez, 2013, 19–34.

Ewalt, Margaret. *Peripheral Wonders: Nature, Knowledge and Enlightenment in the Eighteenth-Century Orinoco*. Lewisburg, PA: Bucknell University Press, 2008.

Fernández Pérez, Joaquín. 'La ciencia ilustrada y las Sociedades Económicas de Amigos del País'. In Sellés, Peset, and Lafuente, eds. *Carlos III y la ciencia de la Ilustración*, 217–32.

Few, Martha. *For All of Humanity: Mesoamerican and Colonial Medicine in Enlightenment Guatemala*. Tucson, AZ: University of Arizona Press, 2015.

'Killing Locusts in Colonial Guatemala'. In Martha Few and Zeb Tortorici, eds. *Centering Animals in Latin American History*. Durham, NC: Duke University Press, 2013, 62–92.

Fiehrer, Thomas. 'Slaves and Freedmen in Colonial Central America: Rediscovering a Forgotten Black Past', *The Journal of Negro History* 64, no. 1 (1979): 39–57.

Fisher, Andrew B. and Matthew D. O'Hara, eds. *Imperial Subjects. Race and Identity in Colonial Latin America*. Durham, NC and London: Duke University Press, 2009.

Firestone, Janet. *The Spanish Royal Corps of Engineers in the Western Borderlands: Instrument of Bourbon Reform, 1764 to 1815*. Glendale, CA: A.H. Clark Co., 1977.

Floyd, Troy S. 'The Guatemalan Merchants, the Government, and the Provincianos, 1750–1800', *The Hispanic American Historical Review*, 41, no. 1 (1961): 90–110.

'The Indigo Merchant: Promoter of Central American Economic Development, 1750–1808', *Business History Review* (1965): 476–9.

Foucault, Michel. *The Order of Things: An Archaeology of the Human Sciences*. London and New York: Routledge Classics, 2002 [1970].

Security, Territory, Population: Lectures at the Collège de France, 1977–1978. New York: Picador, 2009.

Fraser, Nancy. 'Rethinking the Public Sphere: a Contribution to the Critique of Actually Existing Democracy.' In Craig Calhoun, eds. *Habermas and the Public Sphere*. Cambridge, MA and London: MIT Press, 1992, 109–42.

Fróes da Fonseca, Maria Rachel. 'La construcción de la patria por el discurso científico: Mexico y Brasil (1770–1830)', *Secuencia* 45 (1999): 5–26.

Gándara Chacana, Natalia. 'Representaciones de un territorio. La frontera mapuche en los proyectos ilustrados del Reino de Chile en la segunda mitad del siglo XVIII', *Historia Crítica* 59 (2016): 61–80.

García Laguardia, Jorge Mario. *José del Valle, ilustración y liberalismo en Centroamérica*. Tegucigalpa: Departamento Editorial de la UNAH, 1982.

Gerbi, Antonello. *The Dispute of the New World: The History of a Polemic, 1750–1900.* University of Pittsburgh Press, 1973.

Glacken, Clarence. *Traces on the Rhodian Shore: Nature and Culture in Western Thought from Ancient Times to the End of the Eighteenth Century.* Berkeley, CA: University of California Press, 1976.

Glick, Thomas. 'Science and Independence in Latin America', *Hispanic American Historical Review* 71 (1991): 307–34.

Godlewska, Anne. *Geography Unbound: French Geographic Science from Cassini to Humboldt.* University of Chicago Press, 1999.

Gómez, Alejandro. *José del Valle: el político de la independencia centroamericana.* Guatemala: Universidad Francisco Marroquín, 2011.

Gonzáles Alzate, Jorge. 'Hidalgo, José Domingo.' In *Diccionario AFEHC.* Online publication, http://afehc-historia-centroamericana.org/index.php?action=fi_aff& id=3685

'State Reform, Popular Resistance, and Negotiation of Rule in Late Bourbon Guatemala: The Quetzaltenango Aguardiente Monopoly, 1785–1807'. In Dym and Belaubre, *Politics, Economy and Society in Bourbon Central America,* 129–55.

Gonzáles Bueno, Antonio, and Raúl Rodríguez Nozal. *Plantas americanas para la españa ilustrada. Génesis, desarrollo y ocaso del proyecto español de expediciones botánicas.* Madrid: Editorial Complutense, 2000.

Gonzáles Tascón, Ignacio. *Ingeniería española en ultramar: siglos XVI–XIX.* 2 vols, Madrid: Colegio de Ingenieros de Caminos, Canales y Puertos, 1992.

González, Nancie L. *Sojourners of the Caribbean: Ethnogenesis and Ethnohistory of the Garifuna.* Urbana, IL: University of Illinois Press, 1988.

Goodman, Dena. *The Republic of Letters: A Cultural History of the French Enlightenment.* Ithaca, NY and London: Cornell University Press, 1994.

Graham, Ian. 'Juan Galindo, Enthusiast', *Estudios de Cultura Maya* 3 (1963): 11–35.

Grafe, Regina, and Alejandra Irigoin. 'A Stakeholder Empire: The Political Economy of Spanish Imperial Rule in America', *The Economic History Review* 65, no. 2 (2012), 609–51.

Griffith, William. *Empires in the Wilderness: Foreign Colonization and Development in Guatemala, 1834–1844.* Chapel Hill, NC: University of North Carolina Press, 1965.

'Juan Galindo, Central American Chauvinist', *The Hispanic American Historical Review* 40, no. 1 (1960): 25–52.

Grove, Richard, *Green Imperialism: Colonial Expansion, Tropical Island Edens and the Origins of Environmentalism, 1600–1860.* Cambridge University Press, 1995.

Guerra, François-Xavier, and Annick Lempérière, eds. *Los espacios públicos en Iberoamérica: ambigüedades y problemas, siglos xviii–xix.* México: Fondo de Cultura Económica, 1998.

Guimerá, Agustín, ed. *El reformismo borbónico. Una visión interdisciplinar.* Madrid: Consejo Superior de Investigaciones Científicas, Alianza Editorial, 1996.

Gutiérrez Álvarez, Coralia. 'Racismo y sociedad en la crisis del Imperio español: El caso de los pueblos del Altiplano Occidental de Guatemala'. In Herrera and Webre, eds. *La época coloial,* 249–77.

Habermas, Jürgen. *The Structural Transformation of the Public Sphere.* Cambridge, MA: MIT Press, 1989.

Hall, Carolyn, and Héctor Pérez Brignoli. *Historical Atlas of Central America.* Paperback ed., Norman, OK: University of Oklahoma Press, 2003.

Harley, Brian. *The New Nature of Maps.* Baltimore, MD: Johns Hopkins University Press, 2001.

Herbert, Francis. 'Guatemala, Colonisation, and the Royal Geographical Society in the 1830s: Some Contemporary Manuscript and Printed Documents in the RGS's Collections', *Journal of the International Map Collectors' Society* 107 (2006): 6–12.

Hernández Pérez, José Santos. 'Medicina y salud pública: su difusión a través de la Gaceta de Guatemala (1797–1804).' *eä* 2 (2010) – electronic publication, www.ea-journal.com.

Herrera, Robinson. '"Por Que No Sabemos Firmar": Black Slaves in Early Guatemala', *The Americas* 57, no. 2 (2000): 247–67.

Herrera, Robinson, and Stephen Webre, eds. *La época colonial en Guatemala: estudios de historia cultural y social.* Guatemala: Editorial Universitaria, Universidad de San Carlos de Guatemala, 2013.

Herzog, Tamar. *Frontiers of Possession: Spain and Portugal in Europe and the Americas.* Cambridge, MA and London: Harvard University Press, 2015.

Hoeg, Jerry. 'Andrés Bello's "Ode to Tropical Agriculture": The Landscape of Independence'. In Beatriz Rivera-Barnes and Jerry Hoeg, eds. *Reading and Writing the Latin American Landscape.* New York: Palgrave Macmillan, 2009, 53–66.

Hopkins, Thomas. 'Adam Smith on American Economic Development and the Future of the European Atlantic Empires'. In Reinert and Røge, eds. *Political Economy*, 53–75.

Hutchison, A. A., K. V. Cashman, C. Williams and A. C. Rust. 'The 1717 eruption of Volcán de Fuego, Guatemala: Cascading Hazards and Societal Response', *Quaternary International* 394 (2016), 69–78.

Jones, Peter M. *Agricultural Enlightenment: Knowledge, Technology, and Nature, 1750–1840.* Oxford University Press, 2016.

Jonsson, Fredrik Albritton. *Enlightenment's Frontier: The Scottish Highlands and the Origins of Environmentalism.* New Haven, CT: Yale University Press, 2013.

'Scottish Tobacco and Rhubarb: The Natural Order of Civil Cameralism in the Scottish Enlightenment', *Eighteenth-Century Studies* 49, no. 2 (2016): 129–47.

Kagan, Richard. *Urban Images of the Hispanic World, 1493–1793.* New Haven, CT and London: Yale University Press, 2000.

Kramer, Wendy, W., George Lovell and Christopher H. Lutz. 'Pillage in the Archives: The Whereabouts of Guatemalan Documentary Treasures', *Latin American Research Review* 48, no. 3 (2013): 153–67.

Koerner, Lisbet. *Linnaeus: Nature and Nation.* Cambridge, MA: Harvard University Press, 1999.

Kohl, Philip, Irina Podgorny, and Stefanie Gänger. *Nature and Antiquities: The Making of Archaeology in the Americas.* Tucson, AZ: The University of Arizona Press, 2014.

Lafuente, Antonio. 'Institucionalización metropolitana de la ciencia española en el siglo XVIII'. In Antonio Lafuente and José Sala Catalá, eds. *Ciencia colonial en América.* Madrid: Alianza, 1992, 91–118.

'Enlightenment in an Imperial Context: Local Science in the Late-Eighteenth-Century Hispanic World', *Osiris* 2nd Series 15 (2000): 155–73.

Lafuente, Antonio and Leoncio López-Ocón. 'Scientific Traditions and Enlightenment Expeditions.' In José Saldaña, ed. *Science in Latin America. A History.* 1st English ed., Austin, TX: University of Texas Press, 2006, 123–50.

Lafuente, Antonio, and Nuria Valverde. 'Linnaean Botany and Spanish Imperial Biopolitics'. In Schiebinger and Swan, *Colonial Botany*, 134–47.

Lanning, John Tate. *The Eighteenth-Century Enlightenment in the University of San Carlos de Guatemala.* Ithaca, NY: Cornell University Press, 1956.

Latour, Bruno. *Science in Action: How to Follow Scientists and Engineers Through Society.* Cambridge, MA: Harvard University Press, 1987.

Lefebvre, Henri. *The Production of Space.* Oxford: Blackwell, 1991.

Leonard, Thomas. 'Central America and the United States: Overlooked Foreign Policy Objectives', *The Americas* 50, no. 1 (1993): 1–30.

Levine, Philippa. *The Amateur and the Professional: Antiquarians, Historians and Archaeologists in Victorian England, 1838–1886.* Cambridge University Press, 1986.

Lindenfeld, David. *The Practical Imagination. The German Sciences of State in the Nineteenth Century.* University of Chicago Press, 1997.

Liss, Robert. 'Frontier Tales: Tokugawa Japan in Translation'. In *The Brokered World: Go-Betweens and Global Intelligence, 1770–1820.* Edited by Simon Schaffer, Lissa Roberts, Kapil Raj, and James Delbourgo, 1–47. Sagamore Beach, MA: Watson Publishing, 2009.

Livingstone, David. *Putting Science in Its Place: Geographies of Scientific Knowledge.* University of Chicago Press, 2003.

Livingstone, David, and Charles Withers, eds. *Geography and Enlightenment.* Chicago, IL and London: The University of Chicago Press, 1999.

Lovell, George. *Conquest and Survival in Colonial Guatemala. A Historical Geography of the Cuchumatán Highlands, 1500–1821.* 4th ed. Montreal and Kingston: McGill-Queen's University Press, 2015 [1985].

Lovell, George, and Christopher Lutz, 'Core and Periphery in Colonial Guatemala.' In Carol A. Smith, *Guatemalan Indians and the State: 1540 to 1988.* Austin, TX: University of Texas Press, 1990), 35–51.

Lowood, Henry. *Patriotism, Profit, and the Promotion of Science in the German Enlightenment: The Economic and Scientific Societies, 1760–1815.* New York: Garland Publishing, 1991.

Lucena Giraldo, Manuel, 'Los experimentos agrícolas en la Guyana española'. In Antonio Lafuente, Alberto Elena, and María Luisa Ortega, eds. *Mundialización de la ciencia y cultura nacional*, Madrid: Ediciones Doce Calles, 1993.

Luján Muñoz, Jorge. *Relaciones geográficas e históricas del siglo XVIII del reino de Guatemala. Vol. I. Relaciones geográficas e históricas de la década de 1740.* Guatemala: Universidad Del Valle, 2006.

Luján Muñoz, Jorge, and Alberto Herrarte, eds., *Historia general de Guatemala*, Vol. IV: *Desde la república federal hasta 1898.* Guatemala: Asociación de Amigos del País, Fundación para la Cultura y el Desarrollo, 1995.

Luján Muñoz, Jorge, and Cristina Zilbermann de Luján, eds. *Historia general de Guatemala*, Vol. III: *Siglo xviii hasta la independencia.* Guatemala City: Asociación de Amigos del País, Fundación para la Cultura y el Desarrollo, 1995.

Luján Muñoz, Jorge, and Marion Popenoe de Hatch, eds., *Historia general de Guatemala*, Vol. 1: *Época precolombiana*. Guatemala City: Asociación de Amigos del País, Fundación para la Cultura y el Desarrollo, 1999.

Luque Alcaide, Elisa. *La Sociedad Económica de Amigos del País de Guatemala*. Seville: Escuela de Estudios Hispano-Americanos, 1962.

Madeira Santos, Catarina. 'Administrative Knowledge in a Colonial Context: Angola in the Eighteenth Century.' *The British Journal for the History of Science* 43, no. 4 (2010): 539–56.

Majzul Lolmay, Filiberto Patal. *Rusoltzik ri Kaqchikel / Diccionario estándar bilingüe Kaqchikel-Español*. 2nd ed. Antigua Guatemala: OKMA, 2013.

Maldonado Polo, José Luis. *Las huellas de la razón: la expedición científica de Centroamérica (1795–1803)*. Madrid: Consejo Superior de Investigaciones Científicas, 2001.

Martín-Valdepeñas Yagüe, Elisa. 'Del amigo del país al ciudadano útil: una aproximación al discurso patriótico en la Real Sociedad Económica Matritense de Amigos del País en el Antiguo Régimen', *Cuadernos de Historia Moderna, Anejo XI: La nación antes del nacionalismo en la Monarquía Hispánica (1777–1824)* (2012): 23–47.

Martín Blasco, Julio, and Jesús María García Añoveros. *El arzobispo de Guatemala don Pedro Cortés y Larraz (Belchite 1712, Zaragoza 1786). Defensor de la justicia y de la verdad*. Belchite: Ayuntamiento, 1992.

Martínez Peláez, Severo. *La patria del criollo. Ensayo de interpretación de la realidad colonial guatemalteca*. Paperback ed., Mexico City: Fondo de Cultura Económica, 2006 (1970).

Martínez Salinas, Maria Luisa. *La colonización de la costa centroamericana de la Mosquitia en el siglo xviii. Familias canarias en el proyecto poblador*. Valladolid: Ediciones Universidad de Valladolid, 2015.

Matthew, Laura E. *Memories of Conquest: Becoming Mexicano in Colonial Guatemala*. Chapel Hill, NC: University of North Carolina Press, 2012.

McAleer, John. '"A Young Slip of Botany": Botanical Networks, the South Atlantic, and Britain's Maritime Worlds, c.1790–1810,' *Journal of Global History* 11, no. 1 (2016): 24–43.

McCook, Stuart. *States of Nature. Science, Agriculture and Environment in the Spanish Caribbean, 1760–1940*. Austin, TX: University of Texas Press, 2002.

McCreery, David. *Rural Guatemala, 1760–1940*. Stanford University Press, 1994.

McFarlane, Anthony. 'Identity, Enlightenment and Political Dissent in Late Colonial Spanish America.' *Transactions of the Royal Historical Society*, 6th Series, 8 (1998): 309–35.

'Science and Sedition in Spanish America: New Granada in the Age of Revolution, 1776–1810.' In Susan Manning and Peter France, eds. *Enlightenment and Emancipation*. Lewisburg, PA: Bucknell University Press, 2006, 97–116.

Medina, José Toribio. *La imprenta en Guatemala. 1660–1821*. 2nd ed. Guatemala City: Imprenta Nacional de Guatemala, 1960.

Meléndez, Mariselle. 'The Cultural Production of Space in Colonial Latin America.' In Barney Warf and Santa Arias, eds. *The Spatial Turn: Interdisciplinary Perspectives*. London: Routledge, 2009, 173–91.

'*Patria, Criollos* and Blacks: Imagining the Nation in the *Mercurio peruano*, 1791–1795.' *Colonial Latin American Review* 15 (2006): 207–27.

'Spanish American Enlightenments: Local Epistemologies and Transnational Exchanges in Eighteenth-Century Newspapers.' *Dieciocho, Anejo 4* (2009): 115–33.

Migletti, Sarah, and John Morgan, eds. *Governing the Environment in the Early Modern World: Theory and Practice.* London and New York: Routledge, 2017.

'Introduction'. In Migletti and Morgan, eds. *Governing the Environment*, 1–17.

Mokyr, Joel. *The Enlightened Economy: An Economic History of Britain, 1700–1850.* New Haven, CT and London: Yale University Press, 2009.

Monguió, Luis. '*Palabras e ideas: "patria"* y "*nación*" en el virreinato del Perú,' *Revista iberoamericana*, 44 (1978): 451–70.

Morán Turina, Miguel. 'Ir interrogando antigüedades'. In Almagro Gorbea and Maier Allende, eds. *Corona y arqueología en el siglo de las luces*, 49–57.

Morato Moreno, Manuel. 'El mapa de la Relación Geográfica de Zapotitlán (1579): una isla de racionalidad en un océano de empirismo'. *Journal of Latin American Geography* 10, no. 2 (2011): 217–29.

Mundy, Barbara. *The Mapping of New Spain: Indigenous Cartography and the Maps of the Relaciones Geográficas.* University of Chicago Press, 1996.

Muñoz Calvo, María Luisa. 'Las actividades de José Mariano Mociño en el Reino de Guatemala (1795–1799).' In José Luis Peset, ed. *Ciencia, vida y espacio en Iberoamérica*, Vol. I. Madrid: Consejo Superior de Investigaciones Científicas, 1989, 3–19.

Muñoz Pérez, José. 'Los proyectos sobre España e Indias en el siglo XVIII. El proyectismo como género', *Revista de estudios políticos* 54 (1955), 169–95.

Muthu, Sankar. *Enlightenment Against Empire*, Princeton University Press, 2003.

Navarrete, Carlos. *Palenque, 1784: el inicio de la aventura arqueológica maya.* Mexico City: UNAM, 2000.

Nieto Olarte, Mauricio. *Remedios para el imperio: historia natural y la apropiación del Nuevo Mundo.* Bogotá: Instituto Colombiano de Antropología e Historia, 2000.

Nieto Olarte, Mauricio, Paola Castaño and Diana Ojeda. 'Ilustración y orden social: el problema de la población en el *Semanario del Nuevo Reyno de Granada* (1808–1810)', *Revista de Indias* 65 (2005), 683–708.

'"El influjo del clima sobre los seres organizados" y la retórica ilustrada en el *Semanario del Nuevo Reyno de Granada*', *Historia Crítica* 30 (2005): 91–114.

Offen, Karl. 'British Logwood Extraction from the Mosquitia: The Origin of a Myth', *The Hispanic American Historical Review* 80, no. 1 (2000): 113–35.

'Edge of Empire'. In Dym and Offen, eds. *Mapping Latin America*, 88–92.

Ortega Martínez, Franciso, and Alexander Chaparro Silva, eds. *Disfraz y pluma de todos. Opinión pública y cultura política, siglos XVIII y XIX.* Bogotá: Universidad Nacional de Colombia, 2012.

Owensby, Brian. *Empire of Law and Indian Justice in Colonial Mexico.* Stanford University Press, 2008.

Padrón, Ricardo. *The Spacious Word: Cartography, Literature and Empire in Early Modern Spain.* University of Chicago Press, 2004.

Pagden, Anthony. 'Identity Formation in Spanish America'. In Nicholas Canny and Anthony Pagden, eds. *Colonial Identity in the Atlantic World, 1500–1800.* Princeton, NJ: Institute for Advanced Study, 1987, 51–94.

Palma Murga, Gustavo. 'Between Fidelity and Pragmatism. Guatemala's Commercial Elite Responds to Bourbon Reforms on Trade and Contraband.' In Dym and Belaubre, *Politics, Economy and Society in Bourbon Central America*, 101–27.

Palma Murga, Gustavo, and Arturo Taracena Arriola. 'Las dinámicas agrarias en Guatemala entre 1524 y 1944'. In Gustavo Palma Murga, Arturo Taracena Arriola, and José Aylwin Oyarzun, eds. *Procesos agrarios desde el siglo XVI a los Acuerdos de Paz*. Guatemala City: FLACSO, MINUGUA, CONTIERRA, 2002, 15–72.

Palti, Elias José. 'Recent Studies on the Emergence of a Public Sphere in Latin America', *Latin American Research Review* 36 (2001): 255–66.

Paquette, Gabriel. 'The Dissolution of the Spanish Atlantic Monarchy.' *The Historical Journal* 52, no. 1 (2009): 175–212.

Enlightenment, Governance, and Reform in Spain and Its Empire, 1759–1808. Basingstoke: Palgrave Macmillan, 2008.

'The Image of Imperial Spain in British Political Thought, 1750–1800', *Bulletin of Spanish Studies* 81, no. 2 (2004), 187–214.

'State-Civil Society Cooperation and Conflict in the Spanish Empire: The Intellectual and Political Activities of the Ultramarine Consulados and Economic Societies, c. 1780–1810', *Journal of Latin American Studies* 39 (2007): 263–98.

Paquette, Gabriel, ed. *Enlightened Reform in Southern Europe and its Atlantic Colonies, c. 1750–1810*. Farnham: Ashgate, 2009.

Pattridge, Blake. *Institution Building and State Formation in Nineteenth-Century Latin America: The University of San Carlos, Guatemala*. New York: Peter Lang, 2004.

Payne Iglesias, Elizet. *El puerto de Truxillo: Un viaje hacia su melancólico abandono*. Tegucigalpa: Editorial Guaymuras, 2007.

Peset, José Luis. 'Ciencia e independencia en la América española.' In Antonio Lafuente, Alberto Elena, and María Luisa Ortega, eds. *Mundialización de la ciencia y cultura nacional*. Madrid: Ediciones Doce Calles, 1993, 195–217.

Ciencia y libertad. El papel del científico ante la independencia americana. Madrid: Consejo Superior de Investigaciones Científicas, 1987.

Phelan, John Leddy. 'Authority and Flexibility in the Spanish Imperial Bureaucracy', *Administrative Science Quarterly* 5, no. 1 (1960), 47–65.

Philip, Kavita. *Civilising Natures: Race, Resources and Modernity in Colonial South India*. Hyderabad: Orient Longman, 2003.

Philipps, Denise. *Acolytes of Nature: Defining Natural Science in Germany, 1770–1850*. University of Chicago Press, 2012.

Pillsbury, Joanne, and Lisa Trever. 'The King, the Bishop, and the Creation of an American Antiquity', *Ñawpa Pacha* 29, no. 1 (2008), 191–219.

Pinto Sória, Julio César. *Centroamérica, de la colonia al estado nacional: 1800–1840*. Guatemala: Editorial Universitaria de Guatemala, 1989.

Podgorny, Irina. 'The Reliability of the Ruins,' *Journal of Spanish Cultural Studies* 8, no. 2 (2007): 213–33.

'"Silent and Alone": How the Ruins of Palenque Were Taught to Speak the Language of Archaeology.' In Ludomir Lozny, ed. *Comparative Archaeologies: A Sociological View of the Science of the Past*. New York: Springer, 2011, 527–54.

Ponce, Pilar. 'Burocracia colonial y territorio americano: Las relaciones de Indias'. In Antonio Lafuente and José Sala Catalá, eds. *Ciencia colonial en América*. Madrid: Alianza, 1992, 29–44.

Porter, Roy, and Mikuláš Teich. *The Enlightenment in National Context*. Cambridge University Press, 1981.

Poupeney Hart, Catherine. 'Entre historia natural y relación geográfica: el discurso sobre la tierra en el reino de Guatemala. Siglo xviii'. In Ignacio Arellano and Fermín del Pino, eds. *Lecturas y ediciones de crónicas de Indias. Una propuesta interdisciplinaria*. Madrid: Iberoamericana, 2004, 441–60.

'Prensa e ilustración: José Rossi y Rubí, del Mercurio Peruano a la Gaceta de Guatemala', *Istmo 13: Documentación VIII Congreso Centroamericano de Historia, Guatemala, julio 2006: Trabajos seleccionados sobre literatura e historia*. Online publication: http://istmo.denison.edu/n13/proyectos/prensa.html.

'Parcours journalistiques en régime colonial: José Rossi y Rubí, Alejandro Ramírez et Simón Bergaño', *El argonauta español* 6 (2009): 6–11. http://argonauta.revues.org/603.

'Entre gaceta y "espectador": avatares de la prensa antigua en América Central', *Cuadernos de ilustración y romanticismo* 16 (2010): 1–22.

Pratt, Marie Louise. *Imperial Eyes. Travel Writing and Transculturation*. Paperback Ed. New York: Routledge, 2008 [1992].

Premo, Bianca. *The Enlightenment on Trial: Ordinary Litigants and Colonialism in the Spanish Empire*. Oxford University Press, 2017.

Puerto Sarmiento, Francisco. *La ilusión quebrada: Botánica, sanidad y política científica en la España ilustrada*. Madrid: Consejo Superior de Investigaciones Científicas, 1988.

'José Antonio de Alzate y Ramírez ante la ciencia española ilustrada'. In Patricia Aceves Pastrana, ed. *Periodismo científico en el siglo xviii*. Mexico City: UNAM, 2001, 79–107.

Raillard, Matthieu. 'The Masson de Morsvilliers Affair Reconsidered: Nation, Hybridism and Spain's Eighteenth-Century Cultural Identity', *Dieciocho* 32 (2009): 31–48.

Raj, Kapil. *Relocating Modern Science: Circulation and the Construction of Knowledge in South Asia and Europe, 1650–1900*. Basingstoke and New York: Palgrave Macmillan, 2007.

Rajan, S. Ravi. *Modernizing Nature: Forestry and Imperial Eco-Development 1800–1950*. Oxford University Press, 2006.

Rama, Ángel. *The Lettered City*. Edited and translated by John Chasteen. Durham, NC and London: Duke University Press, 1996 [Original 1984].

Ramírez, Paul. 'Enlightened Publics for Public Health: Assessing Disease in Colonial Mexico,' *Endeavour* 37 no. 1 (2013): 3–12.

Reinert, Sophus. 'The Empire of Emulation: A Quantitative Analysis of Economic Translations in the European World, 1500–1849'. In Reinert and Røge, eds. *The Political Economy of Empire*, 105–28.

Reinert, Sophus. *Translating Empire: Emulation and the Origins of Political Economy*. Cambridge, MA: Harvard University Press, 2011.

Reinert, Sophus and Pernille Røge, eds. *The Political Economy of Empire in the Early Modern World.* Basingstoke: Palgrave Macmillan, 2013.

Reinhold, Meyer. 'The Quest for "Useful Knowledge" in Eighteenth-Century America', *Proceedings of the American Philosophical Society* 119, no. 2 (1975): 108–32.

Restall, Matthew. 'Imperial Rivalries'. In Dym and Offen, *Mapping Latin America*, 79–83.

Reyes, José Luis. *Apuntes para una monografía de la Sociedad Económica de Amigos del País.* Guatemala City: Editorial José de Pineda Ibarra, 1964.

Roberts, Lissa. '*"Le centre de toutes choses"*: Constructing and Managing Centralization on the Isle de France', *History of Science* 52, no. 3 (2014): 319–42.

Rodríguez, Jaime. *"We Are Now the True Spaniards." Sovereignty, Revolution, Independence, and the Emergence of the Federal Republic of Mexico, 1808–1824.* Stanford University Press, 2012.

Rodríguez, Mario. *A Palmerstonian Diplomat in Central America. Frederick Chatfield, Esq.* Tucson, AZ: The University of Arizona Press, 1964.

The Cádiz Experiment in Central America, 1808 to 1826. Berkeley, CA: University of California Press, 1978.

Rojas Lima, Flavio, ed. *Diccionario histórico-biográfico de Guatemala.* Guatemala City: Asociación de Amigos del País, 2004.

Romero Sandóval, Roberto. 'Viajeros en Palenque, siglos XVIII y XIX: un estudio histórico a través de su bibliografía', *Boletín del Instituto de Investigaciones Bibliográficas. Nueva época* 2, no. 1 (1997): 9–40.

Rubio Sánchez, Manuel. *Historia de la Sociedad Económica de Amigos del País.* Guatemala City: Editorial Académica Centroamericana, 1981.

Historia del Puerto de Trujillo, Vol. II Edición del 25 Aniversario. Banco Central de Honduras: Tegucigalpa, 2000.

'Puerto de Iztapa o de la Independencia. primera parte', *Antropología e Historia de Guatemala* 8, no. 2 (1956): 24–49.

Safier, Neil. *Measuring the New World: Enlightenment Science and South America.* University of Chicago Press, 2008.

Sagastume Paiz, Tania, 'Las propuestas ilustradas sobre la propiedad corporativa, 1750–1811', *Estudios* 59, tercera época (2014), 41–70.

Saint-Lu, André. *Condición colonial y conciencia criolla en Guatemala (1524–1821).* Guatemala: Editorial Universitaria, 1978.

Saladino García, Alberto. *Ciencia y prensa durante la ilustración latinoamericana.* Mexico: Universidad Autónoma del Estado de México, 1996.

Saldaña, Juan José, ed. *Science in Latin America. A History.* 1st English ed. Austin, TX: University of Texas Press, 2006.

Samayoa Guevara, Héctor Humberto. *Ensayos sobre la independencia de Centroamérica.* Guatemala: Ibarra, 1972.

Sánchez-Blanco, Francisco. *El Absolutismo y las Luces en el reinado de Carlos III.* Madrid: Marcial Pons, 2002.

La mentalidad ilustrada. Madrid: Taurus, 1999.

Sandman, Alison. 'Controlling Knowledge. Navigation, Cartography, and Secrecy in the Early Modern Spanish Atlantic'. In Nicholas Dew and James Delbourgo, eds. *Science and Empire in the Atlantic World.* New York: Routledge, 2008, 31–51.

Schabas, Margaret. *The Natural Origins of Economics*. Chicago, IL and London: University of Chicago Press, 2005.

Schaffer, Simon. 'Measuring Virtue: Eudiometry, Enlightenment, and Pneumatic Medicine'. In Andrew Cunningham and Roger French, eds. *The Medical Enlightenment of the Eighteenth Century*. Cambridge University Press, 1990, 281–318.

Schiebinger, Londa, and Claudia Swan, eds. *Colonial Botany*. Paperback ed. Philadelphia, PA: University of Pennsylvania Press, 2007 [2005].

Schofield, James. 'Jeremy Bentham: Legislator of the World', *Current Legal Problems* 51 (1998): 115–47.

Sellers-García, Sylvia. *Distance and Documents at the Spanish Empire's Periphery*. Stanford University Press, 2014.

   'The Mail in Time: Postal Routes and Conceptions of Distance in Colonial Guatemala', *Colonial Latin American Review* 21 (2012): 77–99.

Sellés, Manuel, José Luis Peset, and Antonio Lafuente, eds. *Carlos III y la ciencia de la Ilustración*. Madrid: Alianza, 1988.

Serrano, Elena. 'Making Oeconomic People: The Spanish Magazine of Agriculture and Arts for Parish Rectors (1797–1808)', *History and Technology* 30, no. 3 (2014), 149–76.

Serrera Contreras, Ramón María. *Tráfico terrestre y red vial en las Indias Españolas*. Ministerio del Interior. Madrid and Barcelona: Dirección General de Tráfico, 1992.

Shafer, Robert. *The Economic Societies in the Spanish World, 1763–1821*. Syracuse University Press, 1958.

Shapin, Steven. 'Pump and Circumstance: Robert Boyle's Literary Technology', *Social Studies of Science* 14 (1984): 481–520.

Shapin, Steven and Simon Schaffer. *Leviathan and the Air-Pump*. Princeton University Press, 1989.

Shaw, Michael Crozier. 'El siglo de hazer caminos. Spanish Road Reforms during the Eighteenth Century. A Survey and Assessment', *Dieciocho* 32, no. 2 (2009): 413–34.

Sigurðsson, Haraldur. *Melting the Earth: The History of Ideas on Volcanic Eruptions*. Oxford University Press, 1999.

Silva, Renán. *La ilustración en el virreinato de Nueva Granada: estudios de historia cultural*. Medellín: Carreta Editores, 2005 [1981].

   Los Ilustrados de la Nueva Granada, 1760–1808: genealogía de una comunidad de interpretación. Medellín: Banco de la República; EAFIT, 2002.

   Prensa y revolución a finales del s. xviii: contribución a un análisis de la formación de la ideología de independencia nacional. Bogotá: Banco de la República, 1988.

Sluyter, Andrew. 'The Making of the Myth in Postcolonial Development: Material-Conceptual Landscape Transformation in Sixteenth-Century Veracruz', *Annals of the Association of American Geographers* 89 (1999): 377–401.

Smith, Crosbie, and Jon Agar, eds. *Making Space for Science: Territorial Themes in the Shaping of Knowledge*. Basingstoke, New York, and Manchester: Palgrave Macmillan and Manchester University Press, 1998.

Smith, Carol. *Guatemalan Indians and the State, 1540 to 1988*. Austin, TX: University of Texas Press, 1990.

Smith, Robert S. 'Financing the Central American Federation, 1821–1838', *The Hispanic American Historical Review* 43, no. 4 (1963): 483–510.

'Origins of the Consulado of Guatemala', *The Hispanic American Historical Review* 26, no. 2 (1956): 150–60.

Smith, Theresa Ann. *The Emerging Female Citizen: Gender and Enlightenment in Spain*. Berkeley and Los Angeles, CA: University of California Press, 2006.

Socolow, Susan Migden. *The Bureaucrats of Buenos Aires, 1769–1810: Amor al Real Servicio*. Durham, NC and London: Duke University Press, 1987.

Solano, Francisco de. *Antonio de Ulloa y la Nueva España*. Mexico City: UNAM, 1987.

'El Archivo General de Indias y la promoción del americanismo científico'. In Sellés et al., eds. *Carlos III y la ciencia de la Ilustración*, 277–96.

Relaciones geográficas del arzobispado de Mexico, 1743. Madrid: Consejo Superior de Investigaciones Científicas, 1988.

Spary, Emma. 'Botanical Networks Revisited'. In Regina Dauser, Stefan Hächler, Michael Kempe, Franz, and Martin Stuber, eds. *Wissen im Netz. Botanik und Pflanzentransfer in europäischen Korrespondenznetzen des 18. Jahrhunderts*. Berlin: Akademie Verlag, 2008, 113–34.

'"Peaches Which the Patriarchs Lacked": Natural History, Natural Resources, and the Natural Economy in France', *History of Political Economy* 35: Annual Supplement (2003): 14–41.

Utopia's Garden. French Natural History from Old Regime to Revolution. Chicago, IL and London: University of Chicago Press, 2000.

Stapelbroek, Koen, and Jani Marjanen, eds. *The Rise of Economic Societies in the Eighteenth Century*. Basingstoke and New York: Palgrave Macmillan, 2012.

Stein, Barbara and Stanley Stein, *Apogee of Empire: Spain and New Spain in the Age of Charles III, 1759–1789*. Baltimore, MD: The Johns Hopkins University Press, 2003.

Edge of Crisis: War and Trade in the Spanish Atlantic, 1789–1808. Baltimore, MD: The Johns Hopkins University Press, 2009.

Tang, Chenxi. *The Geographic Imagination of Modernity: Geography, Literature, and Philosophy in German Romanticism*. Stanford University Press, 2008.

Taracena Arriola, Arturo. *Etnicidad, estado y nación en Guatemala, 1808–1944*, Vol. I, Antigua Guatemala: CIRMA, 2002.

La expedición científica al reino de Guatemala. Guatemala City: Editorial Universitaria de Guatemala, 1983.

Taracena Arriola, Arturo, Juan, Pablo Piraand Celia Marcos, 'La construcción nacional del territorio de Guatemala', *Revista Historia* 45 (2002), 9–33.

Tarver Denova, Hollis and Emily Slape, eds. *The Spanish Empire: A Historical Encyclopedia*, Vol. I. Santa Barbara: ABC-CLIO, 2016.

Terrall, Mary. 'The Uses of Anonymity in the Age of Reason'. In Mario Biagioli and Peter Galison, eds. *Scientific Authorship: Credit and Intellectual Property in Science*. New York and London: Routledge, 2003, 91–112.

Thompson, Kenneth. 'Trees as a Theme in Medical Geography and Public Health', *Bulletin of the New York Academy of Medicine* 54, no. 5 (1978): 517–31.

Thurner, Mark. *History's Peru: The Poetics of Colonial and Postcolonial Historiography.* Gainesville, FL: University of Florida Press, 2011.

Tompson, Doug. 'The Establecimientos Costeros of Bourbon Central America, 1787–1800'. In Dym and Belaubre, eds. *Politics, Economy and Society in Bourbon Central America*, 157–84.

Torres Santo Domingo, Marta, 'Un bestseller del siglo XVIII: El viaje de George Anson alrededor del mundo', *Biblio 3 W* 9, no. 531 (2004) Online publication: www .ub.edu/geocrit/b3 w-531.htm.

Townsend Ezcurra, Andrés. *Las provincias unidas de Centroamérica: fundación de la república.* San José: Editorial Costa Rica, 1973 [1958].

Twinam, Ann. *Purchasing Whiteness. Pardos, Mulattos, and the Quest for Social Mobility in the Spanish Indies.* Stanford University Press, 2015.

Trever, Lisa, and Pillsbury, Joanne, 'Martínez Compañón and His Illustrated "Museum"'. In Daniela Bleichmar and Peter Mancall, eds. *Collecting Across Cultures* (Philadelphia, PA: University of Pennsylvania Press, 2010), 236–53.

Uribe Uran, Victor. 'The Birth of a Public Sphere in Latin America during the Age of Revolution', *Comparative Studies in Society and History* 42 (2000): 425–57.

Urteaga, Luis. 'La teoría de los climas y los orígenes del ambientalismo', *Geocritica* 99 (1993): 1–55.

Valverde, Nuria, and Antonio Lafuente. 'Space Production and Spanish Imperial Geopolitics'. In Gabriella Bleichmar et al., eds. *Science in the Spanish and Portuguese Empires, 1500–1800.* Stanford University Press, 2009, 198–215.

Varela Fernández, Julia. 'La educación ilustrada o como fabricar sujetos dóciles y útiles', *Revista de educación* (1988), 245–74.

Vinson, Ben. *Bearing Arms for His Majesty: The Free-Colored Militia in Colonial Mexico.* Stanford University Press, 2003.

Wakefield, Andre. 'Cameralism: A German Alternative to Mercantilism.' In Philip J. Stern and Carl Wennerlind, eds. *Mercantilism Reimagined: Political Economy in Early Modern Britain and Its Empire.* Oxford University Press, 2014, 134–50.

Walker, Charles. 'The Upper Classes and their Upper Stories: Architecture and the Aftermath of the Lima Earthquake of 1746', *Hispanic American Historical Review* 83 (2003): 53–82.

Walker, Tamara J. *Exquisite Slaves: Race, Clothing, and Status in Colonial Lima.* Cambridge University Press, 2017.

Warren, Adam. *Medicine and Politics in Colonial Peru. Population Growth and the Bourbon Reforms.* University of Pittsburgh Press, 2010.

Weber, David. *Bárbaros: Spaniards and their Savages in the Age of Enlightenment.* New Haven, CT and London: Yale University Press, 2005.

West, Robert. 'The *Relaciones Geográficas* of Mexico and Central America, 1740–1792'. In Howard F. Cline, ed. *Handbook of Middle American Indians. Guide to Ethnohistorical Sources, Part 1.* Austin, TX: University of Texas Press, 1972, 396–452.

Withers, Charles. *Placing the Enlightenment: Thinking Geographically about the Age of Reason*. Chicago, IL and London: University of Chicago Press, 2007.

Woodward, Ralph Lee. *Class Privilege and Economic Development: The Consulado de Comercio of Guatemala, 1793–1871*. Chapel Hill, NC: University of North Carolina Press, 1966.

'The Aftermath of Independence, 1821-c.1870'. In Leslie Bethell, ed. *Central America since Independence*. Cambridge University Press, 1991, 1–36.

Wortman, Miles, 'Bourbon Reforms in Central America', *The Americas* 32 (1975): 222–38.

Government and Society in Central America, 1680–1840. New York: Columbia University Press, 1982.

Zeta Quinde, Rosa. *El pensamiento ilustrado en el Mercurio Peruano 1791–1794*. Piura: Universidad de Piura, 2000.

Zilbermann de Luján, Cristina. 'Destrucción y traslado de la capital. La Nueva Guatemala de la Asunción'. In Luján Muñoz and Zilbermann de Luján, *Historia general de Guatemala*, Vol. iii, 199–210.

Index

Page numbers in italics denote definitions.

Academy of Sciences (Guatemala), 197–8, 210–11
acclimatisation, 82–6, 112, 118, 180
Achim, Miruna, 32
Afanador Llach, María José, 12, 155
Africans and Afro-Caribbeans
  enslaved. *See* slavery
  settlers, 174, 180
agriculture. *See also* crops, labour, climate, settlements
  and new settlements, 173–8, 216. *See also* settlements
  as priority of reformers, 50, 69, 122, 164
  decline of, 24, 169
  in Economic Society, 64–6, 91, 168
  treatises, 104, 130, 203
aguardiente (liquor), 64, 182
algalia. See medicinal plants
Alvarado, Pedro de, 191
Alzate y Ramírez, José Antonio de, 36, 68, 91, 98, 114, 142, 188
Anson, George, 141
Antigua Guatemala. *See* Santiago de Guatemala
archaeology, 30–55, 208, 226
  antiquities in natural history, 31, 48
  Chan Chan, 52
  Copán, 31, 53, 226
  Creole, 36, 49–50
  Palenque, 29–57
  Pompeii, 42, 46
architecture, 42, 47, 54, 178, 183
archives. *See* bureaucracy, geography
Atlas Guatemalteco. See maps
Audiencia de Guatemala, *1–3*, *See also* periphery

Baily, John (Juan), 213–15, 218
bananas. *See* crops (plantains)
Barrundia, José del, 223

Batres, Ventura, 89
beehives, 89
Belaubre, Christophe, 170
Belize (British Honduras), 15, 173, 209, 216–17, 222, 226
Bello, Andrés, 165
Bentham, Jeremy, 105, 220
Bergaño y Villegas, Simón, 22, 67
Bernasconi, Antonio, 33, 42–9
Berquist Soule, Emily, 52
Beteta, Ignacio, 68, 123, 130, 133, 190
biopower. *See* Foucault
Bogotá. *See* New Granada
borderlands, 173, 181
borders, national
  mapped, 204, 226, 234
  with Belize, 217
Botanical Expedition, Royal, 10, 30, 219
  to New Spain and Guatemala, 74, 80, 87, 114
botany
  botanical garden, 9, 83, 85, 112, 180
  botanical material exchanged, 60, 78, 80, 81, 83, 210, *See also* acclimatisation
  botanical networks. *See* networks
  botanical treatises, 88, 113
  British, 12, 82–5, 102, 210. *See also* British Board of Agriculture
  economic or 'useful', 8, 78, 82, 111, 112, 220, 231. *See also* crops, dyes
  Linnaean taxonomy, 74, 95–9, 113
  medical botany. *See* medicinal plants
  plant equivalents, 112–14
  practical, 95–9, 102, 117, 232
Bourbon Reforms, *9*, 16, 26, 32, 48–51, 173
Britain. *See also* Belize, Jamaica, Sumatra, botany
  British agents, 220. *See also* Baily, Chatfield
  improvement, 8, 12, 65, 83, 167

investment and speculation, 210, 214, 216–17, 223
London, 216, 217, 221
Pacific presence, 141, 173
political relationship with, 24, 211, 216–17
privateers or pirates, 173, 185
Royal Geographical Society (RGS), 165, 211–16
scientific relationship with, 86, 209, 211, 212–13, 214–16, 219, 223
settlements, 151, 167, 173
trading relationship with, 82, 151, 168, 182. *See also* contraband
British Board of Agriculture, 83, 84
Buffon, Comte de, 129
bureaucracy, 33, 55
and enlightened reform, 16, 50, 148, 232
red tape preventing reform, 145

Cabello Carro, Paz, 32
cabinet of natural history. *See* natural history
cameralism. *See* political economy
Campillo y Cosío, José del, 9, 173
Campomanes, Pedro de, 9, 48, 175
Candiani, Vera, 136
Cañizares-Esguerra, Jorge, 32, 48, 195, 218
Caribbean, 81, 84, 150, 152, 173, 177, 188. *See also* Honduras, Jamaica, Cuba, harbours
Carter, Paul, 97
Carvajal, Micaela, 74
Casaús Arzú, Marta Elena, 200
castas, 15, 73, 179
Catholic Church. *See also* parish priests, religious orders
in Bourbon Reforms, 9
cattle, 24, 134, 178, 180, 183
censorship, 22, 68
Central American Federation
Guatemala in, 195, 202
Charles III of Spain, 48, 199
Chatfield, Frederick, 211, 215
Chiapas, 19, 29–55, 61, 72, 75, 117, 136–7, 155, 166, 172, 226
chorography. *See* geography
Ciudad Real. *See* Chiapas
Clavijero, Francisco de, 22
climate. *See* environment, acclimatisation
Colombia, 6, 217, 219, 221, 222. *See also* New Granada
colonies, internal. *See* settlements
Comayagua. *See* Honduras
Concoha Chet, Héctor, 160
conquest, remembered, 51, 53, 191
conservation. *See* resource management
Consulado de Comercio, 6

harbours and rivers, 141, 149–54, 160, 163, 183–6
improvement and reform, 19, 170, 178
monopoly, 150–2
opposed to Economic Society, 149
re-establishment and legacy, 199, 205–7, 212, 225
roads, 143
contraband trade, 151–2, 167, 173, 183, 207, 232
Córdova, Antonio, 101
Córdova, Matías de, 19, 105
Cortes de Cádiz, 22, 23, 105, 199, 221
Cortés y Larraz, Pedro, 25, 46
Costa Rica, 1, 18, 23, 133
cotton. *See* crops
Cozar, Prudencio, 77–9
Craib, Raymond, 135
Creoles, *14*, 18, 49, 50, 52–3, 165
crops. *See also* agriculture, botany (economic), dyes
ackee, 83
avocados, 107
breadfruit, 83, 109
cacao, 50, 63–6, 131–4, 158, 165, 170, 176, 177
cotton, 9, 59, 62–3, 66, 158, 204
flax, 21, 63, 65, 92, 167, 182
linen. *See* flax
maize, 166, 188, 212
planned by reformers, 90, 166–8, 182, 208
plantains, 106–10, 166, 178, 187
potatoes, 178, 182
rice, 83, 85–6, 166, 171, 178, 180
sugar, 15, 64, 133
wheat, 62, 72, 188
Cuba, 86, 133, 185
Havana, 22, 60, 89, 102, 174, 189, 221

Del Río, Andrés, 221
Del Río, Antonio, 33–7, 50–7
Del Valle, José Cecilio. *See* Valle
demography, 170–2
Dighero, Bernardo, 153, 212, 223
Domás y Valle, José, 23, 100, 153, 182
Domás, María Josefa, 74
Drayton, Richard, 167
dyes, 74
cochineal, 50, 80–1
from *rubia* plant, 80
grana. *See* cochineal
indigo production, 50, 80, 87, 103, 111, 150
indigo trade, 24, 141–4, 156, 164, 173
Dym, Jordana, 5, 18, 67, 73, 194, 198, 222

earthquakes, 44–7
  and Economic Society, 23, 58–9
  Lima, 44
  of 1773, 3, 23, 44–5, 148, 156, 186
Economic Society. *See also* agriculture,
    Gazeta, natural history, networks,
    patriotism
  and imperial administration, 6, 16–18, 76,
    145, 148, 173, 176. *See also* bureaucracy
  and Royal Botanical Expedition, 114–16
  formalising knowledge, 115, 134
  juntas de correspondencia, 66, 133, 155, 174
    See also Honduras (Trujillo)
  legacy, 198–9, 207
  membership, 16–18, 60–2,
    73–6, 80, 94
  name, 4, 202
  on Indians and the poor, 63–5, 72, 166, 169
  prizes offered, 63, 72, 105
  purpose, *4*, 6–8, 23, 58
  re-establishment, independence,
    199–204, 223
  suspension, 21, 199
education
  in Economic Society, 59, 100, 105, 125,
    172
  population in need of, 23, 68–9, 127, 164,
    168, 197, 198
  University of San Carlos, 18, 201
El Salvador, 23, 60, 77, 111, 141, 164, 215
  San Salvador, 44, 155
elites, 10, *15*, 71–2
empirical knowledge
  experiments, *78, 95*, 95, 117
emulation, 9, 103
encyclopaedia
  Encyclopédie Méthodique, 105, 128–30,
    154, 220
  genre, 122, 129, 130, 135
engineers, 29, 36, 37, 43, 124, 125, 128, 148,
    186, 206, 215, 221. *See also* Diez Navarro,
    Rivera Maestre, Sierra, Garci Aguirre
Enlightenment, *10–13*, 91
  collaborative, 60, 81
  embodied, *21*, 89, 91, 108, 145, 153, 160,
    200, 201. *See also* patriotism
  French authors, 50, 104, 189, 190. *See also*
    Montesquieu, Voltaire, Moheau
  practical, 1, 21, 89, 100. *See also* networks
    (practical)
  public happiness, *6*, 48, 151
  Scottish, 12, 190. *See also* Smith, Adam
  universal, 11, 92–3, 110, 117, 129. *See also*
    theoretical knowledge
environment. *See also* landscape

climate, 52, 85, 129, 134, *157*, 161, 164–7,
    177–91, 203
  climate, changed by humans, 157, 181–92
  public health, 106–10, 157, 181–8
  trees, 106–10, 187, 188
epidemics, 24, 101
Esparragosa, Narciso, 78, 94

Ferdinand VII of Spain, 22
Few, Martha, 101
Flores, José Felipe, 87, 89, 98, 101, 117
Foucault, Michel
  biopower, 175–6
  on natural history, 95–8
France
  French treatises, 104, 110, 203
  jardin du roi, 201
  Pacific presence, 141
  Société de Géographie 212. *See also*
    geography
Francos y Monroy, Cayetano, 24, 46–7, 58
free trade, 9, 82, 168
Fuentes y Guzmán, Francisco Antonio, 38, 166

Galindo, Juan, 204, 207–212, 216,
    218
Galvez, Mariano, 211, 216, 223, 226
García Granados, María Josefa, 74
García Redondo, Antonio, 67, 116, 170, 197
Garci-Aguirre, Pedro, 148, 183
Garrote Bueno, Manuel, 64, 66, 71
Garrote Bueno, Ponciano, 71, 79, 97
Gazeta de Guatemala, 66–9
  and other newspapers, 68, 83, 86–7, 94,
    102, 168
  as community, 18, 70–2, 78–81, 94–5,
    136–42
  as education, 68–9, 127
  public sphere, 70–3, 76, 79, 87, 92, 109,
    127, 130
geography. *See also* maps, infrastructure
  and history, 44–7, 135, 136
  and natural history. *See* natural history
  British. *See* Britain, maps
  chorography, *122*, 122–6, 142, 149, 155
  French, 37, 46, 212
  German, 46, 135. *See also* Humboldt
  in bureaucracy, 30–5, 126, 127, 141,
    152, 163
  in *Gazeta de Guatemala*, 68, 120–55
  legacy of colonial archive, 206–9, 212–13,
    222–5, 228
  military, 31, 35–9, 127, 140
  route descriptions, 31, 127, 137–42, 163, 204
  using archives, 45, 122, 127, 133–4, 190

geology, 211, 219, 221
German lands, 7, 175, 221
Goicoechea, José Antonio. *See* Liendo
    y Goicoechea, José Antonio
Gonzáles Mollinedo y Saravia, Antonio, 23, 61,
    93, 104, 145, 149, 164, 176, 178
Gonzáles, José Mariano, 198
Goodman, Dena, 71
Grove, Richard, 187
Guatemala City
    construction. *See* earthquake of 1773
    opposition to, 75, 150

Habermas, Jürgen, 70
Haitian revolution, 174
Hales, Stephen, 46, 187
harbours
    Acajutla (Sonsonate), 173, 174, 191
    Golfo Dulce, 152, 208
    Izabal, 152, 177, 183–6, 205, 207
    Iztapa, 137, 191, 205, 223
    Omoa, 15, 143, 149–52, 176–9, 182–3,
        188–90
    San Juan del Sur, 151, 214
    Trujillo. *See* Honduras (Trujillo)
    Veracruz, 67, 152
Havana. *See* Cuba
Hidalgo, José Domingo, 126–7, 159, 166
Highlands, Guatemalan, 38, 49, 115, 137, 144,
    161, 173. *See also* Quetzaltenango
Honduras
    British Honduras. *See* Belize
    Comayagua, 13, 18, 150, 160
    contraband trade, 82, 151
    Copán. *See* archaeology
    Danlí, 75
    Economic Society correspondents, 112, 115,
        133, 165
    Trujillo, 85, 133, 172–92
Humboldt, Alexander von, 46, 88, 187, 218,
    219, 222

improvement. *See also* Britain, improvement,
    proyectismo
    agricultural, 12, 83, 85, 104
    and transport. *See* transport
    as ideology, 13–22, 24, 29, 47–51, 59–60,
        145, 181, 220
    based on knowledge, 59
    of climate. *See* environment
Indians / indigenous. *See also* labour, networks,
    population
    expert knowledge, 77, 81, 96–7, 117, 159
    historical knowledge, 33, 49, 56
    ineffective custodians, 159, 167–8

K'iché, 38, 49
Kakchiquel, 96
labour for projects, 63, 65, 72, 162–3
    language, 49, 55, 96–8
    pre-conquest, imagined, 51–4, 171, 227
    reformers' plans, 72, 170, 172, 173, 200, 203
    secret knowledge, 97, 124, 159, 207
infrastructure
    bridges, 144, 148, 205
    canals, 152, 206, 213–14
    harbours. *See* harbours
    roads. *See* roads
Ingenhousz, Jan, 109
Irisarri, Juan Bautista, 81, 137–42, 150, 155,
    174–6, 191

Jamaica, 82–4
Jesuits. *See* religious orders
Juarros, Domingo, 38–9, 135, 214

labour
    agricultural, 64, 66, 169, 178
    forced and tributary, 15–16, 47, 65, 162–3,
        170, 205, 217, 233
    population as resource, 62, 105, 162,
        169–81. *See also* Foucault
    prisoners, 159, 179, 205
Lake Atitlán, 96, 144, 222
land ownership, 14, 65, 170
Landívar, Rafael, 168
landscape
    as Indian, 158, 207
    bountiful, *111*, 143, 165, 166, 167, 181, 219
    cultural landscape, *4*, 6, 20, 32, 155, 160,
        161, 175, 186, 234
    historicised, 44–7, 52, 56, 133, 134, 135,
        189. *See also* geography (and history)
    pre-conquest, 51, 134, 160, 188
    soil, 52, 85, 111, 166, 180, 188, 218
    wild, 157–61, 166–7, 170, 208
Latour, Bruno, 17, 176
León. *See* Nicaragua
Liendo y Goicoechea, José Antonio de, 18, 61,
    68, 98, 107–9, 169, 172, 197, 198
Lindenfeld, David, 17
linen. *See* crops
Linnaeus, Carl. *See* botany
livestock. *See* cattle
Longinos Martínez, José, 74, 114
Lovell, George, 4
lowlands, Guatemalan, 158–61

mandamiento. *See* labour
maps. *See also* Atlas Guatemalteco
    Atlas Guatemalteco, 37–9, 223–7

maps (cont.)
    British, using Spanish sources, 209,
        212–13, 215
    military surveys, 124
    of archaeological sites, 36–9
    of inland Central America, 131, 136, 145
    scepticism of, 123–6
Martínez Compañón, Baltasar Jaime, 52
Martínez Peláez, Severo, 4, 32, 166
Mayorga, Martín de, 44–6
medicinal plants, 11, 50, 94
    algalia, 76–9, 97, 117
    camacarnata, 95–8
    indigenous names, 96–8
    quina (quinine), 101, 113–14, 117, 209
medicine, formal, 78, 93, 101, 116
Melón, Sebastián, 63, 68, 91, 92
Mexico (colonial). *See* New Spain
Mexico (independent), 203, 206,
    218, 221, 226
military, 125, 177–81 *See also* geography
Mociño, José Mariano, 61, 87–8, 98,
    109–10, 141
Moheau, Jean-Baptiste, 190
Montesquieu, 11, 129, 157, 190
mountains. *See* volcanoes, landscape
Muñoz Pérez, José, 59
Muñoz, Juan Bautista, 33, 54
museum. *See* natural history
Mutis, José Celestino, 106–10, 113, 188

Nahuatl, 96
natural disasters. *See* earthquakes, volcano
natural history, *31*, *See also* botany, Foucault
    and geography, 31, 127, 220
    as useful knowledge, 7–9, 221
    British collectors, 210
    cabinet, 35, 114–16, 126
    in Economic Society, 93–101, 111–16,
        165, 168
    museum, 203, 211, 221
networks, *17*
    centres and peripheries, 83, 142
    geographically specific, 90, 93, 98, 110, 112
    global, 81–9, 98, 101–14, 209–22
    parish priests, 62, 126, 153, 232
    population as informants, 31, 81, 114
    practical, 66, 78–81, 89, 99–100. *See also*
        theoretical knowledge
    regional, 60–6, 72, 93–101
New Granada, 6, 100
    Bogotá, 102, 176, 188
New Spain, 36, 80, 91, 120
    Acapulco, 140–1
    Mexico City, 57, 88, 114, 136

Oaxaca, 50, 79–81, 133, 203
San Blas, 141
newspapers, 75, 83, 86, 94, 102, 103, 168, 198,
    218, 220, *See also Gazeta de Guatemala*
Nicaragua, 103, 113, 133, 150–2, 172, 213–17.
    See also harbours, rivers
    León, 18, 61, 75, 76, 92, 107, 155
Nootka Sound, 88, 141

Panama or Nicaragua canal. *See* infrastructure
Paquette, Gabriel, 6, 170
parish priests. *See* networks
paternalism, 12, 16, 63, 72, 143, 153, 162, 170
patriotism, *5*, *19–21*, *See also* Enlightenment
    (embodied)
    Audiencia as target for projects, 19–20, 76,
        93, 116, 128, 151–5
    Creole, 50
    in independence, 198, 200–2, 225
    territory of intervention, *6*, 119, 150
Pauw, Cornelius de, 129, 130
Payés y Font, Juan, 17, 77
periodical press. *See Gazeta de Guatemala*,
    newspapers
periphery, *13*
    Guatemala as 'miserable' and desolate, 13,
        14, 23, 30, 156, 164
    Guatemala as peripheral, 13–14, 91,
        123, 142
    within Central America, 75, 150, 176, 177,
        181, 185
Persia, 110
Peru, 2, 123
    Andes, 165
    El Callao, 133, 139
    Lima, 44, 68, 75
    Loja, 113
    Mercurio Peruano. *See* newspapers
    Trujillo, 52
Petén, 177, 204, 211, 216
Philip, Kavita, 96, 167
physiocrats. *See* political economy
plantains. *See* crops
Podgorny, Irina, 36
political economy
    and cultural landscape, 51, 107, 149, 155
    cameralism, 8
    German, 17
    physiocrats, 8, 105, 175
    theories, 105
population. *See also* education, Enlightenment,
    labour, networks, public sphere
    beneficiaries of Enlightenment, 94, 97, 104,
        157, 185–8
    called guatemaltecos, 73

demography, 127, 198, 204
  estimates, 14
Porta y Costas, Antonio, 152, 186, 212
Poupeney Hart, Catherine, 67
Priestley, John, 109, 187
proyectismo, *59*, 59, 145
public sphere. *See Gazeta de Guatemala*,
  Habermas

Quentas Zayas, Agustín de las, 52, 136, 148, 155
Quetzaltenango, 23, 77, 112, 115, 126, 127
quina (quinine). *See* medicinal plants

Rajan, Ravi, 189
Ramírez y Blanco, Alejandro, 18, 22, 67, 81–9,
  102, 152, 164
Reinert, Sophus, 103
relaciones geográficas. *See* geography (in
  bureaucracy)
religious orders
  Bethlehemite, 72
  Capuchin convent, 106
  Dominicans, 15, 50, 53, 80, 111, 185
  Franciscans, 111
  Jesuits, 111, 124, 129, 218
  Santa Clara convent, 96
repartimiento. *See* labour
resource management, 188, 189
rice. *See* crops
Rivera Maestre, Miguel, 38–9, 223, 226–7
rivers
  Motagua, 152–4, 159–60, 163, 186, 206, 212
  Polochic, 152
  San Juan, 152, 155, 214, 217
  Tulijá, 136, 155
roads
  construction, 43, 143–8, 161–2, 204, 205,
    208, 209
  mule transport, 137, 144, 152, 208
  reshaping geography, 144–55, 183, 208, 209
  route descriptions, 137
Rodríguez de Zea, Agustín and Blas, 66
Rossi y Rubí, José, 18, 64, 68, 78–9, 117,
  143–9, 150, 160–3, 180
Rozier, François, 104, 203

Santiago de Guatemala (Antigua), 23, 133
scientific expedition. *See also* Botanical
  Expedition
  Geodesic Expedition of 1742, 225
  planned by José del Valle, 219

Sellers-García, Sylvia, 102, 137, 207
settlements
  civilising, 159, 183, 190
  new settlements, 85, *136*, 172, 173,
    176–9, 183
Shapin, Steven, 94–5
Sierra, Josef de la, 36, 126
Sigaud de la Fond, Joseph-Aignan, 107, 110
Silva, Renán, 17
Skinner, George Ure, 210
slavery. *See also* labour
  abolishment, 217
  Adam Smith on, 177
  enslaved black people, 15
Spary, Emma, 201
Suchitepéquez, 66, 77–8, 133, 143–50, 160
  Mazatenango, 62–4, 71
Sumatra, 81–6

Tang, Chenxi, 135
textile industry, 23, 168, 182
  silkworms, 74, 79, 203
theoretical knowledge, rejection of, 17,
  95–100, 123, 126
Thompson, George Alexander, 213, 220, 222
Thurner, Mark, 2, 52
transport. *See* rivers, roads, harbours,
  infrastructure

Ulloa, Antonio de, 10, 48
Unanue, Hipólito, 52, 91, 98
United States, 151, 174, 213, 217
  Philadelphia, 86–7, 98

Valle, José Cecilio del
  geography, 204, 222–7
  Indians, 53, 200
  political economy, 105, 134, 165, 193, 200, 203
  scientific networks, 196–8, 203, 216–22
Venezuela, 165, 187
Villaurrutia, Jacobo de, 17, 18, 20, 65, 68, 88,
  112, 115, 151, 155, 157, 158, 197
volcanoes, 42–7, 112, 115, *120*, 168,
  204, 219
Voltaire, 11, 46

Ward, Bernardo, 9, 105, 175
wheat. *See* crops
women, 73–5

Zea, Francisco Antonio de, 100

Lightning Source UK Ltd.
Milton Keynes UK
UKHW020638191022
410718UK00026B/907